Deep Neural Networks in Medical Imaging

Deep Neural Networks in Medical Imaging

Editors

Lucian Mihai Itu
Constantin Suciu
Anamaria Vizitiu

Basel • Beijing • Wuhan • Barcelona • Belgrade • Novi Sad • Cluj • Manchester

Editors
Lucian Mihai Itu
Transilvania University of
Brașov
Brașov
Romania

Constantin Suciu
Transilvania University of
Brașov
Brașov
Romania

Anamaria Vizitiu
Transilvania University of
Brașov
Brașov
Romania

Editorial Office
MDPI AG
Grosspeteranlage 5
4052 Basel, Switzerland

This is a reprint of articles from the Special Issue published online in the open access journal *Applied Sciences* (ISSN 2076-3417) (available at: https://www.mdpi.com/journal/applsci/special_issues/Deep_Neural_Networks_Medical_Imaging).

For citation purposes, cite each article independently as indicated on the article page online and as indicated below:

Lastname, A.A.; Lastname, B.B. Article Title. *Journal Name* **Year**, *Volume Number*, Page Range.

ISBN 978-3-7258-2525-7 (Hbk)
ISBN 978-3-7258-2526-4 (PDF)
doi.org/10.3390/books978-3-7258-2526-4

© 2024 by the authors. Articles in this book are Open Access and distributed under the Creative Commons Attribution (CC BY) license. The book as a whole is distributed by MDPI under the terms and conditions of the Creative Commons Attribution-NonCommercial-NoDerivs (CC BY-NC-ND) license.

Contents

Diana Ioana Stoian, Horia Andrei Leonte, Anamaria Vizitiu, Constantin Suciu and Lucian Mihai Itu
Deep Neural Networks in Medical Imaging: Privacy Preservation, Image Generation and Applications
Reprinted from: *Appl. Sci.* 2023, 13, 11668, doi:10.3390/app132111668 1

Andreea Bianca Popescu, Ioana Antonia Taca, Anamaria Vizitiu, Cosmin Ioan Nita, Constantin Suciu, Lucian Mihai Itu and Alexandru Scafa-Udriste
Obfuscation Algorithm for Privacy-Preserving Deep Learning-Based Medical Image Analysis
Reprinted from: *Appl. Sci.* 2022, 12, 3997, doi:10.3390/app12083997 7

Muhammad Mateen Yaqoob, Muhammad Nazir, Abdullah Yousafzai, Muhammad Amir Khan, Asad Ali Shaikh, Abeer D. Algarni and Hela Elmannai
Modified Artificial Bee Colony Based Feature Optimized Federated Learning for Heart Disease Diagnosis in Healthcare
Reprinted from: *Appl. Sci.* 2022, 12, 12080, doi:10.3390/app122312080 33

Muhammad Mateen Yaqoob, Muhammad Nazir, Muhammad Amir Khan, Sajida Qureshi and Amal Al-Rasheed
Hybrid Classifier-Based Federated Learning in Health Service Providers for Cardiovascular Disease Prediction
Reprinted from: *Appl. Sci.* 2023, 13, 1911, doi:10.3390/app13031911 49

Dominik F. Bauer, Constantin Ulrich, Tom Russ, Alena-Kathrin Golla, Lothar R. Schad and Frank G. Zöllner
End-to-End Deep Learning CT Image Reconstruction for Metal Artifact Reduction
Reprinted from: *Appl. Sci.* 2022, 12, 404, doi:10.3390/app12010404 66

Andrei Puiu, Sureerat Reaungamornrat, Thomas Pheiffer, Lucian Mihai Itu, Constantin Suciu, Florin Cristian Ghesu and Tommaso Mansi
Generative Adversarial CT Volume Extrapolation for Robust Small-to-Large Field of View Registration
Reprinted from: *Appl. Sci.* 2022, 12, 2944, doi:10.3390/app12062944 81

Muhammad Yaqub, Feng Jinchao, Shahzad Ahmed, Kaleem Arshid, Muhammad Atif Bilal, Muhammad Pervez Akhter and Muhammad Sultan Zia
GAN-TL: Generative Adversarial Networks with Transfer Learning for MRI Reconstruction
Reprinted from: *Appl. Sci.* 2022, 12, 8841, doi:10.3390/app12178841 94

Shunlei Li, Muhammad Adeel Azam, Ajay Gunalan and Leonardo S. Mattos
One-Step Enhancer: Deblurring and Denoising of OCT Images
Reprinted from: *Appl. Sci.* 2022, 12, 10092, doi:10.3390/app121910092 114

Rubén G. Barriada, Olga Simó-Servat, Alejandra Planas, Cristina Hernández, Rafael Simó and David Masip
Deep Learning of Retinal Imaging: A Useful Tool for Coronary Artery Calcium Score Prediction in Diabetic Patients
Reprinted from: *Appl. Sci.* 2022, 12, 1401, doi:10.3390/app12031401 125

Costin Florian Ciușdel, Lucian Mihai Itu, Serkan Cimen, Michael Wels, Chris Schwemmer, Philipp Fortner, et al.
Normalizing Flows for Out-of-Distribution Detection: Application to Coronary Artery Segmentation
Reprinted from: *Appl. Sci.* **2022**, *12*, 3839, doi:10.3390/app12083839 . . . . **135**

Cosmin-Andrei Hatfaludi, Irina-Andra Tache, Costin Florian Ciușdel, Andrei Puiu, Diana Stoian, Lucian Mihai Itu, et al.
Towards a Deep-Learning Approach for Prediction of Fractional Flow Reserve from Optical Coherence Tomography
Reprinted from: *Appl. Sci.* **2022**, *12*, 6964, doi:10.3390/app12146964 . . . . **156**

Zahra Mungloo-Dilmohamud, Maleika Heenaye-Mamode Khan, Khadiime Jhumka, Balkrish N. Beedassy, Noorshad Z. Mungloo and Carlos Peña-Reyes
Balancing Data through Data Augmentation Improves the Generality of Transfer Learning for Diabetic Retinopathy Classification
Reprinted from: *Appl. Sci.* **2022**, *12*, 5363, doi:10.3390/app12115363 . . . . **179**

Tao Yu, Huiyi Hu, Xinsen Zhang, Honglin Lei, Jiquan Liu, Weiling Hu, et al.
Real-Time Multi-Label Upper Gastrointestinal Anatomy Recognition from Gastroscope Videos
Reprinted from: *Appl. Sci.* **2022**, *12*, 3306, doi:10.3390/app12073306 . . . . **196**

Nada Mobark, Safwat Hamad and S. Z. Rida
CoroNet: Deep Neural Network-Based End-to-End Training for Breast Cancer Diagnosis
Reprinted from: *Appl. Sci.* **2022**, *12*, 7080, doi:10.3390/app12147080 . . . . **216**

Dehai Zhang, Anquan Ren, Jiashu Liang, Qing Liu, Haoxing Wang and Yu Ma
Improving Medical X-ray Report Generation by Using Knowledge Graph
Reprinted from: *Appl. Sci.* **2022**, *12*, 11111, doi:10.3390/app122111111 . . . . **228**

Editorial

Deep Neural Networks in Medical Imaging: Privacy Preservation, Image Generation and Applications

Diana Ioana Stoian [1,2], Horia Andrei Leonte [1,2], Anamaria Vizitiu [1,2], Constantin Suciu [1,2] and Lucian Mihai Itu [1,2,*]

1. Advanta, Siemens SRL, 15 Noiembrie Bvd, 500097 Brasov, Romania; diana.stoian@siemens.com (D.I.S.); horia.leonte.ext@siemens.com (H.A.L.); anamaria.vizitiu@siemens.com (A.V.); constantin.suciu@siemens.com (C.S.)
2. Automation and Information Technology, Transilvania University of Brasov, Mihai Viteazu nr. 5, 5000174 Brasov, Romania
* Correspondence: lucian.itu@unitbv.ro

Citation: Stoian, D.I.; Leonte, H.A.; Vizitiu, A.; Suciu, C.; Itu, L.M. Deep Neural Networks in Medical Imaging: Privacy Preservation, Image Generation and Applications. *Appl. Sci.* 2023, *13*, 11668. https://doi.org/10.3390/app132111668

Received: 13 October 2023
Accepted: 23 October 2023
Published: 25 October 2023

Copyright: © 2023 by the authors. Licensee MDPI, Basel, Switzerland. This article is an open access article distributed under the terms and conditions of the Creative Commons Attribution (CC BY) license (https://creativecommons.org/licenses/by/4.0/).

1. Introduction

Medical Imaging plays a key role in disease management, starting from baseline risk assessment, diagnosis, staging, therapy planning, therapy delivery, and follow-up. Each type of disease has led to the development of more advanced imaging methods and modalities to help clinicians address the specific challenges in analyzing the underlying disease mechanisms. Imaging data is one of the most important sources of evidence for clinical analysis and medical intervention as it accounts for about 90% of all healthcare data. Researchers have been actively pursuing the development of advanced image analysis algorithms, some of which are routinely used in clinical practice. These developments were driven by the need for a comprehensive quantification of structure and function across several imaging modalities such as Computed Tomography (CT), X-ray Radiography, Magnetic Resonance Imaging (MRI), Ultrasound, Nuclear Medicine Imaging, and Digital Pathology [1].

In the context of the availability of unprecedented data storage capacity and computational power, Deep learning has become the state-of-the-art machine learning technique, providing unprecedented performance at learning patterns in medical images and great promise for helping physicians during clinical decision-making processes. Previously reported deep learning-related studies cover various types of problems, e.g., classification, detection, segmentation, for different types of structures, e.g., landmarks, lesions, organs, in diverse anatomical application areas [2].

The aim of this special issue is to present and highlight novel methods, architectures, techniques, and applications of deep learning in medical imaging. Papers both from theoretical and practical aspects were welcome, including ongoing research projects, experimental results, and recent developments related to, but not limited to, the following topics: image reconstruction; image enhancement; segmentation; registration; computer aided detection; landmark detection; image or view recognition; automated report generation; multi-task learning; transfer learning; generative learning; self-supervised learning; semi-supervised learning; weakly supervised learning; unsupervised learning; federated learning; privacy preserving learning; explainability and interpretability; robustness and out-of-distribution detection; and uncertainty quantification.

2. The Papers

In this Special Issue, we published a total of 14 papers that span across four interesting topics as outlined below.

2.1. Privacy-Preserving Learning

Deep Learning heavily relies on existing and forthcoming patient data to yield precise and dependable outcomes within the realm of healthcare applications. Despite the copiousness of biomedical data, its dissemination and retrieval are hindered by ethical limitations, particularly concerning safeguarded health-related information pertaining to patients. Consequently, the actualization of medical AI systems encounters challenges, as the requisite data for their development and training are ensnared within the confines of hospital security protocols. To engender resilient algorithms, the databases employed for training, validation, and testing must encompass the complete spectrum of pathological deviations and permutations. Additionally, it is imperative to leverage the entirety of accessible information to formulate a more tailored solution. In instances where training datasets lack heterogeneity, there is a propensity for algorithms to exhibit partiality or inclination towards specific patient profiles [2].

In the field of privacy-preserving learning three papers are presented.

The first paper presents a novel approach for protecting sensitive data in the healthcare setting, as nowadays the trend is to send it outside the facility for it to be processed, trained on etc. [3]. The researchers designed the solution to make it robust against both human perception, as well as against different software attacks such as AI reconstruction attempts. Therefore, by using the proposed pipeline, the data will be obfuscated before leaving the healthcare facility, and the external processing (such as training of AI models) can be performed with a satisfactory privacy-accuracy trade-off, i.e., without a significant drop in accuracy. The three main objectives of the paper are to hide the content from any person viewing the images, to make it difficult for an AI to reconstruct the image, as well as to facilitate AI model training on such data. Regarding the technical aspects, a Variational Autoencoder (VAE) is used, trained on around 30,000 images from the Medical MNIST dataset. The VAE has two different output channels, one based on the mean of the normal distribution, which offers more privacy, but limits the performance in further ML applications, and a second one which is based on the standard deviation of the normal distribution, which trades privacy for training performance.

The second paper's focus lies on maintaining and improving the training and prediction accuracy of AI based solutions in heart disease diagnosis, while overcoming data privacy issues [4]. The main mechanism is federated learning, which aims to keep the data on a single device while training via a collaborative system of a shared model. The optimizer framework–the Modified Artificial Bee Colony (M-ABC) has been chosen due to its flexibility and user-friendliness, has less parameters than other algorithms and has a fast convergence rate. These two methods work together, as the federated matched averaging (FedMA) is constructing a privacy-aware framework for a global cloud model, and the M-ABC framework serves as the feature selector on the client's side. The pipeline was trained on the heart disease dataset of UCI Cleveland, with 303 records and 76 attributes.

In the third paper, the same approach is used, but instead of employing the M-ABC optimization, a hybrid M-ABC with Support Vector Machine is used [5]. The SVM is less prone to overfitting, works well in high dimensional spaces, and has good handling of non-linear data. Moreover, it is a suitable candidate for classifying multiple classes. The M-ABC acts as the feature extractor and the SVM as the classification algorithm. The dataset combines over eleven common features, such as blood pressure, cholesterol serum, sex, age etc. from the datasets of Cleveland, Stalog, Hungary, Long Beach, and Switzerland. As far as results are concerned, the proposed solution is both more efficient and more accurate than the previous ones.

2.2. Image Generation

Within the domain of medical imaging, generative models pursue two primary trajectories: (i) transformation from noise to image and (ii) transition between images. The former encompasses methodologies focused on artificially augmenting the dataset, often referred to as augmentation, by training a deep learning-grounded model to transmute a noise vector into authentic-looking images. To illustrate, when confronted with an extensive array of breast imaging data, such as mammograms, a generative model endeavors to fabricate novel images resembling constituents of the imaging dataset. In the context of image-to-image generation, generative models undertake the conversion of a given input image into an alternative representation. A diverse array of challenges within the healthcare domain can be effectively addressed through image-to-image generative models. These encompass tasks spanning the enhancement of image fidelity via denoising, amplification of resolution, image inpainting, amalgamation of multi-modal images, along with image reconstruction and alignment.

In the field of reconstruction and image generation four papers are presented. The first one suggests employing an end-to-end deep learning network for the correction of metal artifacts in CT images [6]. The input of the network is represented by metal-affected NMAR sinograms, and the outputs are artifact-free reconstructed images. The architecture consists of three parts: sinogram refinement used to filter the sinogram, back projection used to reconstruct the image into the image domain, and image refinement used to further refine the reconstruction. All parts are trained simultaneously and furthermore, the network performs the complete CT image reconstruction, and does not require a predefined back projection operator or the exact X-ray beam geometry.

A second paper in this area proposes a method to increase the field of view of intraoperative images obtained from Computer Tomographs [7]. This method is used as a prior step to the registration of two volumes: thin intraoperative volume and preoperative volume. The method consists in extrapolating the thin volume by generating additional slices from the existing ones using a GAN architecture. By enhancing the context information required for the matching process, the results appear to be comparable to those obtained after aligning two high-resolution images having the same field of view.

The third paper presents a transfer learning enhanced GAN technique for image reconstruction using under-sampled MR data [8]. The model was tested on an open-source knee dataset, and a private brain dataset with two different acceleration factors: 2 and 4. Both datasets were divided into training and test sets. The training sets were used for finetuning the model after transfer learning. The results indicate that the proposed model outperforms the other reconstruction techniques for both acceleration factors, suggesting that, by using transfer learning, the variation in image contrast, acceleration factor and anatomy between training and test dataset is smaller. Moreover, the distribution of the reconstructed images, produced by transfer learning is more similar to the distribution of the completely sampled image.

The main objective of the fourth paper is OCT image enhancement through denoising and deblurring of the image on a single step process [9]. The applied method is an unsupervised learning technique with unpaired images and disentangled representation, combined with a GAN architecture. The framework consists in encoders (used to extract relevant features from the raw images: image content, image noise, blur features, and blur-noise features), generators (used to generate from the extracted features blurred, noisy, blurred-noisy and clean images) and discriminators (used to discriminate between generated and real images for each feature). The obtained model was compared with state-of-the-art methods for OCT image enhancement, which were outperformed. Also, a quantitative comparison with state-of-the-art methods indicates that the proposed enhancer performs better than all the other methods, with the best processing speed when the computations were run on a GPU.

2.3. Applications–Cardiovascular Diseases

Cardiovascular ailment (CVD) poses a substantial peril to human well-being and stands as the primary global fatality determinant [10]. The incidence of both mortality and morbidity linked to CVD exhibits an escalating trajectory, particularly within burgeoning territories. This malady precipitates considerable financial ramifications, approximated at 351.2 billion USD in the United States, thereby engendering persistent compromise to the quality of life [11]. Within the European Union, the annual expenditure has been assessed at 210 billion euros, apportioned amongst direct healthcare outlays (53%), diminished productivity (26%), and informal caregiving for individuals afflicted with CVD (21%) [12].

In the field of cardiovascular disease applications four papers are presented. The first paper in this field proposes two methods for binary classifying the risk of CAD based on the CAC (CAC > 400 represents high risk of CAD, while CAC < 400 low risk of CAD) in diabetic patients [13]: the first method consists in employing a state-of-the-art CNN architecture for CAD risk assessment, based on the retina images and the second method consists in employing classical machine learning classifiers on the clinical data (age and presence of diabetic retinopathy). The DL algorithm considered therein is a VGG16 architecture trained on ImageNet and finetuned on the available retina images. By using the proposed methods, two protocols were established that target two specific applications. The statistics (accuracy, precision, recall, F1 score, confusion matrix) were computed to evaluate each method and the protocols. Results show acceptable accuracies when evaluating the methods independently, while when combining the methods either the precision or recall improve depending on the protocol (the protocol that is created based on the particular needs of each application).

The second paper focuses on obtaining a smaller processing time when using a semi-automated approach for the task of segmenting coronary artery lumen by pre-selecting vessel locations likely to require manual inspection and editing [14]. The pre-selection step is formulated as an Out-of-Distribution (OoD) detection problem with the task of detecting mismatched pairs of CCTA lumen images and their corresponding lumen segmentations. Two Normalizing Flows architectures are employed and assessed: a Glow-like baseline, and a NF architecture which uses a novel coupling layer which exhibits an inductive bias favoring the exploitation of semantical features instead of local pixel correlations. The models were assessed on both synthetic mask perturbations and expert annotations. On synthetic perturbations, the results indicate a better performance for the proposed model, when compared with the baseline model. The proposed model also outperforms the baseline, having a sensitivity for detecting faulty annotations close to inter-expert agreement.

The main objective in the third paper is to evaluate the feasibility of using neural networks in predicting invasively measured FFR from the radius of the coronary lumen that is extracted along the centerline of the coronary artery from OCT images [15]. Three different approaches were used for solving this task: a regression, a classification and an FSL (few shot learning) approach, where the task was formulated also as a classification problem. For each approach different types of architectures were considered: ANN, CNN and RNN. The evaluation step is performed on ensembles for each architecture type: each proposed architecture is trained 20 times, with different random seeds, and the final prediction is performed by the mean value (for regression)/probabilities (for classification) of all 20 models. The FSL CNN based ensemble shows the best diagnostic performance, while being the most robust approach by having the smallest standard deviation and uncertainty. Moreover, compared with baseline approaches based on MLD or %DS FSL CNN reaches improved results. Also, the authors demonstrated that the dataset size has a significant impact on the accuracy: a linear increase in performance was observed as a function of dataset size.

The aim of the last paper of this section is to classify fundus images into five classes of diabetic retinopathy by using neural networks with transfer learning and data augmentation [16]. Three architectures were employed for this task: VGG19, ResNet50, DenseNet169. All models were first finetuned on a public dataset (APTOS). Since the public dataset was imbalanced the models were enhanced by further finetuning on the augmented public dataset (APTOS augmented). The resulting models were tested on a blinded test dataset. Results indicate that ResNet50 performs better than all the other models on all classes.

2.4. Applications–Other

In the last category we included three other applications based on medical imaging, related to cancer, gastrointestinal disorders, and respectively medical report generation. In the first paper, the authors try to optimize the radiological workload, by using the knowledge graph method, a novel method that enhances search engines in general proposed by

Google in 2012 [17]. Firstly, there is an initial knowledge association between disease labels, that are defined as nodes. This is done in two steps, with the help of CheXpert tagger, that classifies the reports into 14 different categories, and the SentencePiece tagger/detagger tool, from which the nouns with top k occurrences are selected as additional disease categories. Based on this, a graph convolutional neural network is used to aggregate information between nodes, creating prior knowledge. This is done by generating a hybrid image-text feature, with features extracted from X-ray images with the help of a CNN, and text features extracted from the associated clinical reports using transformers. The transformers represent a better option compared to the classic RNN approach, as radiology reports tend to consist of longer sentences. This hybrid pair is sent through the graph convolutional network, and the node features are split into two branches: a linear classifier for disease classification and a generator head for the report itself. The result is fine-tuned by re-running it through the text classifier. The results are evaluated by the quality of the Natural Language Generation and as well as the clinical efficacy.

The second paper uses the DDSM dataset (Digital Database for Mammography Screening), with a total of 2620 films, of which 695 are normal and 1925 are abnormal, to improve the detection of Breast Cancer, a pathology encountered worldwide, which puts at risk many lives [18]. The employed model, unlike the classic CNN architectures such as VGG, ResNet, MobileNet, etc. is represented by the novel CoroNet: based on the Xception CNN architecture, which consists of a 71-layer deep CNN architecture, pretrained on an ImageNet dataset. The key efficiency improvement of this architecture is the depth wise separable convolution layers with residual connections, which enable a decrease in the number of operations. Separable convolution replaces the classical $n \times n \times k$ convolution with a $1 \times 1 \times k$ point-wise convolution, followed by a channel-wise $n \times n$ spatial convolution, and the residual connections represent "skip connections" which enable the flow of gradients without the need for non-linear functions of activation. This mitigates the disappearing gradient issue. As per the results, this solution outperforms alternative networks.

Finally, in the third paper, a CNN backbone, ResNet-50 pre-trained on ImageNet, is used to extract features from static images, and a GCN (Graph Convolutional Network) is employed for classifying the relationship between labels [19]. An LSTM architecture was used for the temporal association between subsequent frames in the gastroscopy. Those form the proposed GL-Net architecture, which combines the label extraction feature and temporal correlation and dependencies, for real-time predictions in gastroscopy videos. The dataset consists of 49 videos and after video processing and 5 Hz sampling, 23,471 training images and 5798 test images are obtained, with multi-label annotations.

Conflicts of Interest: The authors declare no conflict of interest.

References

1. Sharma, P.; Suehling, M.; Flohr, T.; Comaniciu, D. Artificial intelligence in diagnostic imaging: Status quo, challenges, and future opportunities. *J. Thorac. Imaging* **2020**, *35*, S11–S16. [CrossRef] [PubMed]
2. Puiu, A.; Vizitiu, A.; Nita, C.; Itu, L.; Sharma, P.; Comaniciu, D. Privacy-preserving and explainable AI for cardiovascular imaging. *Stud. Inform. Control.* **2021**, *30*, 21–32. [CrossRef]
3. Popescu, A.; Taca, I.; Vizitiu, A.; Nita, C.; Suciu, C.; Itu, L.; Scafa-Udriste, A. Obfuscation Algorithm for Privacy-Preserving Deep Learning-Based Medical Image Analysis. *Appl. Sci.* **2022**, *12*, 3997. [CrossRef]
4. Yaqoob, M.; Nazir, M.; Yousafzai, A.; Khan, M.; Shaikh, A.; Algarni, A.; Elmannai, H. Modified Artificial Bee Colony Based Feature Optimized Federated Learning for Heart Disease Diagnosis in Healthcare. *Appl. Sci.* **2022**, *12*, 12080. [CrossRef]
5. Yaqoob, M.; Nazir, M.; Khan, M.; Qureshi, S.; Al-Rasheed, A. A Hybrid Classifier-Based Federated Learning in Health Service Providers for Cardiovascular Disease Prediction. *Appl. Sci.* **2023**, *13*, 1911. [CrossRef]
6. Bauer, D.; Ulrich, C.; Russ, T.; Golla, A.; Schad, L.; Zöllner, F. End-to-End Deep Learning CT Image Reconstruction for Metal Artifact Reduction. *Appl. Sci.* **2022**, *12*, 404. [CrossRef]
7. Puiu, A.; Reaungamornrat, S.; Pheiffer, T.; Itu, L.; Suciu, C.; Ghesu, F.; Mansi, T. Generative Adversarial CT Volume Extrapolation for Robust Small-to-Large Field of View Registration. *Appl. Sci.* **2022**, *12*, 2944. [CrossRef]
8. Yaqub, M.; Jinchao, F.; Ahmed, S.; Arshid, K.; Bilal, M.; Akhter, M.; Zia, M. GAN-TL: Generative Adversarial Networks with Transfer Learning for MRI Reconstruction. *Appl. Sci.* **2022**, *12*, 8841. [CrossRef]

9. Li, S.; Azam, M.; Gunalan, A.; Mattos, L. One-Step Enhancer: Deblurring and Denoising of OCT Images. *Appl. Sci.* **2022**, *12*, 10092. [CrossRef]
10. Thomas, H.; Diamond, J.; Vieco, A.; Chaudhuri, S.; Shinnar, E.; Cromer, S.; Perel, P.; Mensah, G.A.; Narula, J.; Johnson, C.O.; et al. Global Atlas of Cardiovascular Disease 2000–2016: The Path to Prevention and Control. *Glob. Heart* **2018**, *13*, 143–163. [CrossRef] [PubMed]
11. Virani, S.S.; Alonso, A.; Benjamin, E.J.; Bittencourt, M.S.; Callaway, C.W.; Carson, A.P.; Chamberlain, A.M.; Chang, A.R.; Cheng, S.; Delling, F.N.; et al. Heart Disease and Stroke Statistics—2020 Update: A Report from the American Heart AsSoCiation. *Circulation* **2020**, *141*, e139–e596. [CrossRef] [PubMed]
12. Timmis, A.; Townsend, N.; Gale, C.; Grobbee, R.; Maniadakis, N.; Flather, M.; Wilkins, E.; Wright, L.; Vos, R.; Bax, J.J.; et al. European SoCiety of Cardiology: Cardiovascular Disease Statistics 2017. *Eur. Heart J.* **2018**, *39*, 508–579. [CrossRef] [PubMed]
13. Barriada, R.; Simó-Servat, O.; Planas, A.; Hernández, C.; Simó, R.; Masip, D. Deep Learning of Retinal Imaging: A Useful Tool for Coronary Artery Calcium Score Prediction in Diabetic Patients. *Appl. Sci.* **2022**, *12*, 1401. [CrossRef]
14. Ciușdel, C.; Itu, L.; Cimen, S.; Wels, M.; Schwemmer, C.; Fortner, P.; Seitz, S.; Andre, F.; Buß, S.; Sharma, P.; et al. Normalizing Flows for Out-of-Distribution Detection: Application to Coronary Artery Segmentation. *Appl. Sci.* **2022**, *12*, 3839. [CrossRef]
15. Hatfaludi, C.; Tache, I.; Ciușdel, C.; Puiu, A.; Stoian, D.; Itu, L.; Calmac, L.; Popa-Fotea, N.; Bataila, V.; Scafa-Udriste, A. Towards a Deep-Learning Approach for Prediction of Fractional Flow Reserve from Optical Coherence Tomography. *Appl. Sci.* **2022**, *12*, 6964. [CrossRef]
16. Mungloo-Dilmohamud, Z.; Heenaye-Mamode Khan, M.; Jhumka, K.; Beedassy, B.; Mungloo, N.; Peña-Reyes, C. Balancing Data through Data Augmentation Improves the Generality of Transfer Learning for Diabetic Retinopathy Classification. *Appl. Sci.* **2022**, *12*, 5363. [CrossRef]
17. Yu, T.; Hu, H.; Zhang, X.; Lei, H.; Liu, J.; Hu, W.; Duan, H.; Si, J. Real-Time Multi-Label Upper Gastrointestinal Anatomy Recognition from Gastroscope Videos. *Appl. Sci.* **2022**, *12*, 3306. [CrossRef]
18. Mobark, N.; Hamad, S.; Rida, S. CoroNet: Deep Neural Network-Based End-to-End Training for Breast Cancer Diagnosis. *Appl. Sci.* **2022**, *12*, 7080. [CrossRef]
19. Zhang, D.; Ren, A.; Liang, J.; Liu, Q.; Wang, H.; Ma, Y. Improving Medical X-ray Report Generation by Using Knowledge Graph. *Appl. Sci.* **2022**, *12*, 11111. [CrossRef]

Disclaimer/Publisher's Note: The statements, opinions and data contained in all publications are solely those of the individual author(s) and contributor(s) and not of MDPI and/or the editor(s). MDPI and/or the editor(s) disclaim responsibility for any injury to people or property resulting from any ideas, methods, instructions or products referred to in the content.

Article

Obfuscation Algorithm for Privacy-Preserving Deep Learning-Based Medical Image Analysis

Andreea Bianca Popescu [1,2,*], Ioana Antonia Taca [1,3], Anamaria Vizitiu [1,2], Cosmin Ioan Nita [2], Constantin Suciu [1,2], Lucian Mihai Itu [1,2] and Alexandru Scafa-Udriste [4,5]

1. Advanta, Siemens SRL, 500097 Brasov, Romania; ioana_antonia29@yahoo.com (I.A.T.); anamaria.vizitiu@siemens.com (A.V.); suciu.constantin@siemens.com (C.S.); lucian.itu@siemens.com (L.M.I.)
2. Department of Automation and Information Technology, Transilvania University of Brașov, 500174 Brasov, Romania; nita.cosmin.ioan@unitbv.ro
3. Department of Mathematics and Computer Science, Transilvania University of Brașov, 500091 Brasov, Romania
4. Department of Cardiology, Emergency Clinical Hospital, 8 Calea Floreasca, 014461 Bucharest, Romania; alexscafa@yahoo.com
5. Department Cardio-Thoracic, University of Medicine and Pharmacy "Carol Davila", 8 Eroii Sanitari, 050474 Bucharest, Romania
* Correspondence: andreea.popescu.ext@siemens.com

Citation: Popescu, A.B.; Taca, I.A.; Vizitiu, A.; Nita, C.I.; Suciu, C.; Itu, L.M.; Scafa-Udriste, A. Obfuscation Algorithm for Privacy-Preserving Deep Learning-Based Medical Image Analysis. *Appl. Sci.* **2022**, *12*, 3997. https://doi.org/10.3390/app12083997

Academic Editor: Jan Egger

Received: 24 March 2022
Accepted: 13 April 2022
Published: 14 April 2022

Publisher's Note: MDPI stays neutral with regard to jurisdictional claims in published maps and institutional affiliations.

Copyright: © 2022 by the authors. Licensee MDPI, Basel, Switzerland. This article is an open access article distributed under the terms and conditions of the Creative Commons Attribution (CC BY) license (https://creativecommons.org/licenses/by/4.0/).

Abstract: Deep learning (DL)-based algorithms have demonstrated remarkable results in potentially improving the performance and the efficiency of healthcare applications. Since the data typically needs to leave the healthcare facility for performing model training and inference, e.g., in a cloud based solution, privacy concerns have been raised. As a result, the demand for privacy-preserving techniques that enable DL model training and inference on secured data has significantly grown. We propose an image obfuscation algorithm that combines a variational autoencoder (VAE) with random non-bijective pixel intensity mapping to protect the content of medical images, which are subsequently employed in the development of DL-based solutions. A binary classifier is trained on secured coronary angiographic frames to evaluate the utility of obfuscated images in the context of model training. Two possible attack configurations are considered to assess the security level against artificial intelligence (AI)-based reconstruction attempts. Similarity metrics are employed to quantify the security against human perception (structural similarity index measure and peak signal-to-noise-ratio). Furthermore, expert readers performed a visual assessment to determine to what extent the reconstructed images are protected against human perception. The proposed algorithm successfully enables DL model training on obfuscated images with no significant computational overhead while ensuring protection against human eye perception and AI-based reconstruction attacks. Regardless of the threat actor's prior knowledge of the target content, the coronary vessels cannot be entirely recovered through an AI-based attack. Although a drop in accuracy can be observed when the classifier is trained on obfuscated images, the performance is deemed satisfactory in the context of a privacy–accuracy trade-off.

Keywords: image obfuscation; deep learning; medical imaging; privacy preserving classification

1. Introduction

In the last decade, machine learning (ML) algorithms have demonstrated remarkable results in potentially improving the performance and the efficiency of healthcare applications. A recent study [1] provides an overview of the benefits that machine learning brings in healthcare, including aiding doctors in their decision making, and decreasing the cost and time it takes to reach a diagnosis. Even though such solutions allow for better resource allocation and treatment selection, they are challenging to implement in real-world circumstances due to several obstacles. The same study emphasizes that one of the most significant problems is the massive amount of high-quality data that are frequently necessary to create and evaluate machine learning models.

A related issue is the ethical aspect of data collection, which necessitates data sourcing for ML, to comply with personal information protection and privacy regulations [2]. The GDPR establishes precise permission standards for data uses in Europe, whereas the HIPAA regulates healthcare data from patient records in the United States. These laws are considerably more challenging to fulfill when clinical users prefer to delegate ML model development and deployment to third parties, and use them via cloud services, e.g., due to a lack of hardware capabilities. According to a recent survey [3], the Machine Learning as a Service (MLaaS) paradigm has appeared as a highly scalable approach for remotely running predictive models, raising at the same time increased security and privacy concerns. The same paper highlights that fully homomorphic encryption (HE) could be a straightforward approach that allows a third party to process encrypted data without knowing its content.

An early effort that combined HE with neural networks, involving the communication between the model owner and the data provider, is described in [4]. CryptoNets [5] eliminates this interaction, but it has the drawback that the encryption technique does not process real numbers. CryptoDL [6] approximates nonlinear functions with low-degree polynomials to overcome model complexity restrictions. However, the use of estimated activation functions reduces the prediction accuracy of the model. More recent studies propose different approaches to increase the classification accuracy at the inference phase in AI-based models employing homomorphic encryption. In [7], adopting a polynomial approximation of Google's Swish activation function, and applying batch normalization, enhanced classification performance on the MNIST and CIFAR-10 datasets. Additional optimizations are performed to reduce the consumption level. J.W. Lee et al. [8] emphasize that the most common activation functions are non-arithmetic functions (ReLU, sigmoid, leaky ReLU), which are not suited for homomorphic computing, because most HE schemes only enable addition and multiplication. They evaluate these non-arithmetic functions with adequate precision using approximation methods. In combination with multiple methods for reducing rescaling and relinearization errors, the bootstrapping strategy enables a deep learning model to be evaluated on encrypted data. According to the numerical verification, the ResNet-20 model produced equivalent results on the CIFAR-10 dataset for both encrypted and unencrypted data. The efficiency of MLaaS is drastically improved in [9], where GPUs acceleration is used to evaluate a pre-trained CNN on encrypted images from MNIST and CIFAR-10 datasets. None of the above-mentioned methods addresses the training phase of models on encrypted data due to the increased number of operations and the longer runtime, this being regarded as an open problem, especially in the case of image-based datasets. For privacy-preserving computations within deep learning models, we suggested a variant of a noise-free matrix-based homomorphic encryption method (MORE [10]) in our earlier work [11]. We validated the methodology using two medical data collections in addition to the MNIST dataset. The encryption step is employed during both training and inference. The experiments showed that the method provides comparable results to those obtained by unencrypted models, while having a low computational overhead. However, the changes made to the original HE scheme to allow computations on rational numbers come at a cost in terms of privacy, as it provides lower security than standard schemes. This method was further used in [12] to design a cloud-based platform for deploying ML algorithms for wearable sensor data, focused on data privacy. We have further addressed the security compromise in [13], where we combined a HE scheme based on modulo operations over integers [14], an encoding scheme that enables computations on rational numbers, and a numerical optimization strategy that facilitates training with a fixed number of operations. Nevertheless, the computational overhead introduced through encoding and encryption represents a significant drawback of the method.

The comprehensive survey [3] includes theoretical concepts, state-of-the-art capabilities, limits, and possible applications for more privacy-preserving machine learning (PPML) solutions based on HE. An overview of techniques based on other privacy-preserving primitives such as multi-party computation (MPC), differential privacy (DP) and federated learning (FL) is provided in [15]. The authors underline that a hybrid PPML system could feasibly imply a trade-off between ML performance and computational overhead.

Another privacy-preserving approach that has received increasing interest is image obfuscation. In the context of PPML, it entails modifying the image so that the content becomes unintelligible while retaining the underlying information to some extent. Obfuscation methods such as mosaicing, blurring and P3 are analyzed in [16]. Mosaicing is used to alter parts of an image inside a window whose size is inversely related to obfuscated image resolution. Blurring applies a Gaussian filter that removes details from images. Despite the fact that mosaicing and blurring make it impossible for the human eye to detect faces or digits in obfuscated images, the authors show that standard image recognition models can extract useful information from the transformed data. The strategy suggested in [17] uses Gaussian noise to obscure only a few images in the dataset (which are considered to have a sensitive content). The authors emphasize that this method could affect the model performance if too many frames require protection.

The obfuscation techniques described in [18] are variations on the mixup approach, which entails creating convex combinations of pairs of samples. The proposed approaches aim to improve the privacy of the training data, while optimizing the model accuracy without increasing the computational cost of the training process. The presented methods are variants of the mixup technique, which entails creating convex combinations of pairs of samples. After mixing, the newly created sample is further obfuscated through pixel grafting, pixel shuffling, noise addition or blurring. In the same research, authors demonstrate that metrics like SSIM (structural similarity index measure) and HaarPSI (Haar wavelet-based perceptual similarity index), which accord with human perception on picture degradation, may be used for privacy assessment. Two datasets that contain images depicting animals were used to validate the methods. The results highlight that a compromise between obfuscation and learning capabilities must always be considered. The Google Vision AI image classifier was queried with obfuscated images, and its recognition performance was lower than that of the human evaluators. Kim et al. [19] performed an interesting study focused on privacy-preservation for medical image analysis. They proposed a client-server system in which the client protects the patient identity by deforming the input image using a system comprising a transformation generator, a segmentation network, and a discriminator. The system is trained in an end-to-end adversarial manner to solve the task of MRI brain segmentation. Being focused on enabling protection against facial recognition, the approaches presented in [20,21] leverage generative adversarial networks to produce more visually pleasing outputs, while providing a solid defense against deep learning-based recognition systems. In [21], for the analyzed scenarios, the trade-off is formulated based on the privacy against face recognition versus the utility in terms of face detection.

Herein, we propose an obfuscation technique that combines variational autoencoders with non-bijective functions. The aim is to achieve a method that enables accurate model training, while ensuring privacy against human eye perception and AI-based reconstruction attacks. The experiments are constructed to reflect the perspective of a clinical user (e.g., hospital) in a specific use case (coronary angiography view classification), and the perspective of a threat actor. Because the hospital lacks the physical resources and the expertise to develop a DL classification model, the inference is performed by a third party, which is considered untrustworthy. In this scenario, this external party is a Machine Learning as a Service (MLaaS) provider who can train a DL model using the clinical data, and then make it available as a cloud service for inference. Since the patient data is considered to be sensitive and private, every angiographic frame used for training or inference is obfuscated to protect data privacy outside of the clinical environment. Conversely, a potential threat actor, that could be the MLaaS provider or an interceptor, may try to acquire illegal access to the clinical data. The considered attack strategy is based on the training of a reconstruction model on original-obfuscated pairs of samples from a public dataset. Because the obfuscation method is considered publicly available as a black-box tool for collaborative purposes, any external entity can use the tool to obfuscate images and obtain a dataset of corresponding image pairs. Two possible attack configurations are formulated. In the first one, the threat actor knows the data source (i.e., hospital) but is unaware of its specific type (coronary angiography, in our case), hence the training is performed on a public dataset containing medical-related samples. Another possibility is that the attacker is a collaborative hospital which knows that the target dataset consists of coronary angiographies, and which trains the reconstruction model on its own angiographic data.

All parties other than the hospital are regarded as untrustworthy in terms of data security, and, in consequence, every externalized angiographic frame is, in fact, an obfuscated image. Even the rightful receiver, in this case the MLaaS provider, is not considered honest regarding data confidentiality, which is why the proposed obfuscation method aims to be irreversible. The goal is to protect the medical images from a highly resourceful entity (in terms of both computer power and data), while allowing for the training of the desired deep learning model directly on the altered images.

The remainder of the paper is organized as follows. The obfuscation techniques, as well as the network architectures, datasets, and procedures for the suggested use case, are presented in Section 2. Section 3 describes the experiments performed from the perspectives of the clinical user and the threat actor, along with the findings. In Section 4, we iterate through the unique characteristics of the proposed technique, present remarks regarding its usefulness in deep learning-based applications, and finally draw the conclusions.

2. Methods and Materials

In the following, we propose a novel strategy that combines two obfuscation approaches to:
1. Hide the content of a sensitive image from the human eye;
2. Make AI-based image reconstruction challenging;
3. Facilitate DL model training using obfuscated images.

The first stage is to train a variational autoencoder, which uses the original (non-obfuscated) dataset as both input and target, and provides an obfuscated counterpart for each sample at the bottleneck. A detailed description of the VAE architecture, training and obfuscation process is presented in Section 2.1. The next step is also described as a stand-alone method in Section 2.2, where every pixel intensity value is randomly translated to another intensity value in a non-bijective manner, to alter the visual information. When the techniques are used in conjunction, the image encoded with the VAE is further obfuscated through pixel substitution, according to a non-bijective mapping function. The entire workflow is detailed in Section 2.3. The clinical usage scenario, the dataset, and the architecture used to solve the classification task are presented in Section 2.4. Section 2.5 describes the procedures employed to evaluate the privacy level provided by the proposed approach against human perception and against AI-based reconstruction attacks.

2.1. Obfuscation Method Based on a Variational Autoencoder

The Variational Autoencoder [22] considered herein is a generative model based on the work of Kingman et al. [23]. It consists of two models that support each other: an encoder (recognition model) and a decoder (generative model). The difference between VAEs and other AEs is that the input is not encoded as a single point, but as a distribution over the latent space, from which the decoder draws random samples. Due to the reparameterization trick, which allows for backpropagation through the layers, the two components of the VAE can be chosen to be (deep) neural networks.

The autoencoders, and by extension VAEs, generate an encoding of the inputs that allow for an accurate reconstruction. This property also ensures that the encoding contains useful information extracted from the input, and, hence, it can be employed in further DL-based analysis or model training, e.g., within an obfuscation method based on VAE.

From a probabilistic perspective, a VAE implies approximate inference in a latent Gaussian model, where the model likelihood and the approximate posterior are parameterized by neural networks. The recognition model compresses the input data x into a dimensionally reduced latent space χ, while the generative model reconstructs the data given the hidden representation $z \in \chi$. Let us denote the encoder $q_\theta(z|x)$ and the decoder $p_\phi(x|z)$, where θ and ϕ represent the neural network parameters.

The latent variables $z \in \chi$ are considered to be drawn from a simple distribution: $p(z) = \mathcal{N}(0, I)$, named prior (here, I denotes the identity matrix). The input data x have a likelihood $p(x|z)$ that is conditioned on z. As a result, a joint probability distribution over data and latent variables can be defined:

$$p(x, z) = p(x|z)p(z). \quad (1)$$

The aim is to calculate the posterior distribution $p(z|x)$. This can be achieved by applying Bayes' rule:

$$p(z|x) = \frac{p(x|z)p(z)}{p(x)}, \quad (2)$$

where $p(x)$ can be obtained by marginalizing out z: $p(x) = \int p(x|z)p(z)dz$. Unfortunately, the integral is usually intractable [24]. As a consequence, an approximation of this posterior distribution is required.

There are two main ways for posterior approximation: applying Markov Chain Monte Carlo (MCMC) methods such as the Metropolis–Hastings algorithm [25] or Gibbs sampling [26], and variational inference (VI) [27]. VAE uses the latter because the sampling methods converge slower [28]. This approach implies approximating the posterior with a family of Gaussian distributions $q_\lambda(z|x)$, where parameters λ represent the mean and the variance of each hidden representation. As a result, the encoder parameterizes the approximate posterior $q_\theta(z|x, \lambda)$, taking x as input data, and parameters λ as outputs. On the other hand, the decoder parameterizes the likelihood $p(x|z)$, having the latent variables as input and the parameters to distribution $p_\phi(x|z)$ as output. The approximation is penalized by computing the Kullback–Leibler (KL) divergence that measures the distance between $q_\theta(z|x, \lambda)$ and $p(z)$.

Hereupon, the loss function which is minimized during training is composed of two terms: (i) the reconstruction error between input data x and output data x', and (ii) the KL divergence between the approximate posterior and $p(z)$, chosen to be a normal distribution:

$$Loss = \mathcal{L}(x, x') + KL(q_\theta(z|x, \lambda) || p(z)). \quad (3)$$

The first step of our method is to train a convolutional VAE on another dataset from the same domain as the working dataset. Additionally, one of the layers is used for noise addition. At the bottleneck, the information is divided between two channels to obtain an encoded version of the input. Those channels correspond to the mean (channel 1) and standard deviation (channel 2) of the normal distribution obtained from the encoder. Any of the channels can then be used for a subsequent DL model training on obfuscated images. From the trained VAE, only the encoder is retained as a black-box obfuscation tool. As there is no need for a reconstruction once an image has been obfuscated, the decoder is discarded. Figure 1 displays the workflow described for the obfuscation method based on a VAE.

For our experiments, the VAE is trained on the Medical MNIST dataset [29]. The dataset contains 6 classes of X-ray images, that are randomly distributed for training (30,000 images) and validation (12,000). More details about the Medical MNIST dataset are presented in Section 2.5.

During training, the 64 × 64 images are passed through three convolutional layers of 32, 8, and 4 filters, respectively, with a 3 × 3 receptive field. ReLU is the activation function chosen for each layer. The architecture of the decoder consists of three convolutional, ReLU activated layers of 4, 8, and 32 filters, followed by one dense layer. The VAE is trained for 10 epochs.

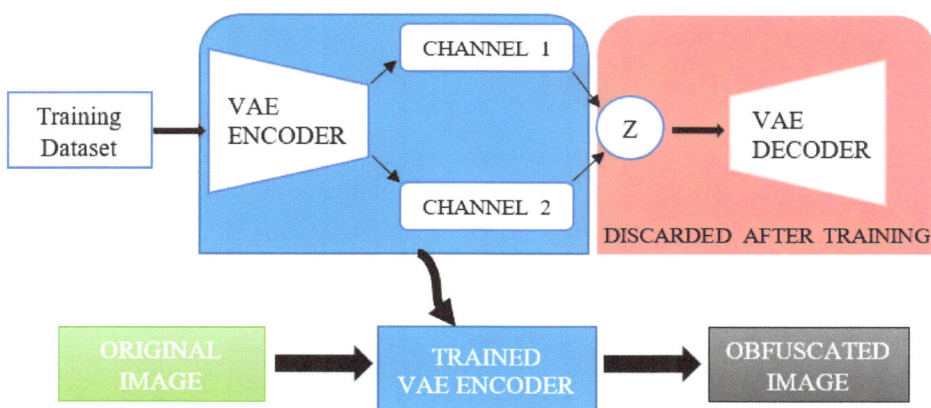

Figure 1. Schematic representation of the obfuscation technique based on a VAE.

The trained encoder can be used for obfuscating medical images. A channel option must be selected, depending on the desired result. The first channel, corresponding to the mean of the normal distribution, usually assures a better privacy level than the second channel, as it does not preserve as much detailed information from the input. This limits, though, its usefulness in further AI-algorithms. The channel corresponding to the standard deviation of the normal distribution tends to preserve more useful information from the original image. As a result, it is preferred in cases where the obfuscated images would be used in machine learning tasks. This channel, although depending on the initial structure of the original image, may or may not ensure the imposed or desired level of privacy. For example, in the encoding of an image with a monochromatic background, most probably sensitive details will be visible, which could uncover the nature of the original image. Such an example is shown in Figure 2, where the original image, representing a coronary angiography, has an almost monochromatic background. As a result, in the image obtained from channel 2, the main vessel can be seen.

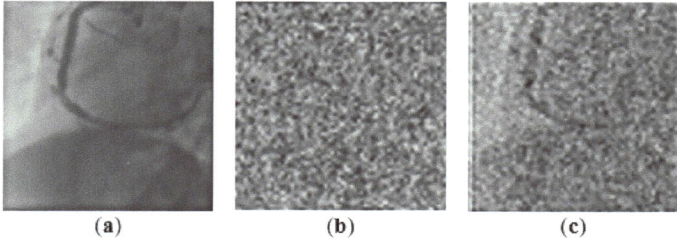

Figure 2. Comparison between the original frame (**a**) and the obfuscated counterparts when channel 1 (**b**) and channel 2 (**c**) are chosen.

2.2. Obfuscation Based on Non-Bijective Pixel Intensity Shuffling

This approach starts with a simple obfuscation technique—random pixel intensity shuffling. Every pixel intensity is randomly associated with another value from the same interval as described by Equation (4), where $range(a, b)$ is a function that returns all integer numbers between a and b, including the interval's endpoints, and $shuffle(x)$ is a function that randomly interchanges the positions of the elements of a list x inside the returned array. We call this array a map because it creates connections between each possible pixel intensity embodied in the list of indexes of the array and a new random value contained in the array at the corresponding position.

$$intensityMap = shuffle(range(0, 255))\qquad(4)$$

This association is a bijective function because for each domain component there is only one corresponding element in the codomain. Although this operation preserves the underlying information of the images, while making them unrecognizable for the human eye, the approach is still susceptible to AI-based attacks, statistical or even reverse engineering attacks. Presuming that an external party has access to the obfuscation algorithm in a black-box form, an unlimited number of new images can be obfuscated, and a statistical evaluation should reveal that a one-to-one mapping was used. By reversing this mapping, a potential attacker can obtain the original images with no information loss. Training a deep learning model to reconstruct the obfuscated images is another attack approach. In anticipation of this kind of attack, a second step is proposed for this obfuscation method. The bijective function is modified so that the injectivity property is lost. In other words, multiple elements of the domain will correspond to the same element of the codomain. This effect is achieved by applying the same $mod\ N$ operation on each value of the previously obtained map. Hence, the obfuscation method can be defined by a function $f : A \to B$, where $A = [0, 255]$ and $B = [0, N)$. When obfuscating an image, an iteration across all pixels must be performed. In Equation (5), pv denotes the intensity of the pixel found at the (i, j) coordinates in the $image$ matrix.

$$pv = image_{i,j}\qquad(5)$$

This value is modified according to Equation (6), where the mod function represents the typical modulo operation and the pv value is used as an index.

$$image_{i,j} = intensityMap_{pv}\ mod\ N\qquad(6)$$

Figure 3 synthesizes the steps proposed for this obfuscation technique. The key concept is that applying a $mod\ N$ operation limits the range of possible values to N elements. However, this is not equivalent to filtering the highest intensities due to the previously performed random associations. Thus, more details are preserved in images by arbitrary but consistent replacement of $256 - N$ pixel intensities. Since the obfuscation function is represented by a many-to-one mapping, the task of reconstructing unseen images becomes more complex and more uncertain, even for an AI-based model trained on original-obfuscated image pairs.

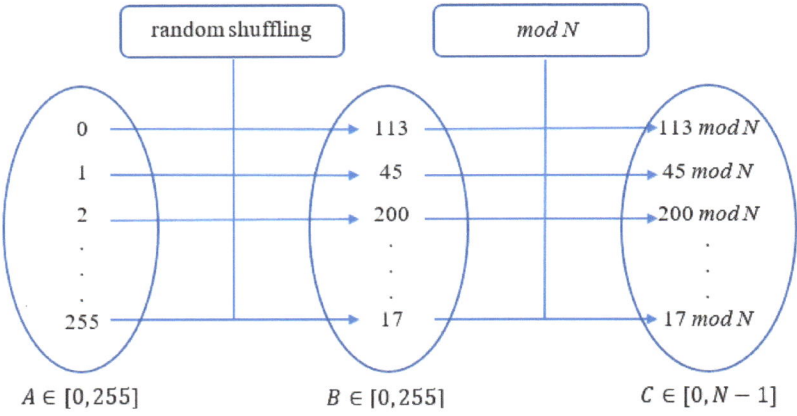

Figure 3. Schematic representation of the obfuscation technique based on pixel intensity shuffling.

The N value is an adjustable parameter that improves security when being set to lower values. As a function of this parameter, the underlying information is preserved in different degrees, presumably retaining enough details in the images for DL-based applications. Figure 4 displays a comparison between an angiographic frame Figure 4a and the obfuscated counterparts when bijective Figure 4b or non-bijectiveFigure 4c–e mapping is applied. The obfuscated samples are rescaled in $[0, 255]$ interval to allow a better visual comparison.

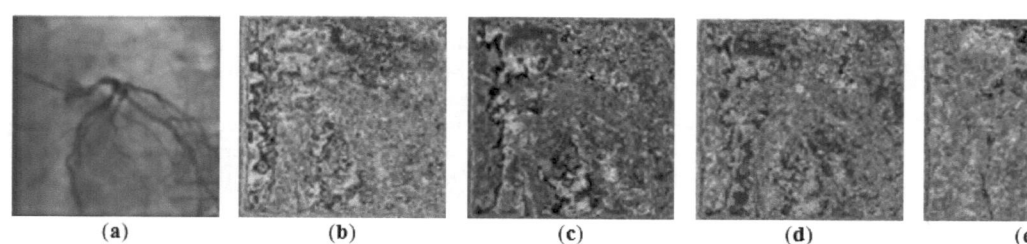

(a)   (b)   (c)   (d)   (e)

Figure 4. Comparison between the original frame (**a**) and the obfuscated counterparts when $N = 256$ (**b**), $N = 156$ (**c**), $N = 50$ (**d**) and $N = 45$ (**e**).

2.3. Secure Obfuscation Algorithm

As previously explained, the security of VAE obfuscation also depends on the image itself. For images with a uniform distribution of pixel intensities, the method will not only protect the content from the human eye perception but, due to the additional noise, also make it more difficult for an AI-based model to reconstruct the original image. In contrast, the human eye would be able to discern the environment from the main structures, or even details of the structures, in a dichromatic image where two predominant intensities describe the object and the background. The noise level can vary, but this would also affect the utility of the image. Using a non-bijective function to substitute the intensities makes the obfuscated images unrecognizable by the human eye. Although the modulo operation is meant to protect against more sophisticated attacks, the success rate of an AI-based reconstruction attack depends on the value of N. The smaller this parameter is, the more difficult the reconstruction becomes. However, this implies a trade-off between privacy and utility. We integrate the strengths of each method into a new obfuscation algorithm to maximize their effectiveness. The steps are as follows, in the order in which they should be performed :

1. The VAE model is trained on images similar to those that will be obfuscated in the clinical use case.
2. All pixel intensities are randomly shuffled, and a *modulo N* operation is performed on each resulting value leading to a non-bijective mapping between different intensities.
3. The original image is encoded using the VAE encoder.
4. Each pixel value of the encoded image is substituted with the corresponding value in the non-bijective map.

As a result, an obfuscated image is created, which retains the original image's underlying relevant information and can be used for further analysis and processing (e.g., image classification). Regardless of the initial structure of an image, combining the techniques improves privacy. First, the eye perception is affected by the intensity shuffling even if, after encoding, the sensitive content is still distinguishable. Then, the protection against AI-based reconstruction attacks is ensured by the conjunction of noise and non-bijectivity. The entire obfuscation workflow is schematically depicted in Figure 5.

Although the underlying information of an image is preserved using this technique, an essential requirement that must be met to use multiple images in the same application (e.g., training a classifier on obfuscated images) is that the same encoder and the same shuffling map should be applied on all images (both for training and inference). The trade-off between privacy and utility can be managed by tuning certain method-specific parameters according to the needs of the use case. For the technique based on non-bijective intensity mapping, the choice of parameter N may influence the image utility. Regarding confidentiality, a higher N implies less information retained in the obfuscated image and, thus, a more difficult to perform image reconstruction. Figure 6 displays an original angiographic frame and the obfuscated counterparts for each obfuscation approach. The chosen value for the modulo operator N in Figure 6c is 96. More examples are included in Appendix A, Figure A1.

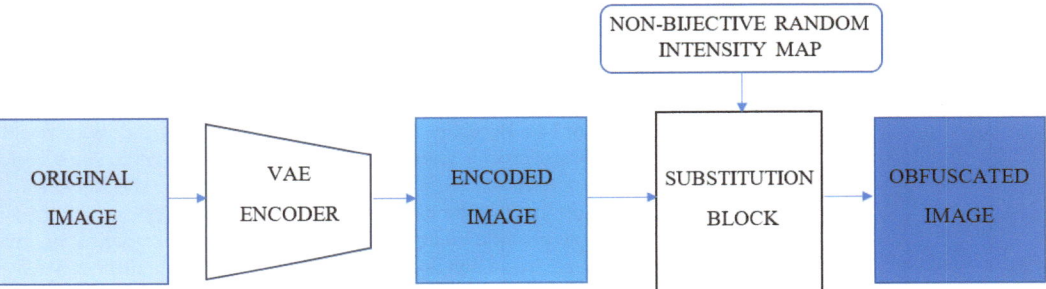

Figure 5. Schematic representation of the secure obfuscation algorithm.

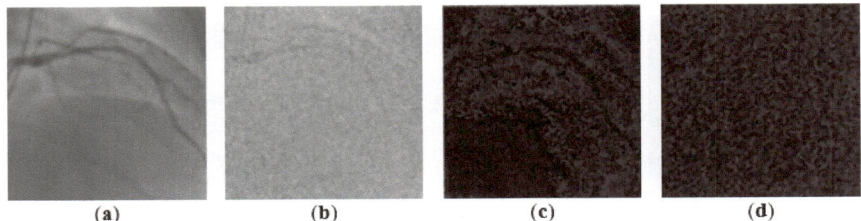

Figure 6. Comparison between an original angiographic frame (**a**) and the obfuscated counterparts when using (**b**) encoding, (**c**) non-bijective intensity mapping, and (**d**) combined algorithm.

2.4. Utility Level Evaluation

As the methods described above rely on reducing, to a certain degree, the information from the original images, their utility after the obfuscation must be evaluated. To perform this analysis, the same DL model is trained for multiple levels of obfuscation, including no obfuscation. The methods presented in Sections 2.1 and 2.2 are employed separately and in conjunction, as described in Section 2.3, to obfuscate an in-house dataset consisting of coronary angiography frames. The same experiment is run for multiple values of N, ranging between 1 and 255. The utility of obfuscated images is determined by comparing the accuracy achieved on a testing dataset for different degrees of obfuscation.

The task is to train a binary classifier to distinguish between RCA and LCA views in angiographic frames. Figure 7 depicts one sample of each category. The dataset contains 3280 coronary angiographies, balanced between the two classes. A subset of 600 images is used for validation, and another subset of 700 images is retained for evaluation purposes. The rest of the 1980 angiographic frames are used for training. Augmentation techniques such as shifting, flipping, zooming and rotation are applied. The original size of the frames is 512×512 pixels, but experiments with different input shapes have shown that a size of 128×128 ensures almost no loss in classification performance with a lower computational time. The pixels values are normalized through min-max scaling in the $[0, 1]$ range.

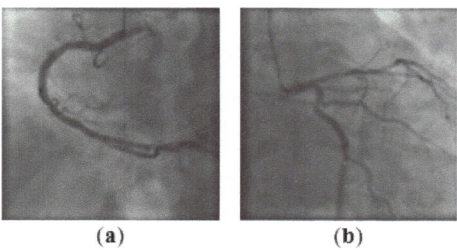

Figure 7. RCA—right coronary artery (**a**) and LCA—left coronary artery (**b**).

The images (obfuscated or not) are passed through four convolutional layers of 16 and 32 filters with a 3 × 3 receptive field during training. The pooling layers downsample the images by a factor of two by using the maximum value of a window. After the last convolutional layer, a flatting layer is added to convert the features matrix into a vector. The fully connected layers contain 512, 1024 and 2 nodes, respectively. The ReLU function is employed as an activation function for all layers, except for the last one where the softmax activation is used. Each convolutional layer is followed by a local normalization layer [30] to make the model more robust to image degradation. To limit the overfitting, between 25% and 50% of the connections of the neurons are dropped through dropout layers. Furthermore, although the maximum number of epochs is set to 30, early stopping is employed when the validation loss is not decreasing within 10 consecutive epochs. A learning rate scheduler is used to achieve good convergence, starting from 1×10^{-3}, and diminishing the value with every epoch. The workflow of an inference step using the obfuscation algorithm is depicted in Figure 8.

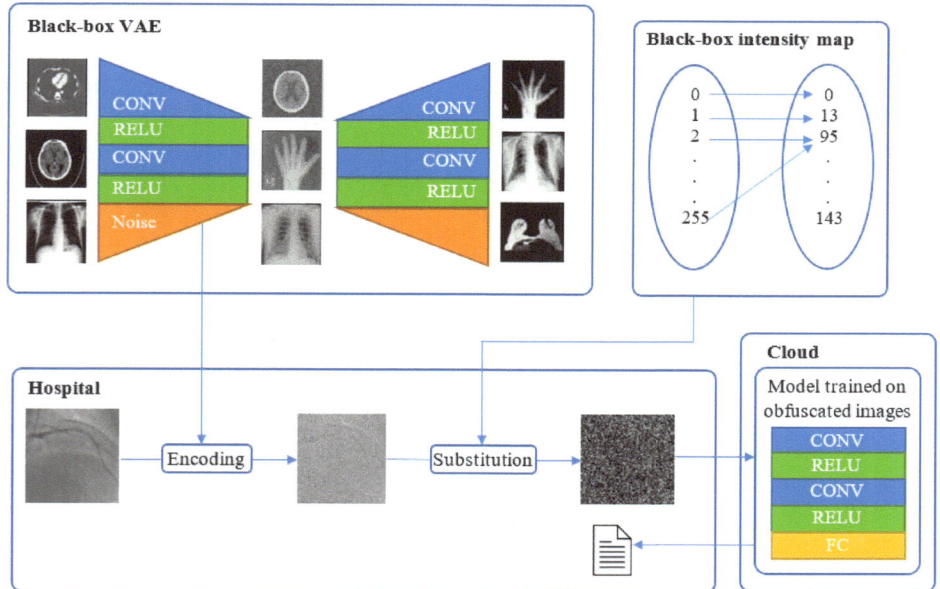

Figure 8. Detailed workflow of inference using the secure obfuscation algorithm.

The Keras framework [31] was used to build the convolutional neural network, and the local normalization layer is based on [30]. The experiments were run on a computer equipped with an Intel i7 CPU (Intel, Santa Clara, CA, USA) at 4.2 GHz, 32 GB RAM and an NVIDIA GeForce GTX 1050 Ti GPU (Nvidia, Santa Clara, CA, USA) with 4 GB of dedicated memory.

2.5. Privacy Level Evaluation

To compare the degree of privacy provided by each proposed technique, we employ similarity metrics such as SSIM and PSNR (peak signal-to-noise-ratio) assessed between the original and the corresponding obfuscated images. As stated in [18], SSIM is an image quality metric that can quantify image privacy. It considers perceptual phenomena like brightness and contrast, as well as structural information changes. SSIM can take values between 0 and 1, where 0 means no structural similarity, and 1 indicates identical images. Therefore, lower values correspond to an increased security. PSNR is expressed using the decibel scale, and typical values for good quality images (with a bit depth of 8) are between 30 and 50 dB. As a result, values below the lower threshold indicate that the image is protected against human perception. The entire testing subset owned by the hypothetical clinical user is employed for this evaluation. The averaged results are presented in Section 3.3.

Two possible attack configurations are considered to assess the level of security against AI-based reconstruction. The considered scenario is that of an external party willing to access the original data sent by the hospital or by a specific patient. The general assumption is that the obfuscation algorithm used by the hospital is publicly available as a black-box tool. The privacy parameter N is also presumed to be known. This means that another clinical user or an MLaaS provider, or even an external interceptor can use the tool to obfuscate images and obtain a dataset of corresponding image pairs. Moreover, because the data source is known, the threat actor might guess that the dataset consists of medical images. The workflow of an entity willing to gain unauthorized access to the data has the following steps: obfuscating a dataset of medical images using the same obfuscation tool as the hospital, training a deep learning model to reconstruct the original frames from the obfuscated images, intercepting obfuscated images, and reconstructing the original images using the previously trained model.

In the first attack configuration, the interceptor assumes that the targeted data contains medical images, but is unaware of their type (E_1); therefore, the malicious actor trains the reconstruction model using a publicly available dataset with different medical-related classes. In the following experiments (see Section 3.3), the reconstruction model is trained using the Medical MNIST dataset [29]. It contains six classes of X-ray images (abdomen CT, breast MRI, CXR, chest CT, hand radiography, head CT), each class totalling around 7000 samples. All 40,954 medical images are used for training, and the evaluation is performed on the intercepted obfuscated dataset. The Medical MNIST images have a size of 64×64 pixels, but they are resized to 128×128, the dimensions of the frames sent by the hospital. Figure 9 depicts a sample of each category of the Medical MNIST dataset.

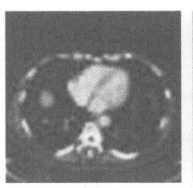
(a)
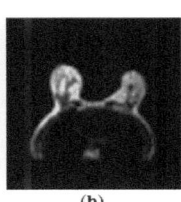
(b)
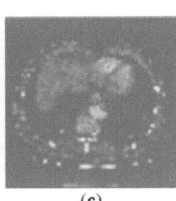
(c)
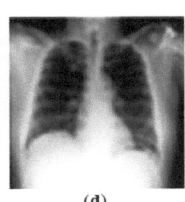
(d)
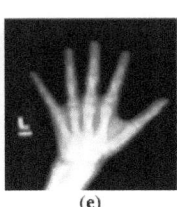
(e)
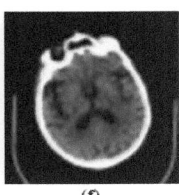
(f)

Figure 9. Medical MNIST samples: abdomen CT (**a**), breast MRI (**b**), CXR (**c**), chest CT (**d**), hand radiography (**e**), head CT (**f**).

Another possibility is that the type of the medical images is well known, so a similar dataset is used to train the reconstruction model (E_2). For example, two clinical partners want to create an aggregated dataset containing coronary angiographies for training a view classification model, but they both wish to keep their data confidential. However, one of the partners is willing to obtain the content provided by the other. As they both use the same obfuscation tool, the threat actor obfuscates his angiographic dataset, and uses it to train a reconstruction model. Then, the malicious actor intercepts the obfuscated frames of the victim, and tries to undo the obfuscation. The (in-house) dataset used in these experiments contains 8365 angiographies (5779 LCA and 2586 RCA), all employed for training. Their original size (512 × 512) is modified to 128 × 128.

Before training, both the inputs (obfuscated images) and the targets (original images) are normalized through min-max scaling in the [0, 1] interval. The U-Net architecture introduced in [32] is employed for reconstruction. The first half of the network, which behaves like an encoder, consists of convolutional and pooling layers that perform downsampling. Each decoder block combines its input with information from the corresponding encoder block, and performs convolutional and upsampling operations. The same activation function, number of filters, kernel size, pooling window and stride as in the original paper were used. The batch size and momentum values were set to 1 and 0.99, respectively. The model was trained for 30 epochs with a learning rate of 0.001. The architecture was implemented in the PyTorch framework [33], and the models were trained on a machine equipped with 128 GB RAM and NVIDIA GeForce GTX 1080 Ti GPU with 11 GB of dedicated memory.

The reconstruction network was trained on images obfuscated using the methods described in Sections 2.1 and 2.2, and the algorithm presented in Section 2.3 for multiple values of the parameter N. To determine the degree of similarity between the reconstructed images and the original counterparts, SSIM and PSNR are computed across all frames sent by the victim (the training dataset of the classifier). Considering the threshold values of SSIM, in the results presented in Section 3.3, a lower SSIM value denotes a poor reconstruction performance and a high privacy level. Regarding the interpretation of PSNR, in the following experiments, values under 30 indicate inaccurate reconstruction and high security. The scikit-image library [34] was employed for computing the similarity metrics.

Expert readers manually performed a visual assessment to determine to what extent the reconstructed images are protected against human perception. The assessment was performed on 50 frames (25 LCA, 25 RCA). Since in most cases the background was reconstructed more accurately than the arteries, two separate scores were assigned for each image. A scale from 1 to 5 was chosen, where 1 indicates that the object was not reconstructed at all and 5 denotes a visual similarity larger than 95%. Some scoring guidelines were formulated to limit the evaluation bias. Tables 1 and 2 synthesize the links between scores and image descriptions.

Figures 10 and 11 display for each score an evaluation example corresponding to the scoring guidelines. The mean scores are computed for all evaluations of all frames. The LCA and RCA frames were also considered separately to determine if reconstruction performs better on a specific class.

Table 1. Scoring guidelines concerning the vessels' accurateness.

Score	Vessel Tree Description
1	No vessel is visible in the image.
2	There are some fine lines in the background, but it is hard to distinguish whether they are blood vessels or to identify the angiographic view.
3	The main vessel is visible, but there are many missing details, and additional artifacts are present.
4	All branches are visible but not with the same clarity as in the original image. Enough details are present to be able to distinguish the angiographic view.
5	The reconstruction is more than 95% similar to the original image. Some portions might be unclear, or some additional artifacts might be present, but the main arteries are well visible.

Table 2. Scoring guidelines concerning the background accurateness.

Score	Background Description
1	The background is almost monochromatic.
2	The prominent shadows are vaguely captured.
3	More accurate intensities are captured, but the background is still diffused overall.
4	The background is close to the original one in shape and pixel intensities. Some diffused areas or additional artifacts might be present.
5	The reconstructed background is more than 95% similar to the original one. The same shapes and shadows are depicted, but might differ in pixel intensity in specific regions.

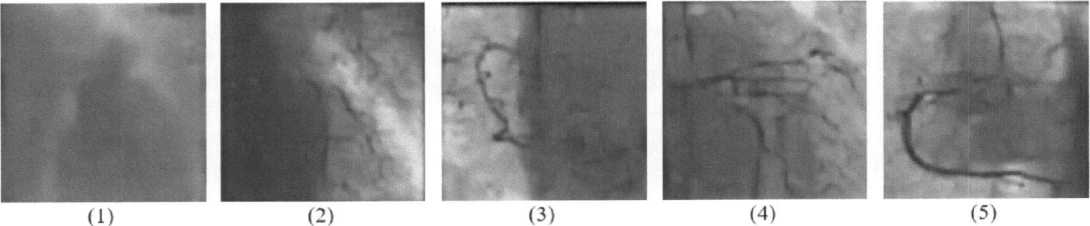

Figure 10. Examples of reconstructed angiographies and the scores assigned concerning the vessels' accurateness: (**1**) no visible vessels; (**2–4**) intermediate scores; (**5**) accurate vessels reconstruction.

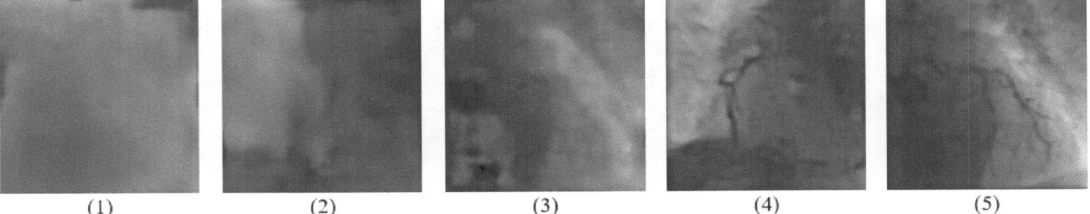

Figure 11. Examples of reconstructed angiographies and the scores assigned concerning the background accurateness: (**1**) monochromatic background; (**2–4**) intermediate scores; (**5**) accurate background reconstruction.

3. Experiments and Results

3.1. Angiographic View Classification

To evaluate the utility of angiographic frames after obfuscation, we formulate four experiments in which convolutional neural networks are trained to solve the angiographic view classification task:

- C_1—original images are used (no obfuscation);
- C_2—images are obfuscated using only the VAE encoder;
- C_3—images are obfuscated only through intensity substitution according to a non-bijective map;
- C_4—images are obfuscated using both methods, as described in Section 2.3.

The details regarding these experiments are presented in Section 2.4. The accuracy obtained by the DL model for each configuration on a testing subset is reported in Table 3.

Table 3. Comparison between DL-model performance when trained on original and obfuscated images, respectively.

	C_1	C_2	C_3	C_4	[11]
Test Accuracy	97.57%	93.71%	88.57%	82.71%	96.20%

After altering the angiographic frames using the VAE encoder, the performance drops by approximately 4%. The method based on a non-bijective map applied to pixel intensities ($N = 96$) leads to a decrease in accuracy of 9%. Although using both techniques causes a significant performance drop compared to the model trained on original images, the accuracy value remains above 80% and may be considered satisfactory in the context of a privacy--accuracy trade-off. The purpose of these experiments is not to achieve state-of-the-art performance on obfuscated images but to compare the results when the same architecture and different obfuscation techniques are employed.

The last column of the table displays the performance previously achieved on the same dataset, using a different DL model and employing the MORE [10] homomorphic encryption scheme as a privacy-preserving technique. The accuracy was identical for the encrypted and the unencrypted model, but the computational time was around 32 times larger when encrypted data was used. In the experiments C_1–C_4 both training and inference were performed with the same runtime since the complexity of the data is not increased by the obfuscation method. Although the MORE encryption scheme provides some advantages in terms of simplicity, clarity, and practicability, when adapted for PPML its linear structure can raise security concerns [11]. By having access to a large enough number of pairings of encrypted and unencrypted data, and by formulating the key search attack as an optimization problem, this linearity may allow one to find the secret key. Furthermore, the fact that the message to be encrypted will always be found among the eigenvalues of the ciphertext matrix is a benefit in terms of utility but also represents privacy-related disadvantages. The obfuscation method overcomes these limitations, since it is highly non-linear and no decryption key is involved.

The C_4 experiment was run multiple times for different values of parameter N. The results achieved on a testing subset are depicted in Figure 12. As expected, for $N = 1$, the accuracy drops to 50% (random guess) because all images become monochromatic. For the other values of N, no monotonous tendency can be observed, suggesting that even for smaller values, enough details are preserved for the classification to be successfully performed.

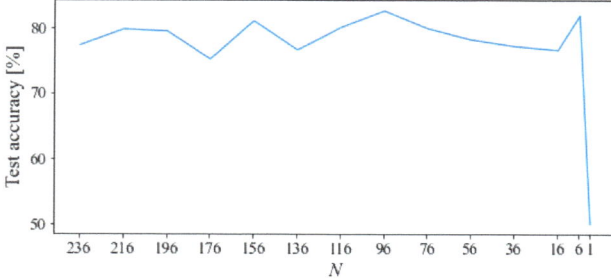

Figure 12. Influence of parameter N on the test accuracy in C_4 configuration experiments.

3.2. Privacy Level of Obfuscated Images

A comparison of the similarity metrics derived for all angiographic frames in the test subset for each of the three procedures is presented in Table 4.

Table 4. Similarity between the original frames and the obfuscated images.

	Encoding	Non-Bijective Mapping	Combined Techniques
SSIM	0.1999	0.0602	0.0512
PSNR [dB]	19.18	9.96	9.92

When employing the VAE encoder to obfuscate images, the results show low similarity and poor quality compared to the original ones. However, applying a non-bijective random mapping leads to an SSIM value below 0.1, corresponding to almost no structural similarity. The PSNR also indicates that applying non-bijective function features increases privacy. Nevertheless, the metrics decrease even further when the methods are simultaneously employed. Thus, the results support the initial hypothesis and motivate the usage in conjunction with the proposed techniques.

3.3. AI-Based Reconstruction Attack

To evaluate the security of obfuscated frames against AI-based reconstruction attacks, we formulate six experiments:

- A_1—E_1 attack configuration when images are obfuscated using only the VAE encoder;
- A_2—E_1 attack configuration when images are obfuscated using only the non-bijective mapping;
- A_3—E_1 attack configuration when images are obfuscated using both encoding and non-bijective mapping;
- A_4—E_2 attack configuration when images are obfuscated using only the VAE encoder;
- A_5—E_2 attack configuration when images are obfuscated using only the non-bijective mapping;
- A_6—E_2 attack configuration when images are obfuscated using both encoding and non-bijective mapping.

Figures 13 and 14 display an example of a reconstructed angiography for each attack configuration. More angiographic samples and their recovered counterparts are presented in Appendix A, Figures A2 and A3. A visual comparison provides the first intuition on the reconstruction capabilities of the AI model in different scenarios. As expected, the performance of the reconstruction model is improved when the training dataset is similar to the targeted dataset, but, even so, all it can restore is the background of the angiographic frames. Because it is typically not possible to identify a patient based on the background of an angiography, this information is not considered sensitive. Even if the background can be recreated through AI-based methods, the obfuscation techniques are deemed secure against AI-based attacks as long as the object of interest (in this case, the coronary vessels) remains unrecognizable after the reconstruction.

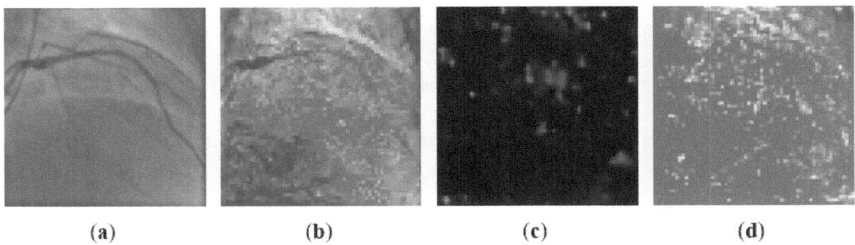

(a) (b) (c) (d)

Figure 13. Comparison between (**a**) an original angiographic frame and the reconstructions obtained with the attack configurations (**b**) A_1, (**c**) A_2, and (**d**) A_3.

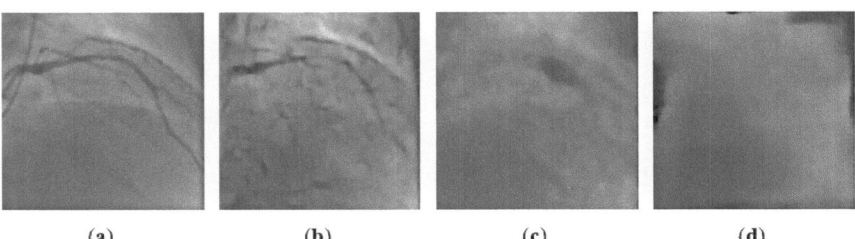

(a) (b) (c) (d)

Figure 14. Comparison between (**a**) an original angiographic frame and the reconstructions obtained with the attack configurations (**b**) A_4, (**c**) A_5, and (**d**) A_6.

For a quantitative analysis, the similarity metrics discussed in Section 2.5 are computed. Higher scores for both SSIM and PSNR indicate better reconstructions. The average values for each attack configuration are presented in Table 5.

Table 5. Similarity between the original frames and the reconstructed images.

Attack Configuration	Experiment	SSIM	PSNR [dB]
E_1	A_1	0.6120	25.23
	A_2	0.3079	9.96
	A_3	0.5100	22.30
E_2	A_4	0.8173	29.54
	A_5	0.7593	26.91
	A_6	0.6855	23.63

These results support the conclusions drawn from the visual inspection. The VAE-based technique enables a certain degree of reconstruction but applying the non-bijective intensity mapping eliminates this shortcoming. Although for the E_1 attack configuration, the best privacy is achieved in experiment A_2 (only non-bijective intensity mapping), this result is not confirmed when a similar dataset is used for training the reconstruction model. For the E_2 setup, the results from A_4, A_5 and A_6 experiments show that, when the methods are used in conjunction, the quality of recreated frames is significantly affected: the vessels are no longer visible, and the background is diffuse.

Experiments A_3 and A_6 were repeatedly run for different values of N. Figures 15 and 16 show how SSIM and PSNR vary as a function of parameter N for the attack configurations E_1 and E_2. As expected, the reconstruction is impossible for $N = 1$ (all information is removed) and is slightly better in E_2 than in E_1. However, there is no monotonous tendency in any configuration. Conversely, in the case of PSNR, although the metrics for $N = 1$ are higher due to the background similarity, there is an oscillating downward trend suggesting that a smaller N implies an increased security level.

The results of the manual evaluation for all three obfuscation approaches are depicted in Figure 17. The mean scores regarding vessels and background reconstruction quality are displayed for each evaluator, alongside the average value. While similar scores were attributed to both vessels and background reconstructions when only encoding was used, for the other two obfuscation approaches, the recovered background presents a higher quality compared to the reconstructed vessels. However, overall, we observe a decreasing trend when comparing the three employed techniques. Even if applying the non-bijective intensity mapping results in a significant privacy improvement, a further decrease in reconstruction quality is noticed when the techniques are used in conjunction.

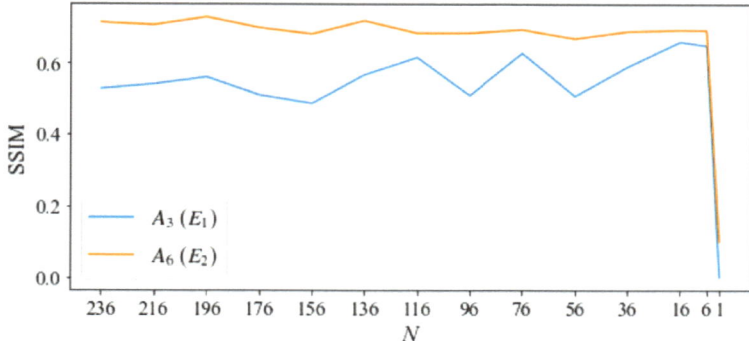

Figure 15. Influence of parameter N on SSIM in A_3 (E_1) and A_6 (E_2), respectively, configuration experiments.

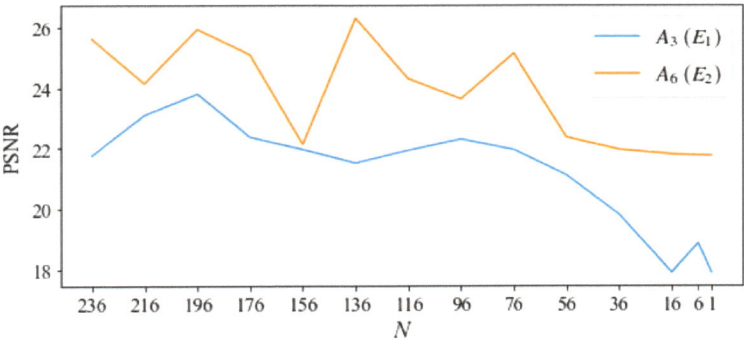

Figure 16. Influence of parameter N on PSNR in A_3 (E_1) and A_6 (E_2), respectively, configuration experiments.

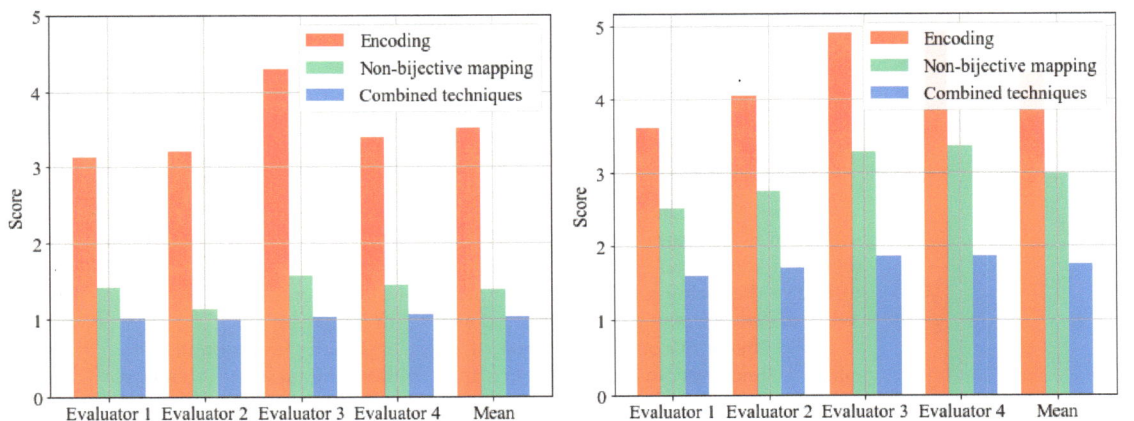

Figure 17. Mean scores of manual evaluation for vessels (**left**) and background (**right**) reconstructions.

Table 6 presents a numerical synthesis of the results. The mean and standard deviation of vessels and background evaluations are displayed for each obfuscation method. The standard deviation is smaller than 1 for each evaluation case, indicating low inter-user variability. The fact that the vessels were mainly evaluated with a score of 1 for the combined procedures strengthens the idea that this strategy provides robust security against recovery attempts.

Table 6. Mean scores regarding the quality of the reconstructed images.

	Encoding		Non-Bijective Mapping		Combined Techniques	
	Vessels	Background	Vessels	Background	Vessels	Background
Mean score	3.50 ± 0.91	4.37 ± 0.70	1.40 ± 0.51	2.97 ± 0.79	1.03 ± 0.17	1.75 ± 0.61

4. Discussion and Conclusions

4.1. Advantageous Properties and Limitations

A first key feature of the proposed obfuscation algorithm is its irreversibility. It is impossible to undo the encoding stage, since performing the decoding without having access to the trained decoder is impossible. Furthermore, reversing the non-bijective mapping is also impossible since it establishes many-to-one relationships, resulting in a large number of alternative substitutes. When the data is sensitive but the partners are untrustworthy, this property addresses the challenge of encryption key management generally involved in collaborative model training. Because there is no inverse function, this obfuscation algorithm can be used by multiple entities to create a common dataset and train a more robust DL model without exposing the original data to each other. Even if the receiver is considered trustworthy, the simple fact of externalizing the data exposes it to the risk of being accessed by an unauthorized party. Thus, an essential requirement that the approach must meet is the preservation of the image utility in the altered state. Once this is achieved, the collaborative work can be carried out based on a zero-trust architecture where the original data can be accessed only by authorized personnel inside the hospital environment.

Another benefit of this strategy is that the VAE does not have to be trained on the same or similar dataset as the one being obfuscated. Even if the encoder is not trained on the target dataset, the underlying information is preserved, and the privacy level is unaffected. As a result, there is no need to disclose sensitive data when training the encoder because any publicly available dataset can be used. Furthermore, it does not need to be trained in the clinical environment; it may be provided as a black-box tool. Furthermore, an outside party can not exploit the decoder to reverse the encoding if it is combined with non-bijective intensity shuffling.

The main drawback of the method is the drop in accuracy for the model that uses obfuscated images to perform a specific task. Such solutions, particularly in the medical domain, should provide performance comparable to that of an expert, to be adopted in the clinical decision-making process. Thus, a path for further development of the method would consist of integrating different strategies for enhancing the performance and robustness of models trained on secured images. An interesting research idea in this direction is to assess how training a DL-based model on a mixed dataset (containing both obfuscated and non-obfuscated synthetic images) would improve the performance. The usage of denoising modules directly in the classification network to increase accuracy without compromising privacy should also be investigated.

Another shortcoming is the lack of a precise security level quantification that would allow a clinical user to choose a particular algorithm configuration for a specific use case. To achieve a rigorous separation of the privacy and accuracy levels according to the obfuscation technique specifications, we intend to conduct within future work additional experiments which solve other medical tasks and employ different datasets and DL solutions.

4.2. Privacy-Utility Trade-off Considerations

While parameter N influences the confidentiality level of the obfuscation method that uses non-bijective functions, the degree of privacy provided by the VAE-based approach is dependent on the level of noise added during encoding, and the number of channels obtained at the bottleneck. When several channels are employed, the information is shared between them, resulting in fewer details being preserved in one channel and in more robust security. Furthermore, the valuable information is not evenly dispersed across the different channels, and one may select a particular representation to fulfill a specific requirement. Hence, the clinical user may select between different options for the trade-off between accuracy and privacy (e.g., categorical choice: very high accuracy, high accuracy, balanced, high privacy, very high privacy). For example, a very high privacy requirement may be chosen if easily recognizable patient features are present in the images (MRI data [35]).

Regarding the classification accuracy, although its value is still above 80% when the obfuscation approaches are combined, which is acceptable in the context of a privacy-accuracy trade-off, there is still room for improvement. The purpose of the classification experiments is not to achieve state-of-the-art performance on obfuscated images but to compare the results when the same architecture and different obfuscation techniques are employed. Because the classification task can be successfully performed even when using small values of N, we sought to explore the existence of structural dissimilarities between the LCA and the RCA, which could allow for a superior reconstruction for one of the classes. Figure 18 depicts two different samples (one LCA and one RCA angiographic acquisition) and their dichromatic obfuscated counterparts. Although assigning the proper category solely by visually inspecting the binary images is difficult, it is clear that some characteristics are preserved even when $N = 2$, which allows for a relatively accurate DL-based classification. The scores given by the expert readers for LCA and RCA reconstructions are recorded separately, to see if such differentiating details allow for a more qualitative recovery for specific samples. In terms of vessel scores, there is no substantial difference between the two categories, according to the visual inspection. However, it appears that the background can be better reconstructed for RCA views.

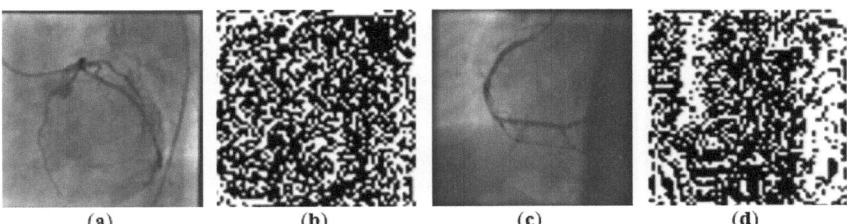

Figure 18. Comparison between LCA (**a**) and RCA (**c**) samples and their obfuscated counterparts when $N = 2$: LCA (**b**) and RCA (**d**).

To demonstrate that there is no statistically significant difference between the two groups, we compute the p-value. The scores are first standardized into the t-score. The p-value is calculated by considering a two-tailed hypothesis. A comparison between the results obtained for the three obfuscation approaches, where the vessels and the background are separately assessed, is presented in Table 7. As the significance level is set to 0.05 and all computed p-values exceed this threshold, we can confirm that the difference between reconstructed RCA and LCA frames is not statistically significant.

Table 7. Statistical significance assessment regarding the reconstruction difference between LCA and RCA views.

	Encoding		Non-Bijective Mapping		Combined Techniques	
	Vessels	Background	Vessels	Background	Vessels	Background
p-value	1	0.765	0.337	0.183	0.678	0.076

4.3. Final Conclusions

In this paper, we present an obfuscation approach that protects the privacy of medical images while allowing for DL model training. Although obfuscation techniques have been previously researched, integrating them into medical applications might be challenging due to the strict privacy and performance requirements. Mosaicing and blurring can be used to make faces and digits unrecognizable to the human eye, as shown in [16]. According to the authors, the obfuscation methods that were evaluated preserve enough information correlated to the original images. Thus, an accurate reconstruction is possible using AI-based models. The approaches proposed in [17] assume that only a part of the images from the dataset contains sensitive information, and these will be obfuscated. However, this is not the case when training models in medical DL-based applications, where the same level of confidentiality is required for all employed data. The method also implies the risk of affecting model accuracy if too many samples need to be secured, which again is not acceptable in a medical application where both privacy and accuracy are crucial.

A promising technique is presented in [18], where images are obfuscated by mixing their pixels with the pixels of another image. Other obfuscation methods were combined with the proposed technique to enhance security, and the experiments showed that the images are protected both from human perception and artificial recognition systems. The performance of models trained on obfuscated images varies with the level of privacy. The loss in accuracy significantly increases when methods are combined, and when privacy parameters are tuned for better security. Another aspect to consider is that the model is trained to perform the cat vs dog classification task, hence the properties of the classes are well defined, and there are many training samples available. However, since the differences between images are very subtle in specific medical imaging applications, and the available data is limited, it is unlikely that training a model on mixed images would achieve high accuracy. In the approach presented by Kim et al. [19], the patient identity is protected by transforming the brain MRI into a proxy image that is sent to the server for segmentation. The altered segmentation mask is then sent to the client, who restores it to the useful version. Compared to our method, this approach differs from the initial requirements perspective, as it is designed to allow for an accurate reconstruction of the processed image. To achieve this, an identity obfuscation loss and a transformation invertibility loss based on SSIM are minimized. The mean average precision and the F1-score are used to assess the re-identification accuracy in the case of an attacker attempting to match an encoded image or segmentation against an existing database. In [20,21], generative models (GANs) were used to create visually appealing images similar to the original ones in terms of basic shape, but distinct in terms of details. Applying this method to X-ray coronary angiographies, for example, might result in synthetic angiographic frames with characteristics which are significantly different from those in the original images (possible stenoses might be excluded, vessel ramifications might be modified, etc). This method is particularly challenging to apply in personalized medicine since the details of each image are required for a proper assessment, but the entire content is confidential. Furthermore, unlike the techniques discussed above, GAN-based methods do not secure information regarding the target objects or the objective of model training (in our use case, the MLaaS provider, or an interceptor who visualizes the obfuscated images, could tell that they are angiographic frames).

The proposed obfuscation algorithm was created with the requirements of a medical use case in mind. Only the computational overhead associated with the obfuscation phase is introduced. Once the data have been secured, training and inference are carried out as if plain data were used. Because the result of the obfuscation is still an image, there is no need for special deep learning libraries or frameworks. Although the privacy-accuracy trade-off must be considered, applying the obfuscation algorithm on medical images successfully hides the sensitive content from human perception and protects it against AI-based reconstruction attacks, while allowing for DL model training with satisfactory performance.

Author Contributions: Conceptualization, A.B.P. and C.I.N.; methodology, A.B.P., I.A.T., A.V., C.S. and L.M.I.; software, A.B.P. and I.A.T.; validation, A.V., C.I.N., A.S.-U. and L.M.I.; formal analysis, A.S.-U., C.I.N. and A.V.; investigation, A.B.P., I.A.T., C.I.N., A.V. and L.M.I.; resources, L.M.I. and A.S.-U.; data curation, A.B.P. and I.A.T.; writing—original draft preparation, A.B.P. and I.A.T.; writing—review and editing, C.I.N., A.V., C.S. and L.M.I.; visualization, A.B.P. and I.A.T.; supervision, C.S. and L.M.I.; project administration, L.M.I.; funding acquisition, L.M.I. All authors have read and agreed to the published version of the manuscript.

Funding: This work was supported by a grant of the Romanian Ministry of Education and Research, CCCDI—UEFISCDI, project number PN-III-P2-2.1-PED-2019-2415, within PNCDI III. This work was also supported by a grant of the Ministry of Education and Research, CNCS/CCCDI— UEFISCDI, project number PN-III-P3-3.6-H2020-2020-0145/2021. This project received funding from the European Union's Horizon 2020 research and innovation program under grant agreement No. 875351.

Institutional Review Board Statement: Not applicable.

Informed Consent Statement: Not applicable.

Data Availability Statement: The Medical MNIST dataset is available at https://github.com/apolanco3225/Medical-MNIST-Classification (accessed on 23 March 2022). The coronary angiography data used to support the findings of this study have not been made available because they were acquired in a research grant (heart.unitbv.ro) which does not allow the publication of the data.

Conflicts of Interest: Andreea Bianca Popescu, Anamaria Vizitiu, Constantin Suciu, and Lucian Mihai Itu are employees of Siemens SRL, Brasov, Romania. Ioana Antonia Taca received scholarships from Siemens SRL, Brasov, Romania. The other authors declare no conflict of interest. The funders had no role in the design of the study; in the collection, analyses, or interpretation of data; in the writing of the manuscript, or in the decision to publish the results.

Abbreviations

The following abbreviations are used in this manuscript:

DL	Deep Learning
VAE	Variational Autoencoder
AI	Artificial Intelligence
ML	Machine Learning
MLaaS	Machine Learning as a Service
SSIM	Structural Similarity Index Measure
PSNR	Peak Signal-to-Noise-Ratio
ReLU	Rectified Linear Unit
LCA	Left Coronary Artery
RCA	Right Coronary Artery
CT	Computed Tomography
MRI	Magnetic Resonance Imaging
CXR	Chest X-Ray
CNN	Convolutional Neural Network

Appendix A

Multiple examples of angiographic frames and the corresponding obfuscated or reconstructed counterparts are presented in this appendix.

Figure A1 displays for each original sample the obfuscated version obtained when encoding and the non-bijective map are used independently and in conjunction. The value of the parameter N used to attain the images displayed under (c) and (d) is 96.

Figure A2 presents reconstructed images in the E_1 attack configuration, when the malicious actor is aware that the target data are medical images but does not know their specific type. The original angiographies are shown in the first column.

The same frames are displayed in Figure A3 along with the recovered images in the E_2 attack configuration, where the threat actor knows that the targeted dataset contains coronary angiographies, and the reconstruction model is trained on a similar dataset. We observe that the more knowledgable the attacker is, the better the reconstruction performance is when only encoding is employed as a security measure. However, the coronary vessels are difficult to recover in both attack configurations, when the second step of the obfuscation algorithm is included.

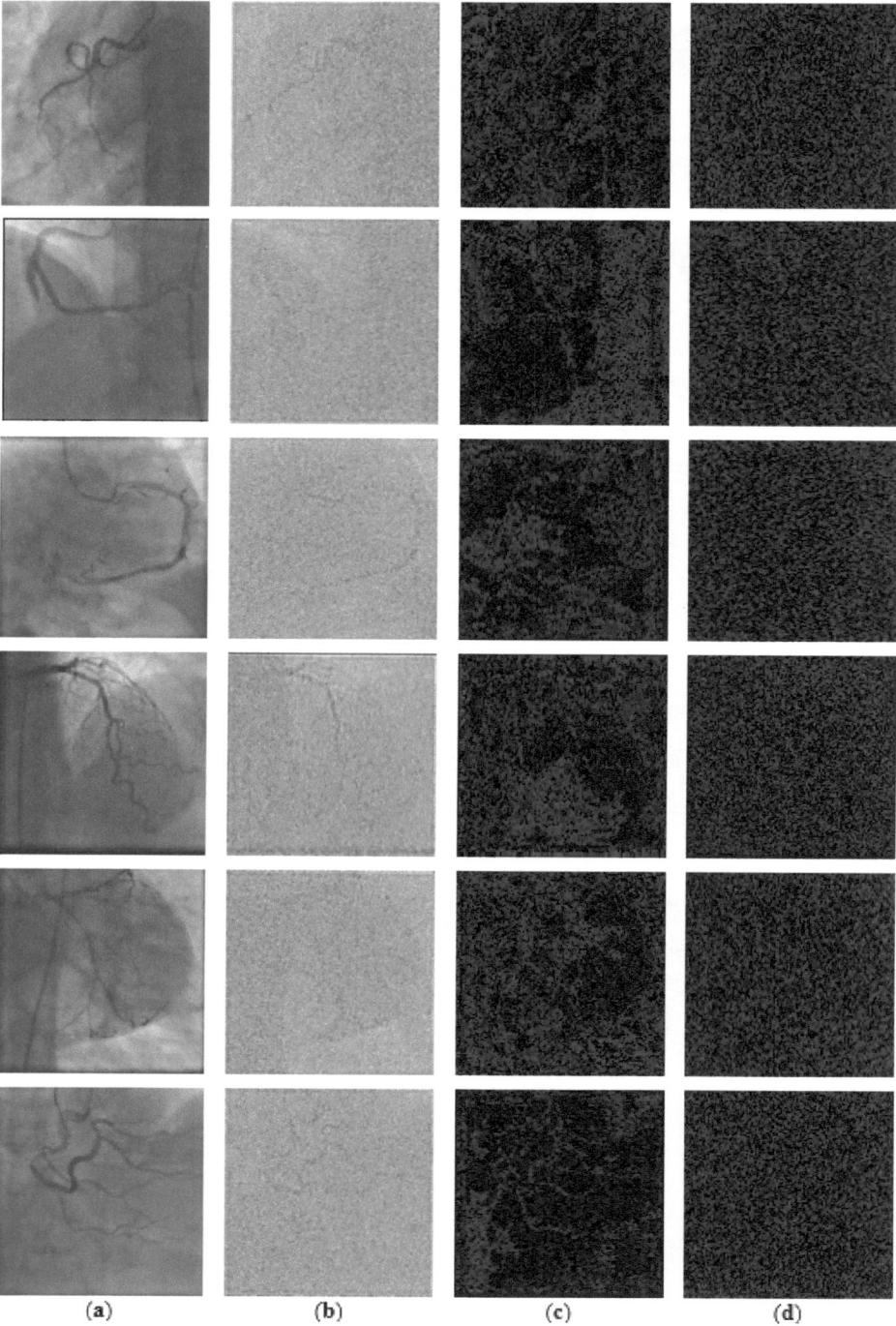

Figure A1. Comparison between (**a**) original and corresponding obfuscated angiographic frames using (**b**) encoding, (**c**) non-bijective intensity mapping, and (**d**) combined algorithm.

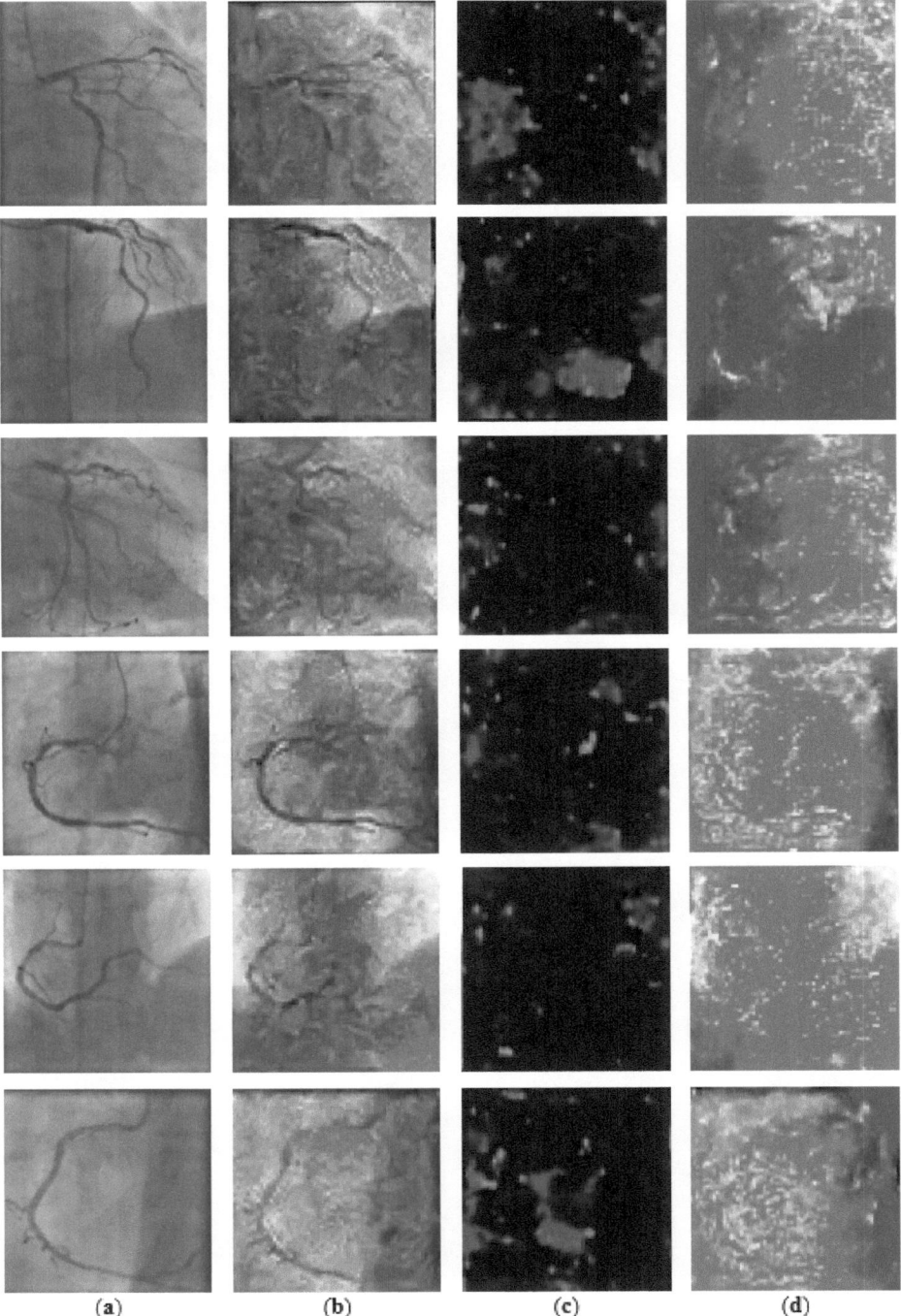

Figure A2. Comparison between (**a**) original angiographic frames and the reconstructions obtained with the attack configurations (**b**) A_1, (**c**) A_2, and (**d**) A_3.

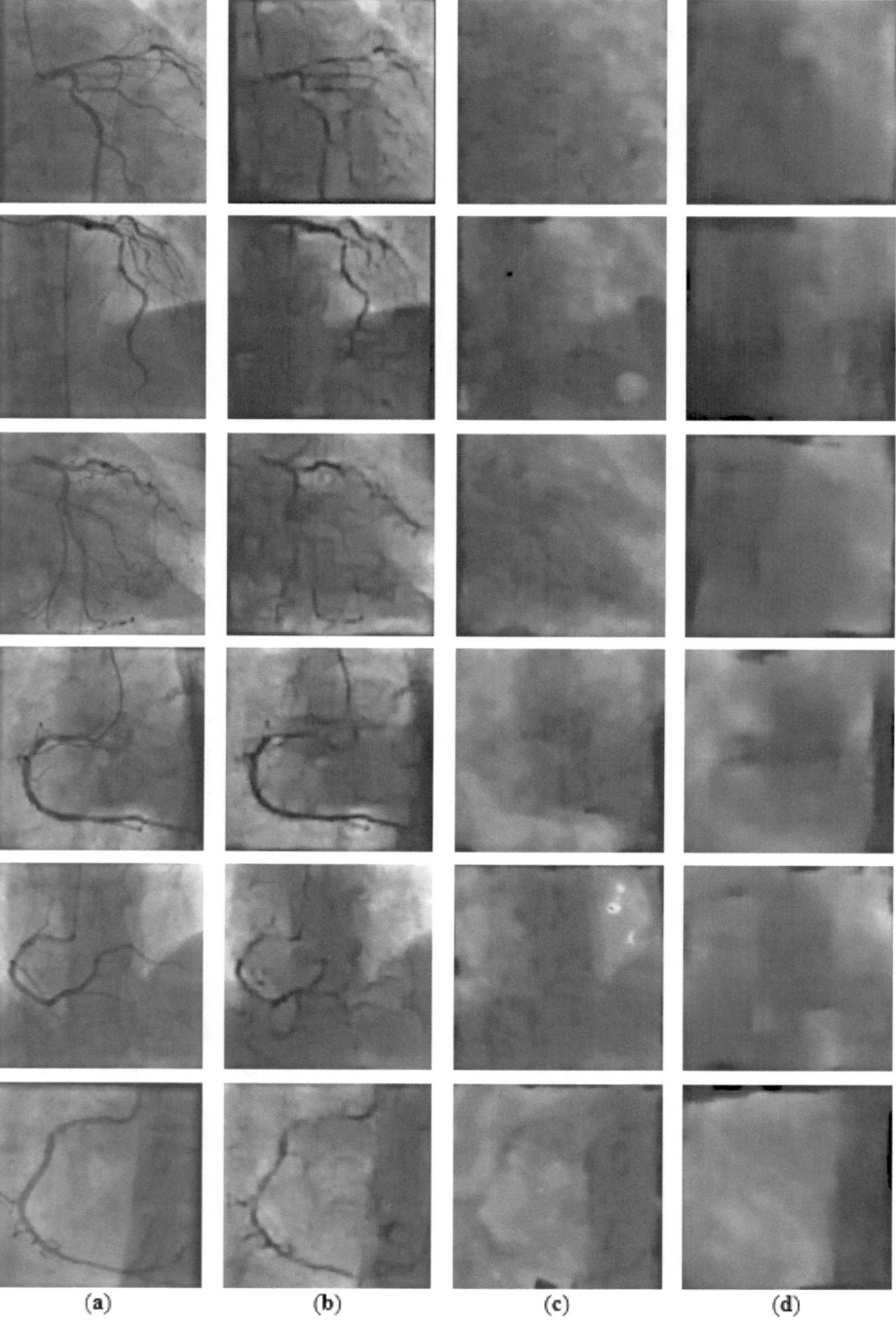

Figure A3. Comparison between (**a**) original angiographic frames and the reconstructions obtained with the attack configurations (**b**) A_4, (**c**) A_5, and (**d**) A_6.

References

1. Gui, C.; Chan, V. Machine learning in medicine. *Univ. West. Ont. Med. J.* **2017**, *86*, 76–78. [CrossRef]
2. Vayena, E.; Blasimme, A.; Cohen, I.G. Machine learning in medicine: Addressing ethical challenges. *PLoS Med.* **2018**, *15*, e1002689. [CrossRef]
3. Pulido-Gaytan, L.B.; Tchernykh, A.; Cortés-Mendoza, J.M.; Babenko, M.; Radchenko, G. A Survey on Privacy-Preserving Machine Learning with Fully Homomorphic Encryption. In *Latin American High Performance Computing Conference*; Springer: Cham, Switzerland, 2020; pp. 115–129.
4. Orlandi, C.; Piva, A.; Barni, M. Oblivious neural network computing via homomorphic encryption. *EURASIP J. Inf. Secur.* **2007**, *2007*, 37343. [CrossRef]
5. Gilad-Bachrach, R.; Dowlin, N.; Laine, K.; Lauter, K.; Naehrig, M.; Wernsing, J. Cryptonets: Applying neural networks to encrypted data with high throughput and accuracy. In Proceedings of the 33rd International Conference on Machine Learning, New York, NY, USA, 20–22 June 2016; pp. 201–210.
6. Hesamifard, E.; Takabi, H.; Ghasemi, M. Cryptodl: Deep neural networks over encrypted data. *arXiv* **2017**, arXiv:1711.05189.
7. Ishiyama, T.; Suzuki, T.; Yamana, H. Highly accurate CNN inference using approximate activation functions over homomorphic encryption. In Proceedings of the 2020 IEEE International Conference on Big Data (Big Data), Atlanta, GA, USA, 10–13 December 2020; pp. 3989–3995.
8. Lee, J.W.; Kang, H.; Lee, Y.; Choi, W.; Eom, J.; Deryabin, M.; Lee, E.; Lee, J.; Yoo, D.; Kim, Y.S.; et al. Privacy-preserving machine learning with fully homomorphic encryption for deep neural network. *arXiv* **2021**, arXiv:2106.07229.
9. Al Badawi, A.; Jin, C.; Lin, J.; Mun, C.F.; Jie, S.J.; Tan, B.H.M.; Nan, X.; Aung, K.M.M.; Chandrasekhar, V.R. Towards the alexnet moment for homomorphic encryption: Hcnn, the first homomorphic cnn on encrypted data with gpus. *IEEE Trans. Emerg. Top. Comput.* **2020**, *9*, 1330–1343. [CrossRef]
10. Kipnis, A.; Hibshoosh, E. Efficient methods for practical fully homomorphic symmetric-key encrypton, randomization and verification. *Cryptology ePrint Archive*. 2012. Available online: https://ia.cr/2012/637 (accessed on 23 March 2022).
11. Vizitiu, A.; Niță, C.I.; Puiu, A.; Suciu, C.; Itu, L.M. Applying Deep Neural Networks over Homomorphic Encrypted Medical Data. *Comput. Math. Methods Med.* **2020**, *2020*, 3910250. [CrossRef]
12. Vizitiu, A.; Nita, C.I.; Toev, R.M.; Suditu, T.; Suciu, C.; Itu, L.M. Framework for Privacy-Preserving Wearable Health Data Analysis: Proof-of-Concept Study for Atrial Fibrillation Detection. *Appl. Sci.* **2021**, *11*, 9049. [CrossRef]
13. Popescu, A.B.; Taca, I.A.; Nita, C.I.; Vizitiu, A.; Demeter, R.; Suciu, C.; Itu, L.M. Privacy preserving classification of eeg data using machine learning and homomorphic encryption. *Appl. Sci.* **2021**, *11*, 7360. [CrossRef]
14. Guang-Li, X.; Xin-Meng, C.; Ping, Z.; Jie, M. A method of homomorphic encryption. *Wuhan Univ. J. Nat. Sci.* **2006**, *11*, 181–184. [CrossRef]
15. Chen, H.; Hussain, S.U.; Boemer, F.; Stapf, E.; Sadeghi, A.R.; Koushanfar, F.; Cammarota, R. Developing privacy-preserving AI systems: The lessons learned. In Proceedings of the 2020 57th ACM/IEEE Design Automation Conference (DAC), San Francisco, CA, USA , 20–24 July 2020; pp. 1–4.
16. McPherson, R.; Shokri, R.; Shmatikov, V. Defeating image obfuscation with deep learning. *arXiv* **2016**, arXiv:1609.00408.
17. Zhang, T.; He, Z.; Lee, R.B. Privacy-preserving machine learning through data obfuscation. *arXiv* **2018**, arXiv:1807.01860.
18. Raynal, M.; Achanta, R.; Humbert, M. Image obfuscation for privacy-preserving machine learning. *arXiv* **2020**, arXiv:2010.10139.
19. Kim, B.N.; Dolz, J.; Desrosiers, C.; Jodoin, P.M. Privacy Preserving for Medical Image Analysis via Non-Linear Deformation Proxy. *arXiv* **2020**, arXiv:2011.12835.
20. Li, T.; Choi, M.S. DeepBlur: A simple and effective method for natural image obfuscation. *arXiv* **2021**, arXiv:2104.02655.
21. Chen, J.W.; Chen, L.J.; Yu, C.M.; Lu, C.S. Perceptual Indistinguishability-Net (PI-Net): Facial image obfuscation with manipulable semantics. In Proceedings of the IEEE/CVF Conference on Computer Vision and Pattern Recognition, Nashville, TN, USA, 20–25 June 2021; pp. 6478–6487.
22. Kingma, D.P.; Welling, M. An introduction to variational autoencoders. *arXiv* **2019**, arXiv:1906.02691.
23. Kingma, D.P.; Welling, M. Auto-encoding variational bayes. *arXiv* **2013**, arXiv:1312.6114.
24. Barber, D. *Bayesian Reasoning and Machine Learning*; Cambridge University Press: Cambridge, UK, 2012.
25. Metropolis, N.; Rosenbluth, A.W.; Rosenbluth, M.N.; Teller, A.H.; Teller, E. Equation of state calculations by fast computing machines. *J. Chem. Phys.* **1953**, *21*, 1087–1092. [CrossRef]
26. Geman, S.; Geman, D. Stochastic relaxation, Gibbs distributions, and the Bayesian restoration of images. *IEEE Trans. Pattern Anal. Mach. Intell.* **1984**, *PAMI-6*, 721–741. [CrossRef]
27. Bernardo, J.; Bayarri, M.; Berger, J.; Dawid, A.; Heckerman, D.; Smith, A.; West, M.; Beal, M.; Ghahramani, Z. The variational Bayesian EM algorithm for incomplete data: With application to scoring graphical model structures. *Bayesian Stat.* **2003**, *7*, 210.
28. Ganguly, A.; Earp, S.W. An Introduction to Variational Inference. *arXiv* **2021**, arXiv:2108.13083.
29. Apolanco3225. Medical MNIST Classification. 2017. Available online: https://github.com/apolanco3225/Medical-MNIST-Classification (accessed on 23 March 2022).
30. Yin, B.; Scholte, H.S.; Bohté, S. LocalNorm: Robust Image Classification Through Dynamically Regularized Normalization. In *International Conference on Artificial Neural Networks*; Springer: Cham, Switzerland, 2021; pp. 240–252.
31. Chollet, F. keras. 2015. Available online: https://github.com/fchollet/keras (accessed on 23 March 2022).

32. Ronneberger, O.; Fischer, P.; Brox, T. U-net: Convolutional networks for biomedical image segmentation. In *International Conference on Medical Image Computing and Computer-Assisted Intervention*; Springer: Cham, Switzerland, 2015; pp. 234–241.
33. Paszke, A.; Gross, S.; Massa, F.; Lerer, A.; Bradbury, J.; Chanan, G.; Killeen, T.; Lin, Z.; Gimelshein, N.; Antiga, L.; et al. PyTorch: An Imperative Style, High-Performance Deep Learning Library. In *Advances in Neural Information Processing Systems 32*; Curran Associates, Inc.: Red Hook, NY, USA, 2019; pp. 8024–8035.
34. Van der Walt, S.; Schönberger, J.L.; Nunez-Iglesias, J.; Boulogne, F.; Warner, J.D.; Yager, N.; Gouillart, E.; Yu, T. Scikit-image: Image processing in Python. *PeerJ* **2014**, *2*, e453. [CrossRef] [PubMed]
35. Schwarz, C.G.; Kremers, W.K.; Therneau, T.M.; Sharp, R.R.; Gunter, J.L.; Vemuri, P.; Arani, A.; Spychalla, A.J.; Kantarci, K.; Knopman, D.S.; et al. Identification of anonymous MRI research participants with face-recognition software. *N. Engl. J. Med.* **2019**, *381*, 1684–1686. [CrossRef] [PubMed]

Article

Modified Artificial Bee Colony Based Feature Optimized Federated Learning for Heart Disease Diagnosis in Healthcare

Muhammad Mateen Yaqoob [1,*], Muhammad Nazir [1], Abdullah Yousafzai [2], Muhammad Amir Khan [3,*], Asad Ali Shaikh [4], Abeer D. Algarni [5] and Hela Elmannai [5]

1. Department of Computer Science, HITEC University Taxila, Taxila 47080, Pakistan
2. Department of Computer Science, University of Central Punjab, Lahore 54782, Pakistan
3. Department of Computer Science, COMSATS University Islamabad Abbottabad Campus, Abbottabad 22060, Pakistan
4. Computer Science Department, Faculty of Computer Sciences, ILMA University, Karachi 75190, Pakistan
5. Department of Information Technology, College of Computer and Information Sciences, Princess Nourah bint Abdulrahman University, P.O. Box 84428, Riyadh 11671, Saudi Arabia
* Correspondence: mateenyaqoob@gmail.com (M.M.Y.); amirkhan@cuiatd.edu.pk (M.A.K.)

Citation: Yaqoob, M.M.; Nazir, M.; Yousafzai, A.; Khan, M.A.; Shaikh, A.A.; Algarni, A.D.; Elmannai, H. Modified Artificial Bee Colony Based Feature Optimized Federated Learning for Heart Disease Diagnosis in Healthcare. *Appl. Sci.* **2022**, *12*, 12080. https://doi.org/10.3390/app122312080

Academic Editors: Lucian Mihai Itu, Constantin Suciu and Anamaria Vizitiu

Received: 5 November 2022
Accepted: 23 November 2022
Published: 25 November 2022

Publisher's Note: MDPI stays neutral with regard to jurisdictional claims in published maps and institutional affiliations.

Copyright: © 2022 by the authors. Licensee MDPI, Basel, Switzerland. This article is an open access article distributed under the terms and conditions of the Creative Commons Attribution (CC BY) license (https://creativecommons.org/licenses/by/4.0/).

Abstract: Heart disease is one of the lethal diseases causing millions of fatalities every year. The Internet of Medical Things (IoMT) based healthcare effectively enables a reduction in death rate by early diagnosis and detection of disease. The biomedical data collected using IoMT contains personalized information about the patient and this data has serious privacy concerns. To overcome data privacy issues, several data protection laws are proposed internationally. These privacy laws created a huge problem for techniques used in traditional machine learning. We propose a framework based on federated matched averaging with a modified Artificial Bee Colony (M-ABC) optimization algorithm to overcome privacy issues and to improve the diagnosis method for the prediction of heart disease in this paper. The proposed technique improves the prediction accuracy, classification error, and communication efficiency as compared to the state-of-the-art federated learning algorithms on the real-world heart disease dataset.

Keywords: privacy aware; federated learning; healthcare; heart disease prediction; feature selection

1. Introduction

Advancement in technologies like the Internet of Things (IoT) and wearable sensing devices enables the storage of records related to the health parameters of patients or people. The IoT in the healthcare environment has led to a new research domain of the Internet of Medical Things (IoMT). The IoMT-based solutions integrated with the healthcare system can enhance care services, and quality of life, and enable cost-effective solutions [1]. Biomedical data related to people like medical records, images, physiological signals, and many other forms are gathered using these technologies. The volume of this biomedical data is huge as it can easily be gathered from a huge number of people using modern technologies [2]. Wearable sensing devices, like smartwatches, wristbands, and many others, enable early detection and warnings of several diseases. The increasing trend in wearable devices helps in efficient data collection and the early detection of diseases. Healthcare is a system that is formed with the intention to prevent, diagnose, and treat various health-related problems in humans. As the advancement and development of healthcare-related technologies take place, data in huge amounts is available from various sources. The development of an efficient healthcare infrastructure system is one of the challenging goals of current modern society.

One of the primary health concerns faced worldwide is cardiovascular disease. According to the World Health Organization (WHO), approximately 18 million deaths occur yearly worldwide due to heart or cardiovascular disease [3]. Heart disease or cardiovascular disease (CVD) is based on various conditions that impact the human heart. Many factors cause heart disease including personal and functional behavior and genetic predisposition. Numerous risk factors include smoking, excessive consumption of caffeine, and alcohol, inactivity, stress, and physical fitness, as high blood pressure, obesity, pre-existing heart disease, and high cholesterol can also be a reason for heart disease. CVD is a serious condition that affects the function of the heart and causes problems such as strokes and reduced blood vessel function. Patients with heart disease do not reach the advanced stages of the disease and it is too late for the damage to be repaired. Early and accurate treatment of heart disease plays a significant role in avoiding death. Machine learning (ML)-based techniques provide a way forward for effective diagnosis of heart disease. A lot of research has been performed and various machine learning models have been used to make classifications and predictions for diagnosing heart disease. A hybrid technique based on random forest and a linear model is suggested in [4] to improve the prediction accuracy of heart disease. For the identification of heart disease in the E-healthcare system and to resolve the problem of feature selection, a system is proposed in [5] based on classification algorithms.

Machine learning (ML) models are frequently trained on sufficient user data in healthcare to track a patient's health status. Regrettably, today's healthcare faces two critical challenges. For starters, real data is frequently found as isolated islands. Even though there is a large amount of data in various organizations, sharing this data is impossible due to concerns about privacy and security. As a result, training powerful models with valuable data is difficult. In addition, the European Union through General Data Protection Regulation (GDPR) [6], China by China through China Cyber Security Law [7], and the United States with the California Consumer Privacy Act (CCPA) of 2018 [8], have recently enforced the protection of user data privacy through these regulatory procedures. Therefore, it is not possible to get huge amounts of user data in real-time healthcare applications. To overcome these challenges, federated learning is proposed recently by Google [9,10]. Recently, some new meta-heuristics techniques are proposed such as monarch butterfly optimization (MBO) [11], slime mold algorithm (SMA) [12], moth search algorithm (MSA) [13], hunger games search (HGS) [14], Runge Kutta method (RUN) [15], and Harris hawks optimization (HHO) [16], to further minimize the fitness function by keeping the size of the population unchanged, to improve the weight adaption rate, to enhance the local searching method, to optimize the dynamic fitness function computation, to avoid the local optimal solutions and increase convergence speed, and to cooperatively search for the optimal local solution, respectively. Several security and privacy challenges in an IoT environment with their use cases are outlined in [17,18].

The aim of federated learning is a privacy-aware collaborative learning mechanism of a shared model by keeping the data on the device. Hence, the users of federated learning will experience personalized machine learning and overcome privacy issues as well. Motivated by these highlighted issues of privacy in healthcare, in this paper, we propose a federated matched averaging with a Modified Artificial Bee Colony (M-ABC) optimization-based framework to overcome privacy issues and to improve the diagnosis method for the prediction of heart disease. The objective of our proposed framework is to develop an overall privacy-aware decentralized learning method for heart disease diagnosis which improves the feature optimization at the client end and the communication efficient global cloud model. We chose M-ABC optimizer because it is highly flexible and user-friendly, uses fewer control parameters than other algorithms such as genetic algorithm (GA) and particle swarm optimization (PSO), is easily hybrid with other optimization algorithms, and possesses strong robustness and a fast convergence rate. In addition, the M-ABC method can also accommodate a random cost objective function. This paper's contributions are as follows:

- We design and propose a privacy-aware framework for the prediction of heart disease in healthcare using an improved federated learning algorithm for cloud and user sites.

- M-ABC optimizer is proposed at the client end for the optimal feature selection of heart disease data. This optimizer enables improved accuracy of prediction and fewer classification errors.
- Federated matched averaging (FedMA)-based algorithm is explored for constructing a privacy-aware framework for a global cloud model.
- We validated and tested the proposed framework with a real-world heart disease dataset. Evaluation of the performance of the proposed framework in terms of prediction accuracy, classification error, and communication efficiency is performed with state-of-the-art federated learning algorithms.

The rest of the paper is organized as follows. Section 2 presents the review of related work. Section 3 explains the materials and the proposed framework. Section 4 is related to the evaluation of performance and results. The last section, Section 5, provides a conclusion and future work of the paper.

2. Literature Review

Privacy and security of data, and data in an isolated form are the two big challenges faced by the current machine learning research domain. Techniques based on machine learning require centralized training data for the model to be trained. Regulations are put into practice for data privacy throughout the world [6–8]. Hence data privacy is a big challenge for traditional machine learning techniques. Federated learning initially proposed by Google, federated stochastic gradient (FedSGD), and averaging (FedAvg) based algorithm brought a ray of hope to overcome these challenges [9]. A technique constructed on federated learning is proposed in [10] to overcome the issue of data isolation and privacy. They proposed a comprehensive framework based on federated learning to tackle the issues related to data security in the traditional artificial intelligence domain. Their proposed solution is categorized into two approaches i.e., horizontal, and vertical federated learning.

Technical aspects such as hardware, platforms, software, protocol, enabling technologies, and other features of the data privacy of federated learning are discussed by the authors in [19]. The authors discussed some of the optimization techniques for federated learning in their article by highlighting their features and performance. They also outlined some of the market implications of federated learning in order to anticipate them. Additionally, some of the advantages, issues, and challenges which refer to the design and deployment of federated learning are presented by the authors. In [20], the authors provide insight into the various machine learning deployment architectures such as centralized, distributed, and federated learning. They have outlined the evolution of machine learning architectures with comprehensive deliberation. Moreover, application areas for federated learning such as the IoT systems, healthcare, Gboard App, edge computing, cybersecurity, and many others were suggested by them.

In the paper [21], the authors developed a model based on federated learning for the prediction of hospitalization of health-related disease patients. They used electronic health records (EHR) data distributed amongst numerous sources or agents. The authors proposed the cluster Primal Dual Splitting (cPDS) algorithm to overcome the problem of large-scale sparse Support Vector Machine (sSVM) using a federated learning technique. Their proposed technique achieves analogous prediction accuracy of the classifier. Authors in [22], tested and evaluated the three federated learning-based algorithms on the MNIST dataset and used a Bayesian correlated t-test. According to their evaluation, FedAvg outperforms CO-OP and FSVRG algorithms when the uploads by clients are limited to 10,000. They have used balanced data distribution in which the clients have the same amount of data. An optimized version of FedAvg is proposed by authors in [23], in which they intend to enhance the accuracy and convergence rate of the state-of-the-art federated learning algorithm. They proposed the Federated Match Averaging (FedMA) algorithm based on the layer-wise federated learning algorithm to adopt Bayesian nonparameterized methods for heterogeneous data. Their proposed FedMA performs better than FedAvg in terms of convergence, and accuracy, and reduces the communication size. To optimize the convergence speed of federated learning, the authors in [24] proposed a fast-convergent algorithm that achieves intelligent selection of each device at every round of the training model. Their algorithm utilizes precise and effective approximation for communication of a near-optimal distribution of device selection to improve the convergence rate.

Authors in [25] have proposed an algorithm that assigns the weights according to the contribution of each class to the local models. The machine learning based algorithms can play their part in the detection of COVID-19 using a dataset of chest X-rays of the patients. A Federated learning-based technique is proposed by the authors in [26] to detect COVID-19 cases with improved model prediction accuracy and loss as compared with the traditional machine learning algorithms. For their work, the authors utilized two datasets which are descriptive datasets with COVID-19-infected cases from Wuhan and patients' chest x-ray radiography images with COVID-19, Pneumonia, and normal images. To resolve the issue of data privacy for the IoMT-based healthcare system, authors in [27] proposed a blockchain-based solution using federated learning. Their proposed algorithm is a hybrid approach based on federated learning and maximization of the Gaussian Mixture Model (FL-EM-GMM) and uses blockchain for model verification, and homomorphic encryption to overcome user data privacy issues. Their proposed method shows that the IoMT data training can be completed using privacy locally to prevent data leakage.

Traditionally, the cloud/server collects sensed data from IoMT devices and then performs the prediction of that sensed data. To develop a privacy-aware heart rate prediction technique, authors in [28] proposed a Bayesian inference federated learning with autoregression with exogenous variable (ARX) model. This FedARX method accomplishes accurate and robust heart rate prediction as compared with the traditional machine learning models. To effectively manage and optimize the computation offloading for IoT-based applications, authors in [29] proposed a meta-heuristic Artificial Bee Colony (ABC) optimization. Their technique intelligently manages the computation workload for resource-constrained IoT applications. Authors in [30] proposed the ABC algorithm for the optimization of numerical problems in a computing environment. For lightweight prediction of computational workload in an IoT-assisted Edge environment, authors in [31] proposed an artificial neural network-based framework. Their proposed multi-objective framework enhances workload management for computationally intensive applications. A long short-term memory (LSTM) based prediction of computational workload technique for offloading in IoT-assisted Mobile Edge Computing is proposed in [32]. A detailed survey of intelligent offloading of computational workload is prepared by authors in [33]. An extensive survey of open-source datasets for the COVID-19 disease is performed by authors in [34]. They categorized the datasets into four classes as the identification of COVID-19 from X-ray images, CT scans, and cough sounds, as well as transmission estimation, case reporting, and diagnosis from demographic, epidemiological, and mobility data.

Other methods were also introduced in the literature for heart disease prediction, such as a hybrid approach of linear discriminant analysis with the modified ant lion optimization for classification [35], a combination of Fuzzy logic algorithm and gradient boosting decision tree (GBDT) [36], a technique based on modified salp swarm optimization (MSSO) and an adaptive neuro-fuzzy inference system (ANFIS) [37], and multi-cost objective function [38]. Heart disease monitoring and prediction based on a hybrid classifier and deep learning centered modified neural network for IoT-assisted healthcare is proposed in [39–42]. Moreover, various methods are proposed for improving the classification error and accuracy, such as the higher-order Boltzmann-based model [43], performance evaluation of classifiers and optimizers for heart disease prediction [44], localization using two-stage classifiers [45], a hybrid classifier based on random forest and naïve bayes [46], hybrid recommender system [47], based on genetic algorithm and hybrid classifiers using the ensembled model with a majority voting technique [48], and Artificial intelligence (AI) based heart disease detection using electrocardiogram (ECG) signals [49].

3. Materials and Methods

A healthcare system built on the Internet of Medical Things (IoMT) makes it possible to collect patient data in real-time for the purposes of early disease diagnosis and treatment. Patients who are diagnosed and treated early have a lower risk of developing heart disease. With the emerging international privacy laws like GDPR [6], China Cyber Security Law [7], and CCPA [8], the traditional machine learning based techniques are unable to overcome the privacy issues as they require user data to be processed for model generation and diagnosis of disease. The IoMT-based sensing devices gather heart disease information from the patients before and after the initiation of heart disease. When it comes to the healthcare system, user data is impossible to share due to privacy and security issues. A federated learning framework for heart disease prediction in the healthcare system is proposed in this paper, which overcomes privacy issues and provides effective heart disease prediction in a privacy-aware healthcare system. The symbols used throughout the study are described in Table 1 below.

Table 1. Description of used symbols.

Used Symbol	Description
X_{ni}	Initialization vector for client sites
C_{nie}	Candidate solution by employed bee
X_{pi}	Random local solution
F_n	Fitness function
C_{nio}	Onlooker bee's candidate solution
C_{nis}	Candidate solution of scout bee
w_{jl}	l^{th} neuron studied on the dataset j
θ_i	Mean Gaussian
$c(w_{jl}, \theta_i)$	Similarity function
K	Number of client sites listed as k
B	Size of local minibatch
η	Learning rate
E	Number of local epochs
ω_o	Initial global cloud model
ω_k	Model of kth client

3.1. Dataset Description

We train and test our proposed framework on the heart disease dataset of UCI Cleveland. This dataset contains 303 records and 76 attributes. A detailed description of the dataset is illustrated in Table 2 below. This table shows the numerous risks of heart disease, their description, and the encoded values of these risks. The encoded values are utilized as the input to our proposed framework.

Table 2. Detailed Description of Dataset.

S#	Risk Name	Description	Encoded Values
1	Age	Age in years	>79 = 2, 61–79 = 1, 51–60 = 0, 35–50 = −1, <35 = −2
2	Sex	Female and Male	Female = 0, Male = 1
3	Blood pressure	In mmHg	Above 139 mmHg = High = 1 120–139 mmHg = Normal = 0 Below 120 mmHg = Low = −1

Table 2. *Cont.*

S#	Risk Name	Description	Encoded Values
4	Serum cholesterol	In mg/dL	>240 mg/dL = High = 1 200–239 mg/dL = Normal = 0 <200 mg/dL = Low = −1
5	Hereditary	Family members diagnosed with heart disease	Yes = 1 No = 0
6	Alcohol	Yes or No	Yes = 1 No = 0
7	Diabetes	Yes or No	Yes = 1 No = 0
8	Resting electrocardiographic	Normal, ST T, or Hypertrophy	Hypertrophy = 2 ST T = 1 Normal = 0
9	Angina induced by exercise	Yes or No	Yes = 1 No = 0
10	Fasting blood sugar	>120 mg/dL	True = 1 False = 0
11	Status of heart (thallium scan)	Reversible defect, Normal, fixed defect	Reversible defect = 7, Normal = 3, fixed defect = 6
12	Smoke	Yes or No	Yes = 1 No = 0
13	Diet	Good, Normal, Poor	Good = 1, Normal = 0, Poor = −1
14	Heart Disease	Yes or No	Yes = 1, No = 0

3.2. Optimal Solution Selection Using M-ABC Algorithm for IoMT Clients

An algorithm based on swarm intelligence, known as the Modified Artificial Bee Colony (M-ABC), has been developed and proposed in [50]. The scout bee, onlooker bee, and employed bee all appear in the M-ABC algorithm. Scout bees are responsible for exploring new food sources, while the onlooker bee chooses a food source based on the dance of an employed bee. As a result, the bees employed are protected from exploitation because they are linked to their food source. Neither the scout bees nor the onlooker bees are associated with any particular food source. They are referred to as "unemployed bees" as a result. The main aim of the fitness function is the optimal selection of classification error and communication efficiency of the received models from the IoMT client sites. The objective of the fitness function is to minimize the classification error and number of rounds consumed to achieve higher accuracy. Algorithm 1 below presents the generalized working of the M-ABC optimizer.

Algorithm 1: Working of Optimizer M-ABC Algorithm

1: IoMT sites initialization phase using Equation (1)
2: **Do Repeat**
3:   Employed bees for new solution using Equation (2)
4:   Onlooker bees candidate solution using Equations (3) and (4)
5:   Phase of Scout bees' candidate solution using Equation (5)
6:   Memorize the best solution you came up with
7: **until** maximum number of cycles reached

3.2.1. Initialization Phase

All the population of healthcare sites is initiated with vector X_{ni}. The initialization of IoMT client sites is done using the below Equation (1) with i ranges from 1 to NP:

$$X_{ni} = l_i + [(rand\,(-1,1) + 2i - 1) * \times (u_i - l_i)]/2NP \qquad (1)$$

The u_i and l_i represent the upper and lower bounds of the parameters, respectively.

3.2.2. Solution Search by Employed Bee

The employed bee scours the neighborhood for new solutions. Using this Equation (2), a new answer can be found. The function τ_{ni} produces a random number in the range of -1 and 1, and X_{pi} is a local random solution. The fitness of the new candidate solution by employed bee C_{nie} is calculated and in case the fitness is high then the solution is memorized. The candidate solution using the below equation of employed bee helps in obtaining an improved feature selection for IoMT client sites.

$$C_{nie} = \begin{cases} \tau_{ni} + rand\,(X_{ni}, -X_{pi}); & \text{if } i = i', \\ X_{pi}, & \text{if } i \neq i'. \end{cases} \qquad (2)$$

3.2.3. Candidate Solution by Onlooker Bee

Employed bees share their candidate solution with onlooker bee and after that, the onlooker bees probabilistically choose their candidate solution C_{nio} using the below Equation (3). To further improve the quality of the candidate solution, the C_{nio} by onlooker bee is utilized as represented by the below equation.

$$C_{nio} = \frac{F_{ni}\,(X_n)}{\sum_{i=1}^{m}(F_m)(X_n)} \qquad (3)$$

The fitness function F_n is computed using the below equation.

$$F_n = \begin{cases} \frac{1}{1+F_{obj}}, & \text{if } F_{obj} \geq 0, \\ 1 + \text{abs}(F_{obj}), & \text{if } F_{obj} < 0. \end{cases} \qquad (4)$$

3.2.4. Scout Bee Phase

The scout bee in M-ABC ensures that the new solution is explored, and it chooses a candidate solution C_{nis} using the firefly algorithm as depicted in below Equation (5), where C_{nis0} is the initial solution. If an employed bee fails to improve its solution within a predetermined time frame, it becomes a scout bee.

$$C_{nis} = C_{nis} + e^{-r_i^2}(C_{nis0} - C_{nis}) + (rand\,(0,1) - 0.5) \qquad (5)$$

3.2.5. Data Collection Using IoMT Clients

The IoMT devices are initially used to collect patient health information, and the connected devices communicate with one another when sending patient data. IoMT devices capture medical information from the patient's body after they are implanted, including the heart rate, blood pressure, glucose level, cholesterol, and pulse rate. Using the proposed M-ABC technique, these details are locally optimized within an IoMT local healthcare site, after which the local model from each IoMT local healthcare site is transferred to the global cloud. Patient data from the UCI repository is also used to assess the efficacy of the proposed technique.

3.3. Design of Proposed Framework

We briefly describe our proposed system model and technique in this section. Additionally, in this section, we provide a comprehensive overview of the federated matched averaging (FedMA) algorithm and M-ABC-based optimization for optimal feature selection and classification. The proposed system model is illustrated in Figure 1 below. We assume that there are five healthcare client sites and one cloud server, this setting can be scaled up for generalization. Our proposed framework consists of heart disease data collection devices that are located inside a healthcare site. Initially, a global model is disseminated by the global cloud towards the healthcare sites, after receiving the model from the cloud, the sites perform feature selection and classification using an M-ABC optimizer, after that perform training on the local data using the received model and then the healthcare sites upload their local model updates to the cloud. On receiving multiple updates of local models, a new global model is computed using FedMA, and this new model is then disseminated among the healthcare sites. In this way, all the training data remains on the device and the privacy concerns are overcome with increased prediction accuracy and less classification errors. The working of the proposed framework is illustrated in Algorithm 2.

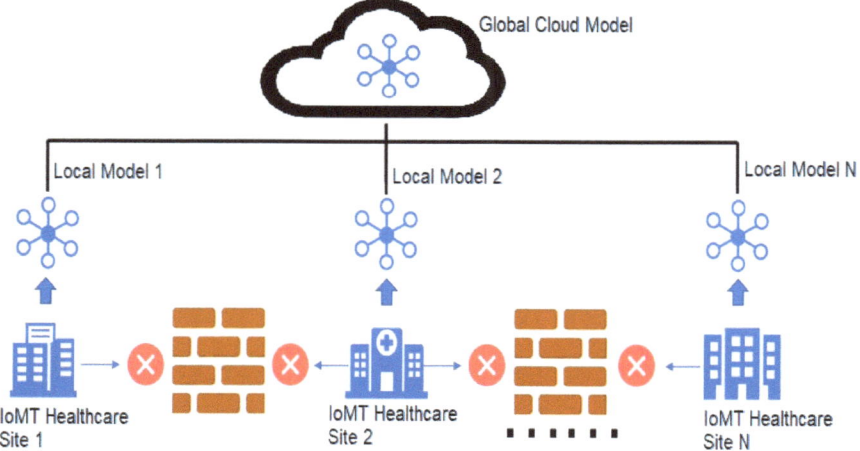

Figure 1. Overview of proposed framework.

FedMA calculates the Maximum Aposteriori Estimate (MAE) of a Bayesian nonparametric model using the Beta Bernoulli process (BBP) using equation 6 below. The w_{jl} be l^{th} neuron studied on the dataset j and $c(..)$ be an appropriate function of similarity. In case the client data sizes are imbalanced, then weighted averaging can be used instead of uniform. The similarity function $c\ (w_{jl},\ \theta_i)$ is the subsequent posterior probability of jth client neuron l generated from a Gaussian with mean θ_i. Due to the nonparametric aspect, their BBP-MAP inference approach allows a number of neurons in the federated model to mildly grow in comparison to the client model sizes. This matched averaging-based global cloud model helps in reducing the communication size to reach the target accuracy and the overall convergence rate of the model is also improved.

$$\min_{\{\pi_{li}^j\}} \sum_{i=1}^{L} \sum_{j,l} \min_{\theta_i} \pi_{li}^j \cdot c(w_{jl}) \text{ s.t. } \sum_i \pi_{li}^j = 1 \forall j, l; \sum_l \pi_{li}^j = 1 \forall\ i, j \qquad (6)$$

Algorithm 2: Learning method of proposed framework for healthcare. The K number of users is listed as k, local minibatch size is shown by β, learning rate is represented by η, and local epochs are represented using E.

Input: Data from various healthcare users $\{U_1, U_2, ---, U_N\}$
Output: Privacy-aware personalized model for each IoMT user ω_k
// **Processing at the global cloud end:**
1: Initialize a global cloud model ω_o
2: **for** every round $r = 1, 2, \ldots$ **do**
(i) $r \leftarrow 2190 \; maximum \; of \; (K, 1)$
(ii) $S_t \leftarrow (r$ is random number of clients)
3: **for** every client $k \in S_r$ **do in parallel**
(i) $\left\{\prod_r^k\right\} \leftarrow$ BBP-MAP $(\{k, C_n, \omega_r\})$ //call BBP-MAE to solve Equation (6)
(ii) $\omega_k \leftarrow \frac{1}{K} \sum_{k=1}^{K} \omega_r^k \prod_r^k$
(iii) $\omega_{r+1} \leftarrow \prod_r^k \omega_k$ //permutate the next weights
4: Distribute ω_k among all users
5: Repeat above steps with every evolving user data
// **Working at Client End** (k, ω):
1: **for each** client in k
(i) $\beta \leftarrow$ (fragment each P_k to groups of β size)
(ii) Compute candidate solution C_n using M-ABC Optimizer using Equations (2), (3), and (5)
2: **for** every local round $i = 1 \ldots E$ **do**
(i) **for** group $b \in \beta$ **do**
(a) $\omega \leftarrow \omega - \eta \nabla l \, (\omega; b)$
3: return ω to the cloud

The proposed framework is devised for both client and cloud ends. This proposed framework is implemented into three stages as described below:

1. Initial Phase: Initially, all the connected IoMT healthcare sites obtain an initial global model ω_o from the cloud and are initiated with vector X_{ni}.
2. Working at Cloud End: To retrieve the weights ω_k of the federated model, the cloud first collects only the weights from the clients and performs matched averaging. The clients then train their local model using their local data while the matching federated is kept frozen once the cloud broadcasts these weights to them. Then, we repeat this process up until the final round of communication.
3. Working at IoMT Client Sites: After data collection using IoMT devices, the collected data is fragmented into local minibatch of size β. The candidate optimal solution C_n for each β is computed using the M-ABC optimizer and the weights of the local computed solution from every IoMT client site are returned to the global cloud.

4. Experimental Evaluation and Results

In this section, we will discuss the simulation process of the proposed framework, simulation environment, and experimental settings for analyzing the efficiency of the proposed framework as a whole and contrast its performance with that of the standard federated learning models.

4.1. Experimental Setup

To evaluate the performance of the proposed framework, we conducted the simulation compromising of 4000 rounds of communication using a python environment using *PyTorch* machine learning libraries on Intel ® Core ™ i7-8550 @ 4GHz system and all the experimentation is performed in this simulated environment. Table 3 below describes the simulation parameters and settings utilized for the experiments.

Table 3. Simulation parameters and settings.

Parameter	Value
Simulation environment	Python
Dataset utilized	UCI Cleveland
Number of communication rounds	4000
Local epochs	{10, 20, 40, 80, 100, 120, 140, 160}
Volume of communication (in GBs)	{0.2, 0.4, 0.6, 0.8, 1.0, 1.2, 1.4, 1.6}
Number of client nodes	5

4.2. Results and Discussion

The performance of the proposed framework for heart disease in terms of prediction accuracy, time to reach the accuracy, communication efficiency, and effect of local epoch on accuracy are measured and compared with state-of-the-art FedSGD, FedAvg, FedMA, and PSO optimizer with FedMA techniques. Figure 2 below shows the comparison of convergence rate with prediction accuracy on the heart disease dataset. The proposed framework achieves 92.89% accuracy on 3000 rounds of communication which is higher than the state-of-the-art FL and FedMA with PSO algorithms. Because our proposed framework utilizes the M-ABC optimizer for healthcare user sites and FedMA for the cloud model, this enables the model to achieve better accuracy faster than existing federated learning algorithms. In FedSGD, FedAvg, and FedMA, the cloud model tends to perform the simple gradient, averaging, and matched averaging, respectively, but their client model does not have any algorithm for feature selection and classification which results in higher convergence time for the cloud model. In PSO with FedMA, the learning rate is improved but the classification and feature selection consume higher convergence, whereas in our proposed framework the learning rate tends to increase faster after every round as compared with FedAvg and FedMA. Therefore, our proposed framework achieves higher accuracy in a lesser number of rounds.

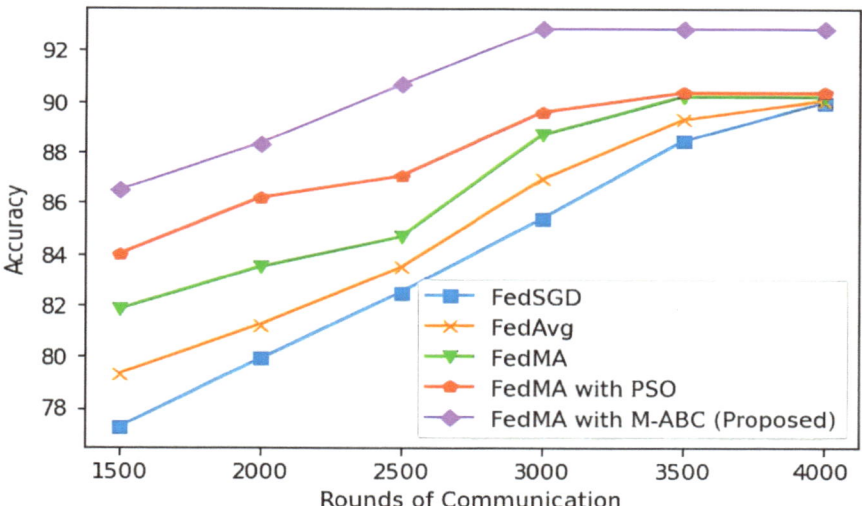

Figure 2. Comparison of communication efficiency.

We have conducted experiments on the effect of local epochs on the accuracy as compared to state-of-the-art FedAvg and FedMA algorithms on the heart disease dataset. We considered the local epochs E to be as {10, 20, 40, 80, 100, 120, 140, 160}. For every E, we evaluated the accuracy test of the proposed framework, FedAvg, and FedMA. The result is illustrated in Figure 3 below. We observed that training our proposed framework for a longer time favors the convergence rate because our proposed framework returns a better global model on the local model with higher model quality as our proposed technique utilizes a modified-ABC optimizer. For FedSGD, FedAvg and FedMA, both did not employ any optimizer, so their accuracy tends to deteriorate as they train for a longer period but in the case of PSO with FedMA, the accuracy remains constant after 80 local epochs which is due to the slow convergence rate of PSO algorithm. This result depicts that user sites can use our proposed framework to continue training their model's local users for as long as they wish.

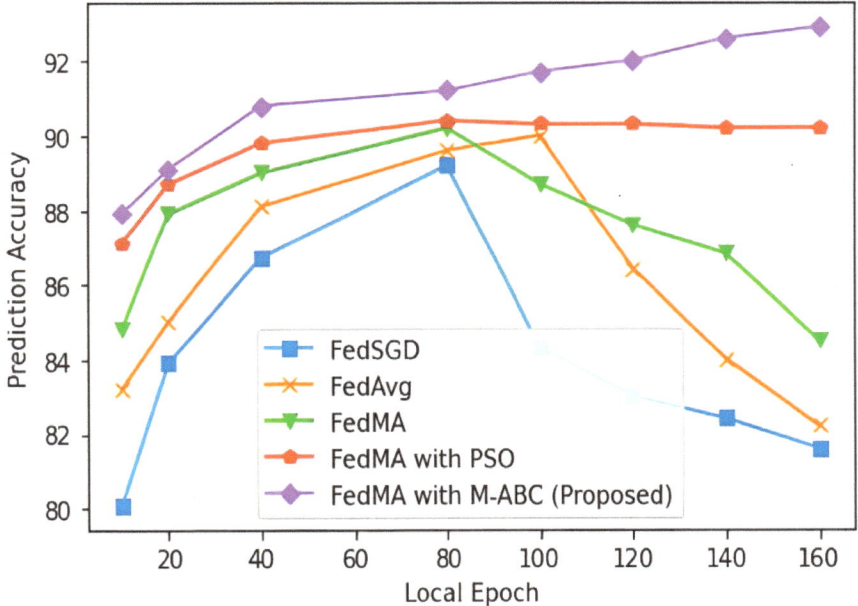

Figure 3. Effect of local epoch on accuracy.

We have evaluated and compared the performance of standard FL, PSO with FedMA and our proposed technique for the effect of prediction accuracy on the volume of communication. For this evaluation, we varied the volume of communication (in Gigabytes) as {0.2, 0.4, 0.6, 0.8, 1.0, 1.2, 1.4, 1.6} and recorded the prediction accuracy of each technique as shown in Figure 4 above. It is observed from the results that the proposed technique achieves better accuracy at both low and high volumes of communication as compared to standard FL and PSO with FedMA. Moreover, in Figure 5 below a comparison of the size of communication used to reach 90% prediction accuracy is illustrated. The proposed technique uses 20% less communication size (in GB) as compared to existing federated learning algorithms.

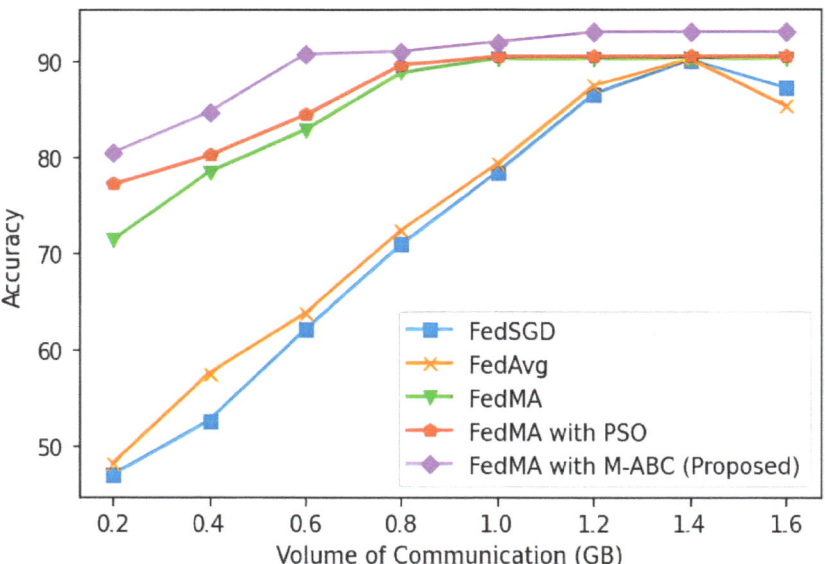

Figure 4. Effect of Accuracy on Amount of Communication.

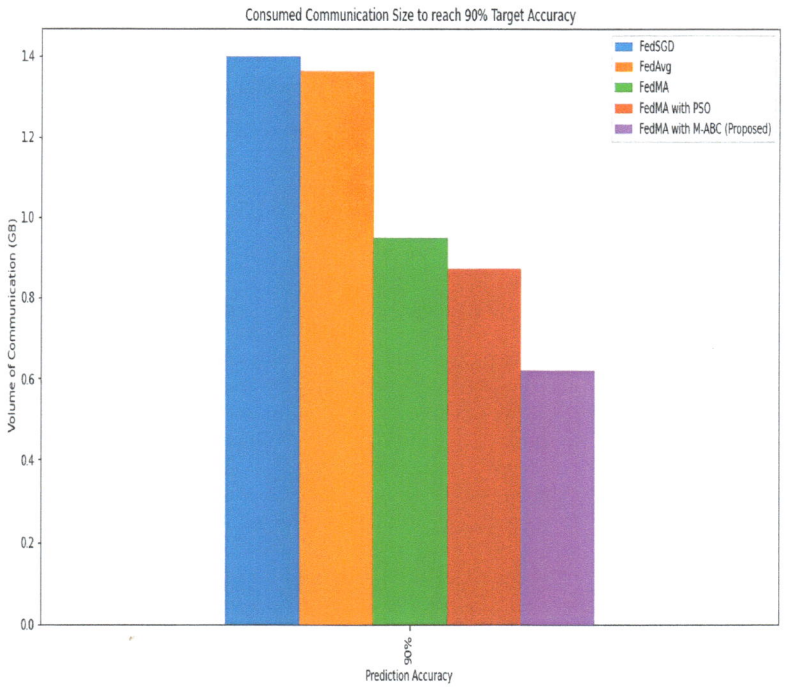

Figure 5. Comparison of communication size consumed to reach 90% prediction accuracy.

The performance metrics such as accuracy, precision, classification error, f-measure, specificity, sensitivity, and the number of rounds consumed to reach the highest accuracy are considered for the performance efficiency comparison of the proposed framework with FedAvg, FedMA, and PSO with FedMA. Accuracy in the context of machine learning means the percentage of all available instances that make the right predictions. Precision is defined as the percentage of correct predictions in the positive instance category. Classification errors are defined as the inaccuracies or percentages of errors available in the case. Three performance measurements are used to identify key features of heart disease. This helps to understand the behavior of different groups for a better selection of features. The results of these parameters are depicted in Tables 4 and 5. Our proposed framework achieves higher target accuracy in a lesser number of rounds as compared to vanilla FL, and PSO with FedMA for the heart disease dataset. As depicted in Table 4, our proposed method delivers a 22% reduction in the number of rounds as compared to FedSGD, FedAvg, and FedMA because the learning rate of our proposed model increases rapidly after every round which results in a 22% less number of rounds. As in Table 5, the proposed framework achieves better scores of prediction accuracy (92.89%), precision (94.2%), sensitivity (96.6), and specificity (81.8) as compared to existing FL algorithms on the heart disease dataset because the learning rate of our proposed model improves after every round of communication with less minibatch size. Hence our proposed framework is best suited for providing better heart disease prediction accuracy with privacy awareness as compared to existing FL algorithms. Moreover, the classification error of our proposed method is 11.8 which is less compared to FedAvg and FedMA due to the M-ABC optimization technique for feature selection and classification used in our proposed framework which results in less classification errors. The optimized features used for the M-ABC optimizer are shown in Table 6 with the details of achieved prediction accuracy. The M-ABC optimizer had 89% accuracy with five functions in the first experiment. Using the same dataset, the M-ABC optimizer with six features yielded 90% accuracy, and eight features achieved 92% accuracy.

Table 4. Time to reach the accuracy of model.

Technique	Accuracy after 4000 Rounds	# of Rounds to Reach 90%	Difference in # of Rounds
FedSGD	90	3988	–
FedAVG	90.07	3871	2.9%
FedMA	90.22	3495	12.4%
FedMA with PSO	90.38	3406	14.6%
FedMA with M-ABC (Proposed)	92.89	3018	24.3%

Table 5. Performance on full features set.

Technique	Accuracy	Precision	Classification Error	F-Measure	Specificity	Sensitivity
FedSGD	90	89.4	22.5	85.1	28.2	83.2
FedAVG	90.07	92.3	20.4	85.8	29.5	85.3
FedMA	90.22	90.1	18.6	86.6	52.5	89.5
FedMA with PSO	90.38	92.5	15.4	86.9	63.8	89.9
FedMA with M-ABC (Proposed)	92.89	94.2	11.8	90.1	81.8	96.6

Table 6. Optimized features with M-ABC optimizer.

Optimized Feature	Accuracy Achieved (in %)
Age, BP, Serum Chol., Rest ECG, Thallium Scan	89.82
Age, BP, Serum Chol., Hereditary, Rest ECG, Thallium Scan	90.72
Age, BP, Serum Chol., Hereditary, Rest ECG, Thallium Scan, Smoke, Diet	92.89

5. Conclusions

We proposed a privacy-aware decentralized federated learning framework for effective heart disease prediction in healthcare in this paper. The proposed framework is a hybrid method of FedMA and M-ABC optimization techniques to improve heart disease prediction while addressing privacy concerns in a healthcare system. The primary goal of this paper is to improve heart disease prediction accuracy as well as training time and communication efficiency. To ensure that our proposed framework is correct and valid, we evaluated and compared the performance in terms of various model prediction-based parameters and communication efficiency with the baseline federated learning FedAvg, FedMA, and with FedMA using PSO optimizer algorithms. The proposed framework indicated improved performance in terms of accuracy, classification error, precision, sensitivity, and communication efficiency. It is observed that the proposed framework provides 2.6% higher accuracy, 7% less classification error, 1.8% more precision, 7.1% higher sensitivity, and 12% fewer rounds are required to achieve the highest level of accuracy.

Our proposed model has some limitations, including the possibility of extending it for scalability in terms of the number of IoMT client sites with the effect of the learning rate on the overall model. In the future, we aim to further improve the privacy-aware healthcare predictive system by using other feature selection algorithms and optimization methods. The diagnosis, treatment, and control of health-related diseases is a major issue due to privacy concerns, hence, in the future, we will work on recovery and treatment of many other critical diseases such as breast cancer, diabetes, skin cancer, and Parkinson's Disease.

Author Contributions: Conceptualization, M.M.Y., M.N. and A.Y.; methodology, M.M.Y., A.D.A. and H.E.; software, M.M.Y. and A.A.S.; validation, M.M.Y., M.N., A.A.S. and M.A.K.; formal analysis, M.M.Y., M.N. and M.A.K.; investigation, M.M.Y.; resources, M.M.Y. and A.D.A.; data curation, M.M.Y. and A.Y.; writing—original draft preparation, M.M.Y., M.N. and A.A.S.; writing—review and editing, M.M.Y., M.N., A.Y., A.D.A. and M.A.K.; visualization, M.M.Y. and M.A.K. supervision, M.N. and A.Y.; project administration, M.N., H.E. and A.D.A. and M.A.K.; funding acquisition, A.D.A. and H.E. All authors have read and agreed to the published version of the manuscript.

Funding: Princess Nourah bint Abdulrahman University Researchers Supporting Project number (PNURSP2022R51), Princess Nourah bint Abdulrahman University, Riyadh, Saudi Arabia.

Data Availability Statement: We ran simulations to see how well the proposed approach performed. Any questions concerning the study in this publication are welcome and can be directed to the lead author (Muhammad Mateen Yaqoob).

Acknowledgments: The authors sincerely appreciate the support from Princess Nourah bint Abdulrahman University Researchers Supporting Project number (PNURSP2022R51), Princess Nourah bint Abdulrahman University, Riyadh, Saudi Arabia.

Conflicts of Interest: The authors declare no conflict of interest.

References

1. Al-Turjman, F.; Nawaz, M.H.; Ulusar, U.D. Intelligence in the Internet of Medical Things era: A systematic review of current and future trends. *Comput. Commun.* **2020**, *150*, 644–660. [CrossRef]
2. Dash, S.; Shakyawar, S.K.; Sharma, M.; Kaushik, S. Big data in healthcare: Management, analysis and future prospects. *J. Big Data* **2019**, *6*, 54. [CrossRef]
3. Watkins, D.A.; Beaton, A.Z.; Carapetis, J.R.; Karthikeyan, G.; Mayosi, B.M.; Wyber, R.; Yacoub, M.H.; Zühlke, L.J. Rheumatic heart disease worldwide: JACC scientific expert panel. *J. Am. Coll. Cardiol.* **2018**, *72*, 1397–1416. [CrossRef]
4. Mohan, S.; Thirumalai, C.; Srivastava, G. Effective Heart Disease Prediction Using Hybrid Machine Learning Techniques. *IEEE Access* **2019**, *7*, 81542–81554. [CrossRef]
5. Li, J.P.; Haq, A.U.; Din, S.U.; Khan, J.; Khan, A.; Saboor, A. Heart Disease Identification Method Using Machine Learning Classification in E-Healthcare. *IEEE Access* **2020**, *8*, 107562–107582. [CrossRef]
6. Voigt, P.; Von dem Bussche, A. Scope of application of the GDPR. In *The EU General Data Protection Regulation*; Springer: Cham, Switzerland, 2017; pp. 9–30.
7. Wagner, J. China's Cybersecurity Law: What you need to know. *The Diplomat*, 1 June 2017. Available online: https://thediplomat.com/2017/06/chinas-cybersecurity-law-what-you-need-to-know/ (accessed on 20 July 2022).

8. De la Torre, L. A Guide to the California Consumer Privacy Act of 2018. SSRN. 2018. Available online: https://papers.ssrn.com/sol3/papers.cfm?abstract_id=3275571 (accessed on 21 July 2022).
9. McMahan, B.; Ramage, D. Federated Learning: Collaborative Machine Learning without Centralized Training Data. Google AI Blog. 2017. Available online: https://ai.googleblog.com/2017/04/federated-learning-collaborative.html (accessed on 2 August 2022).
10. McMahan, B.; Moore, E.; Ramage, D.; Hampson, S.; Aguera y Arcas, B. Communication-Efficient Learning of Deep Networks from Decentralized Data. In Proceedings of the 20th International Conference on Artificial Intelligence and Statistics, Lauderdale, FL, USA, 20–22 April 2017; pp. 1273–1282.
11. Wang, G.G.; Deb, S.; Cui, Z. Monarch butterfly optimization. *Neural Comput. Appl.* **2019**, *31*, 1995–2014. [CrossRef]
12. Li, S.; Chen, H.; Wang, M.; Heidari, A.A.; Mirjalili, S. Slime mould algorithm: A new method for stochastic optimization. *Future Gener. Comput. Syst.* **2020**, *111*, 300–323. [CrossRef]
13. Elaziz, M.A.; Xiong, S.; Jayasena, K.; Li, L. Task scheduling in cloud computing based on hybrid moth search algorithm and differential evolution. *Knowl.-Based Syst.* **2019**, *169*, 39–52. [CrossRef]
14. Yang, Y.; Chen, H.; Heidari, A.A.; Gandomi, A.H. Hunger games search: Visions, conception, implementation, deep analysis, perspectives, and towards performance shifts. *Expert Syst. Appl.* **2021**, *177*, 114864. [CrossRef]
15. Ahmadianfar, I.; Heidari, A.A.; Gandomi, A.H.; Chu, X.; Chen, H. RUN beyond the metaphor: An efficient optimization algorithm based on Runge Kutta method. *Expert Syst. Appl.* **2021**, *181*, 115079. [CrossRef]
16. Heidari, A.A.; Mirjalili, S.; Faris, H.; Aljarah, I.; Mafarja, M.; Chen, H. Harris hawks optimization: Algorithm and applications. *Future Gener. Comput. Syst.* **2019**, *97*, 849–872. [CrossRef]
17. Mousavi, S.K.; Ghaffari, A.; Besharat, S.; Afshari, H. Improving the security of internet of things using cryptographic algorithms: A case of smart irrigation systems. *J. Ambient. Intell. Humaniz. Comput.* **2021**, *12*, 2033–2051. [CrossRef]
18. Mousavi, S.K.; Ghaffari, A.; Besharat, S.; Afshari, H. Security of internet of things based on cryptographic algorithms: A survey. *Wirel. Netw.* **2021**, *27*, 1515–1555. [CrossRef]
19. Aledhari, M.; Razzak, R.; Parizi, R.M.; Saeed, F. Federated Learning: A Survey on Enabling Technologies, Protocols, and Applications. *IEEE Access* **2020**, *8*, 140699–140725. [CrossRef] [PubMed]
20. Rahman, S.A.; Tout, H.; Ould-Slimane, H.; Mourad, A.; Talhi, C.; Guizani, M. A Survey on Federated Learning: The Journey From Centralized to Distributed On-Site Learning and Beyond. *IEEE Internet Things J.* **2021**, *8*, 5476–5497. [CrossRef]
21. Brisimi, T.S.; Chen, R.; Mela, T.; Olshevsky, A.; Paschalidis, I.C.; Shi, W. Federated learning of predictive models from federated Electronic Health Records. *Int. J. Med. Inform.* **2018**, *112*, 59–67. [CrossRef] [PubMed]
22. Nilsson, A.; Smith, S.; Ulm, G.; Gustavsson, E.; Jirstrand, M. A performance evaluation of federated learning algorithms. In Proceedings of the Second Workshop on Distributed Infrastructures for Deep Learning (DIDL), Rennes, France, 10 December 2018; ACM: New York, NY, USA, 2018; pp. 1–8.
23. Wang, H.; Yurochkin, M.; Sun, Y.; Papailiopoulos, D.; Khazaeni, Y. Federated learning with matched averaging. *arXiv* **2020**, arXiv:2002.06440.
24. Nguyen, H.T.; Sehwag, V.; Hosseinalipour, S.; Brinton, C.G.; Chiang, M.; Poor, H.V. Fast-Convergent Federated Learning. *IEEE J. Sel. Areas Commun.* **2021**, *39*, 201–218. [CrossRef]
25. Ma, Z.; Zhao, M.; Cai, X.; Jia, Z. Fast-convergent federated learning with class-weighted aggregation. *J. Syst. Arch.* **2021**, *117*, 102125. [CrossRef]
26. Salam, M.A.; Taha, S.; Ramadan, M. COVID-19 detection using federated machine learning. *PLoS ONE* **2021**, *16*, e0252573. [CrossRef]
27. Cheng, W.; Ou, W.; Yin, X.; Yan, W.; Liu, D.; Liu, C. A Privacy-Protection Model for Patients. *Secur. Commun. Netw.* **2020**, *2020*, 6647562. [CrossRef]
28. Fang, L.; Liu, X.; Su, X.; Ye, J.; Dobson, S.; Hui, P.; Tarkoma, S. Bayesian Inference Federated Learning for Heart Rate Prediction. In Proceedings of the International Conference on Wireless Mobile Communication and Healthcare, Virtual Event, 19 November 2020; Springer: Cham, Switzerland, 2020; pp. 116–130.
29. Babar, M.; Khan, M.S.; Din, A.; Ali, F.; Habib, U.; Kwak, K.S. Intelligent Computation Offloading for IoT Applications in Scalable Edge Computing Using Artificial Bee Colony Optimization. *Complexity* **2021**, *2021*, 5563531. [CrossRef]
30. Karaboga, D. Artificial bee colony algorithm. *Scholarpedia* **2010**, *5*, 6915. [CrossRef]
31. Zaman, S.K.U.; Jehangiri, A.I.; Maqsood, T.; Haq, N.U.; Umar, A.I.; Shuja, J.; Ahmad, Z.; Ben Dhaou, I.; Alsharekh, M.F. LiMPO: Lightweight mobility prediction and offloading framework using machine learning for mobile edge computing. *Clust. Comput.* **2022**, 1–19. [CrossRef]
32. Zaman, S.K.U.; Jehangiri, A.I.; Maqsood, T.; Umar, A.I.; Khan, M.A.; Jhanjhi, N.Z.; Shorfuzzaman, M.; Masud, M. COME-UP: Computation Offloading in Mobile Edge Computing with LSTM Based User Direction Prediction. *Appl. Sci.* **2022**, *12*, 3312. [CrossRef]
33. Zaman, S.K.U.; Jehangiri, A.I.; Maqsood, T.; Ahmad, Z.; Umar, A.I.; Shuja, J.; Alanazi, E.; Alasmary, W. Mobility-aware computational offloading in mobile edge networks: A survey. *Clust. Comput.* **2021**, *24*, 2735–2756. [CrossRef]
34. Shuja, J.; Alanazi, E.; Alasmary, W.; Alashaikh, A. COVID-19 open source data sets: A comprehensive survey. *Appl. Intell.* **2021**, *51*, 1296–1325. [CrossRef]

35. Manimurugan, S.; Almutairi, S.; Aborokbah, M.M.; Narmatha, C.; Ganesan, S.; Chilamkurti, N.; Alzaheb, R.A.; Almoamari, H. Two-Stage Classification Model for the Prediction of Heart Disease Using IoMT and Artificial Intelligence. *Sensors* **2022**, *22*, 476. [CrossRef]
36. Yuan, X.; Chen, J.; Zhang, K.; Wu, Y.; Yang, T. A Stable AI-Based Binary and Multiple Class Heart Disease Prediction Model for IoMT. *IEEE Trans. Ind. Inform.* **2022**, *18*, 2032–2040. [CrossRef]
37. Khan, M.A.; Algarni, F. A Healthcare Monitoring System for the Diagnosis of Heart Disease in the IoMT Cloud Envi-ronment Using MSSO-ANFIS. *IEEE Access* **2020**, *8*, 122259–122269. [CrossRef]
38. Yaqoob, M.M.; Khurshid, W.; Liu, L.; Arif, S.Z.; Khan, I.A.; Khalid, O.; Nawaz, R. Adaptive Multi-Cost Routing Protocol to Enhance Lifetime for Wireless Body Area Network. *Comput. Mater. Contin.* **2022**, *72*, 1089–1103. [CrossRef]
39. Li, C.; Hu, X.; Zhang, L. The IoT-based heart disease monitoring system for pervasive healthcare service. *Procedia Comput. Sci.* **2017**, *112*, 2328–2334. [CrossRef]
40. Khan, M.A. An IoT Framework for Heart Disease Prediction Based on MDCNN Classifier. *IEEE Access* **2020**, *8*, 34717–34727. [CrossRef]
41. Sarmah, S.S. An Efficient IoT-Based Patient Monitoring and Heart Disease Prediction System Using Deep Learning Modified Neural Network. *IEEE Access* **2020**, *8*, 135784–135797. [CrossRef]
42. Makhadmeh, Z.A.; Tolba, A. Utilizing IoT wearable medical device for heart disease prediction using higher order Boltzmann model: A classification approach. *Measurement* **2019**, *147*, 106815. [CrossRef]
43. Ganesan, M.; Sivakumar, N. IoT based heart disease prediction and diagnosis model for healthcare using machine learning models. In Proceedings of the 2019 IEEE International Conference on System, Computation, Automation and Networking (ICSCAN), Pondicherry, India, 29–30 March 2019; pp. 1–5. [CrossRef]
44. Albahri, A.S.; Zaidan, A.A.; Zaidan, B.B.; Alamoodi, A.H.; Shareef, A.H.; Alwan, J.K.; Hamid, R.A.; Aljbory, M.T.; Jasim, A.N.; Baqer, M.J.; et al. Development of IoT-based mhealth framework for various cases of heart disease patients. *Health Technol.* **2021**, *11*, 1013–1033. [CrossRef]
45. Gupta, A.; Yadav, S.; Shahid, S.; Venkanna, U. HeartCare: IoT Based Heart Disease Prediction System. In Proceedings of the 2019 International Conference on Information Technology (ICIT), Bhubaneswar, India, 19–21 December 2019; pp. 88–93.
46. Jabeen, F.; Maqsood, M.; Ghazanfar, M.A.; Aadil, F.; Khan, S.; Khan, M.F.; Mehmood, I. An IoT based efficient hybrid recommender system for cardiovascular disease. *Peer-to-Peer Netw. Appl.* **2019**, *12*, 1263–1276. [CrossRef]
47. Ashri, S.E.A.; El-Gayar, M.M.; El-Daydamony, E.M. HDPF: Heart Disease Prediction Framework Based on Hybrid Classifiers and Genetic Algorithm. *IEEE Access* **2021**, *9*, 146797–146809. [CrossRef]
48. Shin, S.; Kang, M.; Zhang, G.; Jung, J.; Kim, Y.T. Lightweight Ensemble Network for Detecting Heart Disease Using ECG Signals. *Appl. Sci.* **2022**, *12*, 3291. [CrossRef]
49. Ashfaq, Z.; Mumtaz, R.; Rafay, A.; Zaidi, S.M.H.; Saleem, H.; Mumtaz, S.; Shahid, A.; De Poorter, E.; Moerman, I. Embedded AI-Based Digi-Healthcare. *Appl. Sci.* **2022**, *12*, 519. [CrossRef]
50. Panniem, A.; Puphasuk, P. A Modified Artificial Bee Colony Algorithm with Firefly Algorithm Strategy for Continuous Optimization Problems. *J. Appl. Math.* **2018**, *2018*, 1237823. [CrossRef]

Article

Hybrid Classifier-Based Federated Learning in Health Service Providers for Cardiovascular Disease Prediction

Muhammad Mateen Yaqoob [1,*], Muhammad Nazir [1], Muhammad Amir Khan [2,*], Sajida Qureshi [3] and Amal Al-Rasheed [4]

1 Department of Computer Science, HITEC University Taxila, Taxila 47080, Pakistan
2 Department of Computer Science, COMSATS University Islamabad Abbottabad Campus, Abbottabad 22060, Pakistan
3 Computer Science Department, Faculty of Computer Sciences, ILMA University, Karachi 75190, Pakistan
4 Department of Information Systems, College of Computer and Information Sciences, Princess Nourah bint Abdulrahman University, Riyadh 11671, Saudi Arabia
* Correspondence: mateenyaqoob@gmail.com (M.M.Y.); amirkhan@cuiatd.edu.pk (M.A.K.)

Abstract: One of the deadliest diseases, heart disease, claims millions of lives every year worldwide. The biomedical data collected by health service providers (HSPs) contain private information about the patient and are subject to general privacy concerns, and the sharing of the data is restricted under global privacy laws. Furthermore, the sharing and collection of biomedical data have a significant network communication cost and lead to delayed heart disease prediction. To address the training latency, communication cost, and single point of failure, we propose a hybrid framework at the client end of HSP consisting of modified artificial bee colony optimization with support vector machine (MABC-SVM) for optimal feature selection and classification of heart disease. For the HSP server, we proposed federated matched averaging to overcome privacy issues in this paper. We tested and evaluated our proposed technique and compared it with the standard federated learning techniques on the combined cardiovascular disease dataset. Our experimental results show that the proposed hybrid technique improves the prediction accuracy by 1.5%, achieves 1.6% lesser classification error, and utilizes 17.7% lesser rounds to reach the maximum accuracy.

Keywords: heart disease prediction; hybrid technique; ABC-SVM; privacy-aware machine learning; intelligence-based healthcare

1. Introduction

The Internet of Things (IoT) enables the connectivity of physical objects and computational power so they may connect to the Internet. The IoT has the potential to assist in the development of applications that are both adaptable and efficient across a variety of industries, including healthcare, environmental monitoring, and industrial control systems. The IoT in the healthcare environment has led to the establishment of the Internet of Medical Things (IoMT), a cutting-edge area of a cyber physical system for wellness and wellbeing. Integrating these solutions into the HSP system has the potential to improve care services, quality of life, and open the door to cost-effective solutions [1]. For further analysis, the biomedical information pertaining to people is obtained. This information includes medical records, photographs, physiological signals, and many more forms. Given that the IoMT's cyber physical system collects data from several users, the volumetric scale of this biomedical data is enormous [2]. Smartwatches, wristbands, and other wearable sensing devices, among others, help in early illness diagnosis and warning. These wearable devices include strong and application-specific computational architecture that is housed in a distant HSP cloud data center, enhancing their capabilities (for real-time and early detection of health concerns). In IoMT-based healthcare solutions, wearable devices are often collected at the HSP's data center with the goal of preventing, diagnosing, and treating a variety of human health-related issues including cardiovascular illnesses (CVD). The construction of an effective electronic healthcare infrastructure is difficult due to the vast number of data that are gathered from multiple sources, including end users and other stakeholders in the delivery of health services.

According to World Health Organization (WHO) projections, CVD-related mortality accounts for close to 18 million fatalities annually globally [3]. Numerous risk factors, such as a history of heart attack, obesity, stress, high blood pressure, smoking, excessive use of alcohol, and high cholesterol, can all contribute to CVD. CVD impairs heart function and results in issues including strokes and impaired blood vessel function. The ability to treat CVD effectively and quickly is crucial for patient survival. The academic and industry communities are paying close attention to machine learning (ML)-based approaches for the accurate detection and prognosis of cardiac disorders. In [4], for example, a hybrid strategy based on a random forest and linear model was proposed to increase heart disease prediction accuracy. Authors in [5] proposed a feature selection and classification technique for identifying cardiac illness in an e-healthcare system.

To create an effective prediction model for tracking a patient's health state, traditional ML models are trained on vast amounts of user data. Although it is organized by several autonomous HSPs, these healthcare data are available in scattered, isolated silos. Even if there are a lot of aggregate data in different businesses, sharing the data is limited because of worries about security and privacy. Similarly, collected user data from the crowd are too restricted. These restrictions are enforced through regulatory laws such as the European Union's GDPR [6], China Cyber Security Law [7], and the United States' CCPA of 2018 [8]. Hence, it is not trivial to accumulate large amounts of user data in real-time healthcare applications to train powerful predictive models with high-quality training data. On the other hand, if the collection of user data is allowed, it is still not trivial to process these crowd-generated data, since the volume and velocity of the incoming data at the central server of HSP put a lot of burden on the network backhaul, delimited by the processing and storage capabilities of the central server. Indeed, with these restrictions in place, the number of training samples would not be large enough to generalize the model, affecting the performance of the trained model. To overcome these challenges, Google in [9,10] proposed federated learning (FL): a combination of distributed and incremental machine learning. FL is a distributed privacy-preserving machine learning technique that enables the collaborative training of a shared global predictive model without the need of uploading private local data to a central server to overcome the privacy concerns caused by centralized machine learning.

The FL algorithm's efficacy can be further improved by introducing feature selection at the distributed nodes. Feature selection will improve the identification of common features set in the sensory health data and distributed over the healthcare registries. Furthermore, feature selection will also help in dimensionality reduction to lower the computational cost and the model size. In this regard, recently, a feature-optimized federated learning-based technique was proposed in [11]; they addressed the issue of dimensionality reduction and communication efficiency for heart disease by improving the distributed nodes' learning technique. For the security and privacy issues in the cloud computing environment, federated learning incentive-based mechanisms were introduced in [12–14]. Recently, some meta-heuristic techniques have been proposed to further expand the solution search space for cloud-based healthcare systems [15,16]. These techniques also aim to minimize the fitness (objective) function though preserve the size of the population, increase weight adaptation rates, improve local search techniques, offer fitness function-improved computation, provide solutions to avoid local minima, and enhance the convergence rate of the algorithm.

Deep learning (DL) and SVM are both effective methods, although they are made to address distinct challenges. While DL is better suited for big datasets with many features, SVM works well for small to medium-sized datasets with few features. In comparison to

DL, SVM is less prone to overfitting and is known to be successful in high-dimensional space. SVM can be trained using various kernel functions, such as linear, polynomial, and radial basis functions (RBF), which can help to handle non-linear data. In tabular datasets, SVM can be used in conjunction with feature scaling, normalization, and dimensionality reduction techniques to improve the performance. Motivated by these highlighted issues of data privacy, improved feature selection, and classification for heart disease, in this manuscript, we proposed a privacy-aware FL-based framework that utilizes federated matched averaging at the HSPs' cloud end with a hybrid technique of modified artificial bee colony with support vector machine (MABC-SVM) for optimization for effective CVD prediction, respectively, at the client nodes. The M-ABC at the HSP client for optimal solution search works in four phases, i.e., the initialization phase, employed bee phase, onlooker bee phase, and the scout bee phase. These steps are described in detail in Section 4.1. The primary contributions of the proposal are enumerated as follows:

- An FL-based framework is proposed in this paper to overcome the problem of data privacy for HSP systems.
- We utilize the modified version of a federated matched averaging (FedMA) algorithm to preserve the privacy of heart disease data and to address the issues of the HSP's central model updation and communication efficiency.
- A hybrid technique comprised of a modified artificial bee colony and support vector machine (MABC-SVM) is proposed for the prediction of CVD with improved prediction accuracy. This hybrid algorithm is introduced at the client end of HSP.
- Our hybrid method's performance in terms of communication efficiency, classification error, and prediction accuracy is assessed and compared to current FL approaches.

The rest of the paper is organized as follows. Section 2 provides context for federated learning, the MABC method for optimum feature selection, and the SVM classification algorithm. Section 3 is an overview of relevant work. The proposed hybrid FL-based method is explained in Section 4. Section 5 is concerned with performance and the outcomes' evaluation. The conclusion and future work are presented in Section 6.

2. Background

In this section, we provide a brief overview of the methodologies that were used to build the FedMA with MABC-RB-SVM framework for privacy-aware heart disease prediction.

2.1. Basics of Federated Learning

To train a model, one needs access to data, which is the core of the area of artificial intelligence, and it frequently occurs in isolated data islands. The problem of isolated data silos is easily resolved by centralizing data processing. As international privacy protection laws for users strengthen, data collection for training models becomes more challenging. The issue of how to legally address data islands has sparked considerable discussion and research in the field of artificial intelligence. Traditional data analytics approaches are already at capacity due to the many rules that must be adhered to while attempting to address the data silo problem.

By jointly training algorithms without transferring the data, federated learning is a learning paradigm that aims to overcome the issues of data governance and privacy. While data are stored locally, federated learning trains statistical models across data silos. By retaining the data on the device, FL aims to provide a collaborative learning process that is privacy conscious and uses a shared model. As a result, users of FL will benefit from individualized machine learning that also addresses privacy concerns.

2.2. M-ABC-Based Optimization Algorithm

The modified artificial bee colony (M-ABC), a swarm intelligence-based technique, was proposed in [17]. The M-ABC is the upgraded version of artificial bee colony optimization, which is a method that mimics the intelligent foraging behavior of honeybee colonies to find the best solution to a problem. It simulates the bees' actions while they search for nectar, and keeps a group of potential answers, referred to as "bees", that explore the possible solutions and adjust their positions based on the quality of their findings. This method is particularly effective for solving complex optimization issues that are hard to resolve with traditional optimization methods. This M-ABC algorithm has three types of artificial bees such as onlooker, employed, and scout bee. While onlooker bees pick a source constructed on the employed bee's dance, scout bees oversee discovering new food sources. As a result of being connected to their food supply, the employed bees are shielded against exploitation. The observer bees and the scout bees are not connected to any food source. The primary goals of the fitness function are the best possible classification error and communication effectiveness of the models that are obtained from HSP sites. In order to increase accuracy, the fitness function seeks to reduce classification errors and round consumption.

In the M-ABC, the scout bee is combined with the firefly algorithm, and the modified technique is a combination of two different metaheuristic optimization techniques, namely, the artificial bee colony (ABC) and the firefly algorithm. The ABC algorithm generates a population of potential solutions, and the firefly algorithm improves the solutions by simulating the flashing behavior of fireflies. This hybrid algorithm is known for its global optimization abilities and has been applied to various optimization problems, showing better results than the original ABC or firefly algorithm alone. It is useful for solving complex optimization problems that are difficult to solve with traditional techniques.

2.3. SVM Classification Technique

The support vector machine (SVM) is widely used in intelligence-based systems for classification problems. The fundamental principle of the SVM classification algorithm is to discover the negative samples and the optimal selection for dividing positive samples. To attain the best generalization ability while remaining resistant to overfitting, the SVM attempts to determine the trade-off between lowering the training set error and maximizing the margin [18]. In addition, one of the best things about the SVM is that it uses convex quadratic programming, which gives only global minima and keeps the program from getting stuck in local minima.

The data are converted using a method called kernel trick using the SVM. The SVM kernel is a function that converts non-separable problems into separable problems by taking low-dimensional input space and transforming it into higher-dimensional space. Data conversion is used to determine the best splitting line among the expected outcomes. The border can range from a straightforward narrow margin for binary classes to a more challenging splitting including multiple classes [19].

3. Related Work

The contemporary machine learning research field is faced with two major challenges: data isolation and privacy and security issues. In methods utilizing standard ML, centralized training data are necessary. Around the world, laws are implemented to protect the privacy of data [6–8]. Therefore, the main difficulty for conventional machine learning algorithms is data privacy. A federated stochastic gradient descent (FedSGD) and federated averaging (FedAvg)-based technique first developed by Google in [9] offered some hope for overcoming these difficulties. In [10], a method based on FL was suggested to address the problems of data silos and privacy. To address the problems with data security in the conventional artificial intelligence field, they created an extensive architecture based on federated learning. Their suggested solution was divided into two categories: horizontal and vertical FL.

A description of the different machine learning deployment models, including centralized, distributed, and federated learning, was given by the authors in [20]. With careful consideration, they have described how machine learning architectures have developed. The authors of the research in [21] created a federated learning-based model for individuals with diseases that are likely to require hospitalization. They made use of data from electronic health records (EHRs) spread across various sources or agents. To use FL to solve the issue of large-scale sparse computing, the authors presented the clustering-based approach for dual splitting. Their suggested method yielded similar classifier prediction accuracy. The MNIST dataset was utilized by the authors of [22] to test and assess the three FL-based methods. A Bayesian correlated t-test was also employed. When client uploads were restricted to 10,000, FedAvg surpassed CO-OP and FSVRG algorithms, in their assessment. They have employed balanced data distribution, where each customer receives the same volume of information. The authors of [23] suggested a modified version of the standard FL with the aim of improving the algorithm's accuracy and convergence rate. To implement Bayesian non-parameterized approaches for heterogeneous data, they introduced the FedMA algorithm, which is a layer-wise version of the FL algorithm. Their suggested FedMA outperformed in terms of convergence, accuracy, and communication size reduction. The authors of [24] examined technical issues and other factors regarding the data privacy in the distributed implementation environment for FL algorithms. In their study, they outlined the features and results of a few of the optimization strategies for FL implementation. Additionally, they have discussed certain commercial consequences for federated learning that will be expected.

The authors of [25] have suggested an algorithm that distributes weights according to how much each class contributes to the local models. Using patients' chest x-ray data, machine learning-based algorithms can contribute to the identification of COVID-19. In contrast to conventional machine learning techniques, an FL version was suggested in [26] to discover COVID-19 with improved prediction accuracy. A blockchain-based approach based on federated learning was suggested by authors in [27] to address the problem of data privacy for IoMT-based healthcare systems. Their suggested solution was a hybrid strategy built on federated learning and the maximum approximation of the Gaussian mixture model, and it used blockchain to address the issue of user data privacy. Their suggested approach demonstrates that IoMT data training may be carried out utilizing local privacy to stop data leaking.

In the past, researchers gathered sensed data from HSP devices and then utilized that data to predict about several diseases. The authors of [28] suggested a version of FL with a Bayesian inference model to construct a privacy-aware heart rate prediction approach. Comparing this FedARX approach to conventional machine learning models, it achieves accurate and reliable heart rate prediction. A meta-heuristic method called artificial bee colony (ABC) optimization was suggested by the authors in [29] as a way to efficiently manage and optimize the calculation of offloading for IoT-based applications. Their method effectively controls the computing workload for IoT applications with limited resources. The authors in [30] suggested a fast-convergent technique that accomplishes intelligent selection of each device at every round of training the model in order to maximize the convergence speed of federated learning. To increase the convergence rate, their approach employs precise and efficient approximation for the transmission of a nearly optimum distribution of device selection.

Other approaches, such as a hybrid technique combining a linear discriminant analysis with modified ant lion optimization for classification [31], a gradient boosting decision tree with fuzzy logic algorithm [32], a hybrid of modified scalp swarm optimization and adaptive neuro-fuzzy inference system [33], and a multi-objective function using meta-heuristics [34], were also presented in the literature as strategies for predicting heart disease. The use of a hybrid classifier and a modified neural network with a deep learning focus was presented in [35–38] as a method for monitoring and predicting cardiac problems. Methods such as the Boltzmann-based model for higher order [39], modified hybrid method using classifiers and optimizers [40], two-stage-based localization of the classifiers [41], and hybrid classifier based on nave Bayes and random forest [42] were also anticipated to improve classification accuracy with less error.

4. Proposed Hybrid FL-Based Framework

This section suggests a federated learning architecture that addresses privacy concerns and effectively predicts cardiac disease in a healthcare system that is sensitive to privacy concerns. Table 1 provides a description of the symbols used in our suggested framework, where X_{io} represents the initial vector for client sites using the M-ABC algorithm; X_{ri} is the randomly chosen local solution; the candidate solutions of employed, onlooker, and scout bees are represented as C_{en}, C_{on}, and C_{sn}, respectively; Fit[n] is the fitness function; B is the size of the local batch at HSP clients; ω_o is the initial model disseminated by the HSP global orchestrator; similarity function is represented as $c\ (w_{jl},\ \theta_i)$; and the decision function based on RB-SVM for the heart disease dataset d is represented by $D_F\ (d)$.

Table 1. Brief description of the symbols utilized in our proposed framework.

Symbol	Brief Description
X_{io}	Initial vector for MABC at client sites
X_{ri}	Local solution chosen randomly
C_{en}	Employed bee's candidate solution
C_{on}	Candidate solution from onlooker bee
C_{sn}	Candidate solution obtained by scout bee
Fit[n]	Fitness function
N	Number of HSP clients
B	Local minibatch at every HSP client
ω_o	Initial model by HSP global orchestrator
$c\ (w_{jl},\ \theta_i)$	Similarity function
θ_i	Gaussian mean
w_{jl}	Weight of lth neuron on dataset j in MABC
E	Local epochs
η	Learning rate
ω_N	Model of Nth HSP client
d	Input dataset to RB-SVM
K_F	Kernel function
$D_F\ (d)$	Decision function on dataset d in RB-SVM
m_r	Margin function
RB_F	Radial basis function

In the FL-based HSP environment, we have a small number of HSP client devices and the data for independent model training on each device are not sufficient; an FedMA approach is more suitable. Our idea is to train multiple models independently on every HSP client and then average their predictions to produce a final prediction. This can lead to a better performance by reducing overfitting and increasing the diversity of the models. We propose an FL-based framework for privacy-aware prediction of the heart disease. Our proposed framework is constituted for the HSP client- and server-end. For the HSP client, the hybrid model of the RB-SVM with M-ABC for optimal classification and feature selection is proposed. For the HSP server, the FedMA is proposed to overcome the issues of HSP's central model updation and communication efficiency.

4.1. M-ABC-Based Feature Selection

At the HSP client-end, the M-ABC optimizer is used to choose features in the best way possible. The implementation of the M-ABC is completed in the four phases and each phase is described as follows:

4.1.1. Phase-I

Every client site of HSP is initialized by X_{io} vector. This initialization is achieved by using Equation (1):

$$X_{io} = l_i + rand(0,1) * (u_i - l_i) \tag{1}$$

where the u_i and l_i represent the upper and lower bounds of the parameters, respectively.

4.1.2. Phase-II (Searching of Candidate Solution by Employed Bee)

Using Equation (2), the bee searches the local HSP clients for new candidate solutions during this phase. The random integer generated by the function τ_{ni} falls between [−1 and 1], and the local random solution is represented by X_{ri}.

$$C_{en}[i] = X_{io} + \tau_{ni} * (X_{io} - X_{ri}) \tag{2}$$

The fitness function Fit[n] using Equation (3) determines the fitness of a new candidate solution, and if the fitness value is high, the solution is memorized.

$$\text{Fit}[n] = \begin{cases} \frac{1}{1+F_{obj}}, & \text{if } F_{obj} \geq 0, \\ 1 + \text{abs}(F_{obj}), & \text{if } F_{obj} < 0. \end{cases} \tag{3}$$

4.1.3. Phase-III (Onlooker Bee's Candidate Solution)

Employed bees present their potential solution to the onlooker bee, who then makes a probabilistic decision C_{on} using Equation (4).

$$C_{on}[i] = \frac{F_n[i](X_n)}{\sum_{i=1}^{m}(F_m)(X_n)} \tag{4}$$

4.1.4. Phase-IV (Scout Bee with Firefly)

By picking a C_{sn} solution using the firefly process as shown in the equation below, the scout bee ensures that the new solution is evaluated in Equation (5). An employed bee becomes a scout bee if it fails to improve its solution within a defined time range.

$$C_{sn}[i] = C_{sn}[i] + e^{-r_i^2}(C_{sn0}[i] - C_{sn}[i]) + (rand(0,1) - 0.5) \tag{5}$$

4.2. Classification Based on SVM

For a non-linear classification problem with a multidimensional set of features, the decision function using Equation (6) in terms of kernel function $K_F(d, d_j)$ for the input dataset d, m_r as the margin, and weight represented as $ϐj$, can be written as:

$$D_F(d) = \sum_{d=1}^{n} ϐj \cdot K_F(d, d_j) + m_r \tag{6}$$

To solve the heart disease (non-linear discrete) classification problem with a feature set of a high-dimension, kernel function is modified to be the radial basis function (RBF) $RB_F(d, d_j) = e^{(-\gamma|[d-dj]|^2)}$. Therefore, the decision function defined in the above Equation (6) is modified and is computed using Equation (7). In our proposed framework, the RB-SVM classifier is implemented at the HSP client side.

$$D_F(d) = \sum_{d=1}^{n} ϐj \cdot RB_F(d, dj) + m_r \tag{7}$$

4.3. Discussion on Proposed Framework

In this section, we present a full review of our proposed framework comprised of the hybrid MABC-SVM with FedMA-based technique for the effective prediction of CVD. Our suggested system model is depicted in Figures 1 and 2. Our suggested system consists of heart disease data-gathering equipment housed within a healthcare facility. Initially, the HSP global model orchestrator distributes a global model to HSP clients. The HSP clients, upon reception of this model, perform classification and optimal feature selection using our proposed MABC-SVM technique, and then perform the local training. The HSP client nodes send their updated local model towards the HSP central orchestrator. A new global model is generated using the FedMA after receiving repeated updates for local models, and it is distributed among the HSP clients. According to our proposed approach, the privacy issues are addressed as all of the CVD data never left the HSP client node, prediction accuracy is raised, and classification mistakes are decreased. Algorithm 1 below shows how our proposed framework functions.

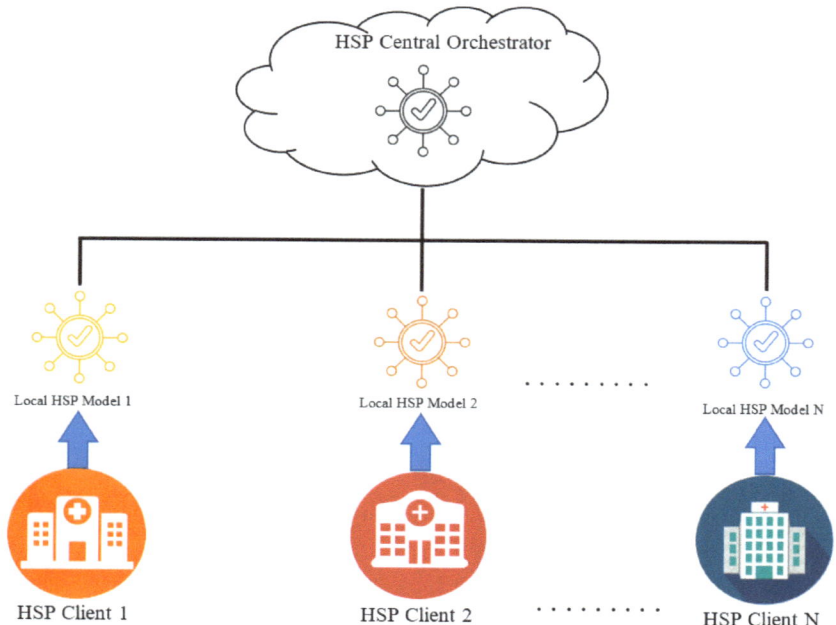

Figure 1. Overview of proposed hybrid federated learning framework.

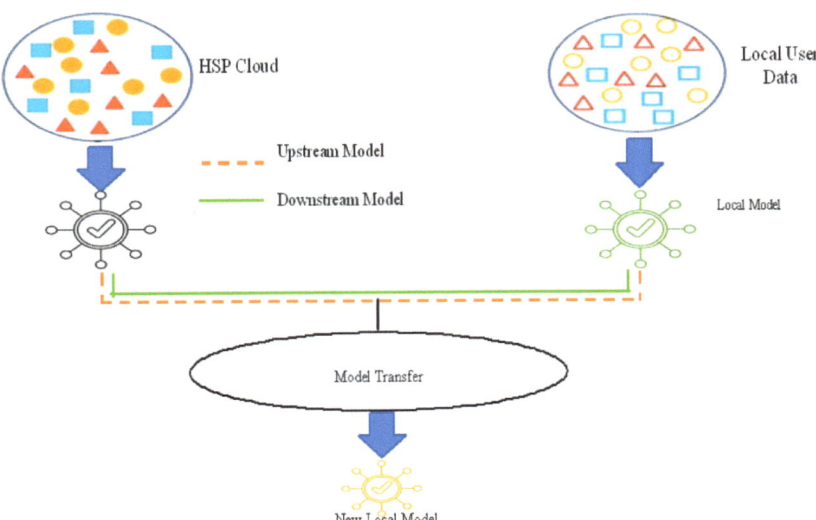

Figure 2. Illustration of model computation at HSP global model orchestrator and user ends.

The proposed FedMA-based framework obtains the maximum aposteriori estimate (MAE) using the Bernoulli process described below:

$$\min_{\{\pi_{li}^j\}} \sum_{i=1}^{L} \sum_{j,l} \min_{\theta_i} \pi_{li}^j \cdot c(w_{jl}) \text{ s.t. } \sum_i \pi_{li}^j = 1 \forall j, l; \sum_l \pi_{li}^j = 1 \forall i, j. \quad (8)$$

where the w_{jl} is lth neuron of the dataset j and an appropriate function of similarity is $c(..)$. The posterior probability of $c(w_{jl}, \theta_i)$ is computed on the jth client neuron l and mean Gaussian θ_i. The total neurons in the federated model can gradually increase in accordance to the sizes of the HSP client models because our suggested inference approach is not reliant on parameters. Our proposed framework is designed for the HSP clients and cloud sites, and it is executed in the following stages:

1. Stage-I (initial): An initial global model ω_o is disseminated to every HSP client user HC_N. After obtaining this initial model, the HSP client is initiated for initial feature selection using X_{i0}.
2. Stage-II (HSP clients): The client nodes will perform feature selection and classification of each fragmented local data of size β using a hybrid MABC with the RB-SVM technique. The updated weights of the local solution are returned to the HSP global orchestrator from every HSP client.
3. Stage-III (HSP global orchestrator): Upon reception of the weights from every HSP client, it performs the matched averaging and obtains an updated weight ω_N for the current round of communication.
4. Stage-IV (finalization at HSP global orchestrator): The updated weights ω_N are computed until there is no evolution in the HSP client models.

Algorithm 1: Proposed hybrid FL-based framework for heart disease prediction

Input: CVD Data from HSP clients $\{HC_1, HC_2, ---, HC_N\}$
Output: Privacy aware model for heart disease at HSP client user ω_N
// **Computation at the HSP global orchestrator:**

1: Initialize with global model ω_o
2: **for** each round $i = 1, 2, \ldots$ **do**
   i) $m \leftarrow max(N, 1)$
   ii) $S[t] \leftarrow$ (m is selected randomly for HSP clients)
3: **do in parallel for** each client $N \in S[t]$
   (i)   Compute inference method using Equation (8) with ($\{N, C_n, \omega_m\}$)
   (ii) $\omega_N \leftarrow \frac{1}{N} \sum_{k=1}^{N} \omega_m^N \Pi_m^N$
   (iii) $\omega_{m+1} \leftarrow \Pi_m^N \omega_N$ //next weights permutation
4: Disseminate ω_N among the HSP clients
5: Repeat until no evolution found in client models

// **Computation at the HSP Client End** (N, ω):

1: **foreach** client in N
   (i) $\beta \leftarrow$ (fragment the local data to β size each into P_N groups)
   (ii) Calculate C_n using MABC with Equations (2), (4) and (5)
   (iii) Perform the decision classification using RB-SVM classifier with Equation (7)
2: **for** every local $i = 1 \ldots E$ epochs **do**
   (i) **for** $b \in \beta$ groups **do**
      (a) Perform gradient descent using $(\omega; b)$
3: send back ω to the HSP global orchestrator

5. Experimental Evaluation and Validation

5.1. Simulation Setup

We use the Python environment (*PyTorch*) on a system with an Intel ® Core TM i7 @ 4 GHz and 16 GB RAM to run a simulation with 5000 communication rounds to evaluate the performance of the proposed framework. For standard FL algorithms at the client end, we implement the SVM classifier. Hence, the standard FL algorithms are versions of the SVM such as FedAvg-SVM, and FedMA-SVM, and we also develop an upgraded version of the vanilla FedMA compromising of a genetic algorithm (GA) and SVM as FedMA with GA-SVM. The effectiveness of our framework for heart disease is assessed and compared with state-of-the-art FedAvg, FedMA, and FedMA with GA-SVM approaches in terms of prediction accuracy, time to attain the accuracy, communication efficiency, and influence of the local epoch on accuracy. We consider the number of HSP client nodes to be five and one HSP server node. However, this proposed framework can be scaled-up for the HSP client nodes.

5.2. Dataset Description

Utilizing the combined dataset of five heart illness datasets, this dataset combines over eleven common features from the datasets of Cleveland, Stalog, Hungary, Long Beach, and Switzerland [43]. This dataset is used for the prediction of CVD, and it consists of various parameters for CVD. The dataset has records of CVD patients recorded using eleven heart disease features. The eleven CVD features of this dataset include resting blood pressure, cholesterol serum, chest pain, max heart rate, depression level, resting electrocardiogram, angina-induced by exercise, fasting blood sugar, ST slope, age, and sex. We train and evaluate our suggested framework on this combined dataset (this dataset is available at https://www.kaggle.com/fedesoriano/heart-failure-prediction, accessed on 8 December

2022). There are 918 entries in the combined dataset collection, along with 76 characteristics in each dataset. Table 2 provides an illustration of the dataset's complete description. The many risks for developing heart disease are included in this table along with their descriptions and encoded values. Our suggested approach uses the encoded values as its input. For the experimentation of this dataset using our proposed framework, the control group refers to the group of patients who do not have heart disease (as determined by the target column of Table 2). The patients with heart failure are considered the experimental group. The target column in the dataset is used to distinguish between the two groups, with 0 indicating no heart failure and 1 indicating heart failure.

Table 2. Thorough description of the combined dataset.

S#	Feature	Explanation	Unit	Coded Values
1	Resting blood pressure (Rt_Bp)	In mmHg	Integer	Low Level = Below 120 = −1 Normal Level = 120–139 = 0 High Level = Above 139 = 1
2	Cholesterol serum (Cl_S)	In mg/dL	Integer	<200 mg/dL = Low = −1 200–239 mg/dL = Normal = 0 >240 mg/dL = High = 1
3	Chest pain (C_P)	Type of chest pain	String	Angina Typical (AT) = 2 Asymptomatic (AS) = 1 Angina Atypical (ATA) = 0 Non-Angina (NA) = −1
4	Max heart rate (MHR)	Maximum achieved heart rate in bpm	Integer	<69 bpm = Low = −1 70–90 bpm = Normal = 0 >91 bpm = High = 1
5	Depression level (Dp_L)	Old peak in ST (numeric value measured for depression level)	Float	<0.5 mm = Normal = 0 >0.5 mm = High = 1
6	Resting electro-cardiogram (Rt_ECG)	Normal, ST T, or LVH	String	LVH = 2 ST T = 1 Normal = 0
7	Angina induced by exercise (AI_bE)	Yes or No	String	Yes = 1 No = 0
8	Fasting blood sugar (F_BS)	>120 mg/dL	Integer	True = 1 False = 0
9	ST slope (ST_S)	Peak exercise slope	String	Up = 2 Flat = 1 Down = 0
10	Age (A)	Age in years	Integer	>77 = 2, 64–77 = 1, 47–63 = 0, 35–46 = −1, <35 = −2
11	Sex (S)	Female and Male	String	Female = 0, Male = 1
12	Target (heart disease)	Yes or No	Integer	Yes = 1, No = 0

5.3. Results and Discussion

In the FL-based HSP environment, the communication rounds refer to the number of times the HSP client model parameters are exchanged during the training process. If the large number of rounds are consumed by an FL model, then it will also increase the communication overhead and computational cost. Therefore, the number of communication rounds is a key aspect in the FL system's overall performance and efficiency. The impact of communication rounds on the algorithm's accuracy in making predictions on the combined dataset is seen in Figure 3. Our proposed framework reaches 93.8% accuracy within the 4500 rounds of communication, which is better than the existing FedAvg-SVM, FedMA-SVM, and FedMA with GA-SVM algorithms. Since the proposed MABC-RB-SVM method employs the hybrid of the MABC optimizer and RB-SVM classifier for optimal feature selection and classification at HSP clients, and for the HSP global orchestrator, we deploy the FedMA which permits our overall model to accomplish better accuracy in a lesser number of communication rounds than the existing FL algorithms. In the FedAvg-SVM and FedMA-SVM, the HSP overall model performs the simple averaging and matched averaging, respectively, on the simple SVM kernel at their client model algorithm which results in consuming higher communication rounds. The learning rate is increased in the GA-SVM with FedMA, but the convergence is consumed more by the classification and feature selection. Consequently, the proposed hybrid framework accomplishes better accuracy and reduces the amount of communication rounds used.

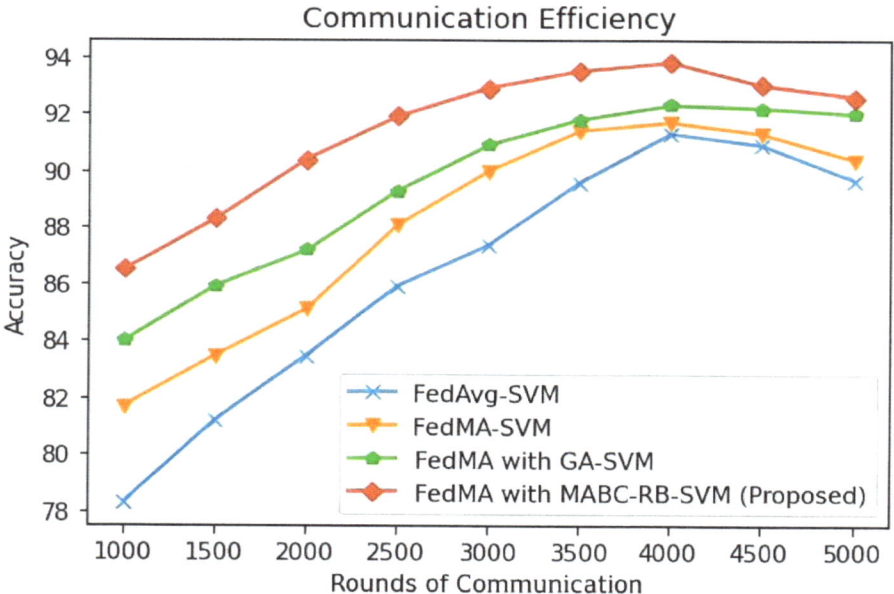

Figure 3. Comparison of convergence rate with prediction accuracy.

We evaluate and vary the local epochs E from 10 to 160, to examine the impact of local epochs on the prediction accuracy of the proposed hybrid technique and existing FL algorithms. The accuracy test on each E of the proposed hybrid framework, FedAvg-SVM, FedMA-SVM, and hybrid of GA-SVM with FedMA, is reviewed and compared. Figure 4 shows the outcome of this test. The findings show that the suggested framework can train for a longer period and supports a higher rate of convergence because it produces a better HSP global model on the local model with a higher model quality. This is because our suggested technique makes use of the RB-SVM and MABC methods at the client side of HSP. The accuracy of traditional FL algorithms, such as FedAvg and FedMA, tends to decrease with time due to the lack of an optimizer. However, in the case of the GA-SVM with FedMA, the accuracy does not decrease much after 100 local epochs, which is attributable to the GA algorithm. This result demonstrates that if user sites implement our recommended structure, they are free to train the local users of their model indefinitely.

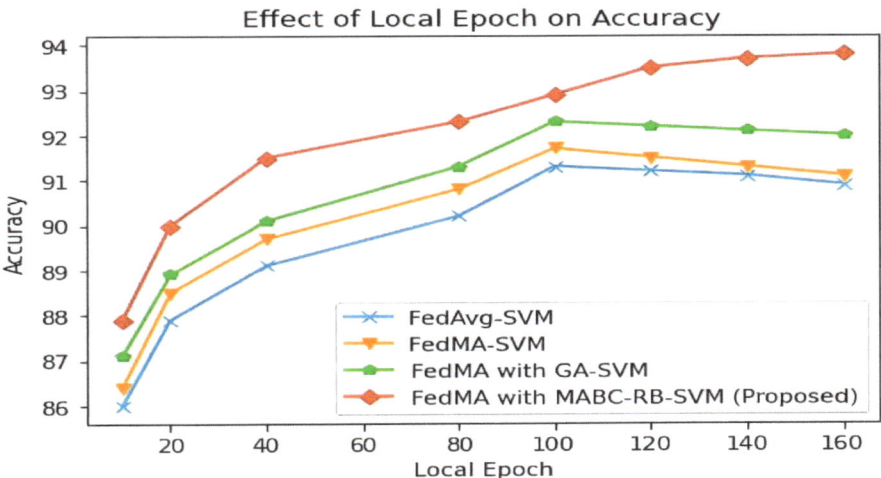

Figure 4. Comparison of the effect of local epoch on the prediction accuracy.

For the influence of prediction accuracy on the utilized communication volume, we examine and compare the performances of the FedAvg-SVM, Fed-MA-SVM, GA-SVM with FedMA, and our suggested approach. We vary the volume of communication (in Gigabytes) for this assessment as {0.6, 1.2, 1.8, 2.4, 3.0, 3.6, 4.2, 4.8, 5.4, 6.0} and record the prediction accuracy of each approach, as shown in Figure 5. The recorded results show that our hybrid approach outperforms traditional FL techniques and FedMA with the GA-SVM in terms of accuracy at both low and large communication volumes. Furthermore, Figure 6 depicts a comparison of the extent of communication necessary to achieve the various target prediction accuracy of the algorithms (70%, 75%, 80%, 85%, and 90%). The GA-SVM outperforms our suggested approach for a lower target accuracy of 70% and 75%. However, when compared to existing FL algorithms, our proposed approach consumes 15–25% less communication size (in GB) for improved target accuracy.

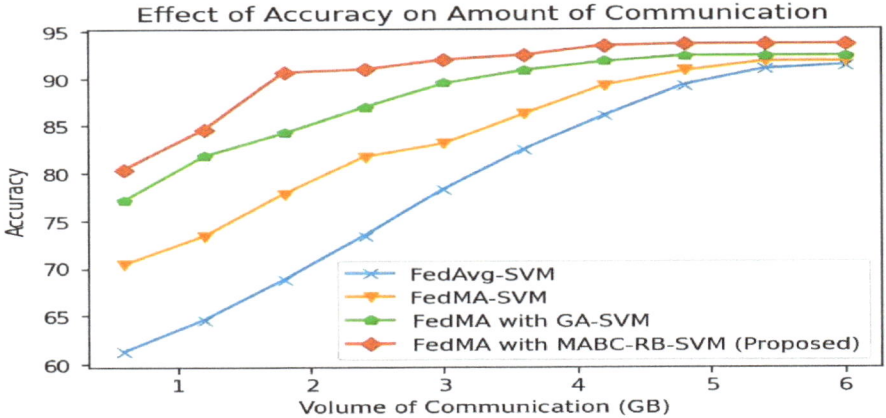

Figure 5. Analysis of prediction accuracy on the volume of communication.

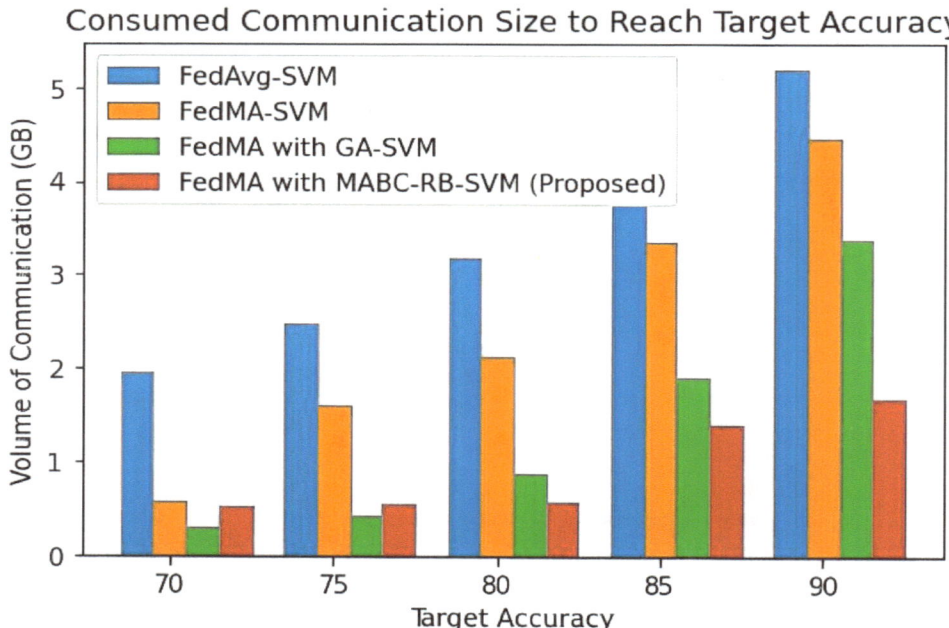

Figure 6. Comparison of consumed communication volume to reach target prediction accuracies.

Performance metrics such as accuracy, precision, classification error, f-measure, specificity, sensitivity, and the number of rounds required to achieve the highest accuracy are assessed for the performance efficiency comparison of the proposed framework with the FedAvg-SVM, FedMA-SVM, and GA-SVM with FedMA. Accuracy in machine learning refers to the proportion of all available examples that yield the right predictions. The fraction of accurate positive instance predictions is what is referred to as precision. Classification errors are defined as the inaccuracies or proportions of mistakes that are readily available in the instance. Three performance indicators are used to identify important heart disease symptoms. This makes it easier to comprehend how different groups behave and enables better feature selection. The results of these parameters are displayed in Tables 3 and 4. The created GA-SVM with FedMA for the heart disease dataset and baseline FL approaches are compared to our proposed framework, which achieves greater target accuracy in less cycles. The number of rounds in our suggested technique is reduced by 37% when compared to existing methods, as shown in Table 3, since our proposed model's learning rate grows quickly after each round, leading to fewer rounds. Table 3 demonstrates that the proposed framework performs better on the heart disease dataset than FL state-of-the-art methods in terms of prediction accuracy (93.8%), precision (94.2%), sensitivity (96.6), and specificity (81.8), because of the proposed model's improved learning rate that increases with each communication round with a smaller minibatch size. Our proposed framework is therefore more equipped to provide increased heart disease prediction accuracy while maintaining privacy when compared to existing baseline FL techniques. In addition, the MABC optimization technique for feature selection and the RB-SVM classification in our proposed framework result in decreased classification errors, resulting in a classification error of 11.9 for our recommended method.

Table 3. Consumed algorithm time for the highest accuracy of the model.

Techniques	Max. Accuracy Achieved	# Of Rounds to Reach 91%	Difference
FedAvg-SVM	91.3	3810	–
FedMA-SVM	91.7	3425	10.1%
FedMA with GA-SVM	92.3	3046	20.1%
FedMA with MABC-RB-SVM (Proposed)	93.8	2408	37.8%

Table 4. Comparison of performance on features of the dataset.

Techniques	Accuracy	F-Measure	Precision	Classification Error	Sensitivity	Specificity
FedAvg-SVM	91.3	87.3	92.3	20.4	85.3	59.5
FedMA-SVM	91.7	88.4	90.1	18.6	89.5	72.5
FedMA with GA-SVM	92.3	89.6	93.7	13.3	91.9	78.8
FedMA with MABC-RB-SVM (Proposed)	93.8	90.1	94.2	11.9	96.6	81.8

6. Conclusions and Future Work

For the objective of early illness detection and treatment, a health service provider (HSP) system to collect patient data in real time has been created. Intelligent healthcare systems can move quickly to save many lives, especially when a patient is in a remote place without access to medical treatment. It is challenging to predict survival in patients with cardiac disease. Due to privacy and security concerns, it is hard to exchange user data when it comes to healthcare systems. In this paper, we proposed a hybrid federated learning framework for improved heart disease prediction and to address privacy issues in the healthcare system. In order to enhance heart disease prediction, the proposed framework combines MABC with RB-SVM feature optimization and classification techniques at the HSP's client node, while FedMA is used at the HSP global orchestrator to solve communication efficacy and privacy problems in the healthcare system. The main goal of this research is to shorten training time and improve communication efficiency while improving the prediction accuracy of heart disease. We evaluated and compared the performance in terms of several model prediction-based metrics and communication efficiency with the baseline FedMA, FedMA, and a developed upgraded version of FedMA using a GA-SVM optimizer and classifier algorithms in order to ensure the accuracy and validity of our proposed framework. Performance metrics including prediction accuracy, classification error, sensitivity, precision, and communication efficiency all showed a considerable improvement under the suggested paradigm. Our findings indicated that the suggested strategies produce outcomes with 1.5% greater accuracy, 1.6% lower classification error, 4.7% higher sensitivity, and 17.7% fewer rounds needed to reach the greatest degree of accuracy. In the future, we will focus on the rehabilitation and treatment of several additional serious illnesses including Parkinson's, diabetes, liver cancer, skin cancer, and breast cancer.

Author Contributions: Conceptualization, M.M.Y. and M.N.; methodology, M.M.Y., M.A.K. and A.A.-R.; software, M.M.Y. and S.Q.; validation, M.M.Y., M.N., A.A.-R and M.A.K.; formal analysis, M.M.Y. and M.N.; investigation, M.M.Y. and S.Q.; resources, M.M.Y. and M.A.K.; data curation, M.M.Y. and M.N.; writing—original draft preparation, M.M.Y., M.N. and M.A.K.; writing—review and editing, M.M.Y., M.N., S.Q. and M.A.K.; visualization, M.M.Y. and M.A.K.; supervision, M.N.; project administration, M.N., S.Q., A.A.-R. and M.A.K.; funding acquisition, A.A.-R. All authors have read and agreed to the published version of the manuscript.

Funding: Princess Nourah bint Abdulrahman University Researchers Supporting Project number (PNURSP2023R235), Princess Nourah bint Abdulrahman University, Riyadh, Saudi Arabia.

Institutional Review Board Statement: Not applicable.

Informed Consent Statement: Not applicable.

Data Availability Statement: We ran simulations to see how well the proposed approach performed. Any questions concerning the study in this publication are welcome and can be directed to the lead author (Muhammad Mateen Yaqoob) upon request.

Acknowledgments: The authors sincerely appreciate the support from Princess Nourah bint Abdulrahman University Researchers Supporting Project number (PNURSP2023R235), Princess Nourah bint Abdulrahman University, Riyadh, Saudi Arabia.

Conflicts of Interest: The authors declare no conflict of interest.

References

1. Turjman, F.A.; Nawaz, M.H.; Uluser, U.D. Intelligence in the Internet of Medical Things era: A systematic review of current and future trends. *Comput. Commun.* **2020**, *150*, 644–660. [CrossRef]
2. Dash, S.; Shakyawar, S.K.; Sharma, M.; Kaushik, S. Big data in healthcare: Management, analysis and future prospects. *J. Big Data* **2019**, *6*, 54. [CrossRef]
3. Watkins, D.A.; Beaton, A.Z.; Carapetis, J.R.; Karthikeyan, G.; Mayosi, B.M.; Wyber, R.; Yacoub, M.H.; Zühlke, L.J. Rheumatic heart disease worldwide: JACC scientific expert panel. *J. Am. Coll. Cardiol.* **2018**, *72*, 1397–1416. [CrossRef] [PubMed]
4. Mohan, S.; Thirumalai, C.; Srivastava, G. Effective Heart Disease Prediction Using Hybrid Machine Learning Techniques. *IEEE Access* **2019**, *7*, 81542–81554. [CrossRef]
5. Li, J.P.; Haq, A.U.; Din, S.U.; Khan, J.; Khan, A.; Saboor, A. Heart Disease Identification Method Using Machine Learning Classification in E-Healthcare. *IEEE Access* **2020**, *8*, 107562–107582. [CrossRef]
6. Voigt, P.; dem Bussche, A.V. Scope of application of the GDPR. In *The EU General Data Protection Regulation*; Springer: Cham, Switzerland, 2017; pp. 9–30.
7. Wagner, J. China's Cybersecurity Law: What You Need to Know. The Diplomat. 2017. Available online: https://thediplomat.com/2017/06/chinas-cybersecurity-law-what-you-need-to-know/ (accessed on 10 October 2022).
8. de la Torre, L. A Guide to the California Consumer Privacy Act of 2018. 2018. Available online: http://dx.doi.org/10.2139/ssrn.3275571 (accessed on 10 October 2022).
9. McMahan, B.; Ramage, D. Federated Learning: Collaborative Machine Learning without Centralized Training Data. Google AI Blog 2017. Available online: https://ai.googleblog.com/2017/04/federated-learning-collaborative.html (accessed on 11 October 2022).
10. McMahan, B.; Moore, E.; Ramage, D.; Hampson, S.; Areas, B.A.Y. Communication-Efficient Learning of Deep Networks from Decentralized Data. In Proceedings of the 20th International Conference on Artificial Intelligence and Statistics, Fort Lauderdale, FL, USA, 20–22 April 2017; pp. 1273–1282.
11. Yaqoob, M.M.; Nazir, M.; Yousafzai, A.; Khan, M.A.; Shaikh, A.A.; Algarni, A.D.; Elmannai, H. Modified Artificial Bee Colony Based Feature Optimized Federated Learning for Heart Disease Diagnosis in Healthcare. *Appl. Sci.* **2022**, *12*, 12080. [CrossRef]
12. Xu, X.; Liu, W.; Zhang, Y.; Zhang, X.; Dou, W.; Qi, L.; Bhuiyan, M.Z.A. Psdf: Privacy-aware iov service deployment with federated learning in cloud-edge computing. *ACM Trans. Intell. Syst. Technol. (TIST)* **2022**, *13*, 70. [CrossRef]
13. Zhang, X.; Hu, M.; Xia, J.; Wei, T.; Chen, M.; Hu, S. Efficient Federated Learning for Cloud-Based AIoT Applications. *IEEE Trans. Comput.-Aided Des. Integr. Circuits Syst.* **2021**, *40*, 2211–2223. [CrossRef]
14. Yang, J.; Zheng, J.; Zhang, Z.; Chen, Q.; Wong, D.S.; Li, Y. Security of federated learning for cloud-edge intelligence collaborative computing. *Int. J. Intell. Syst.* **2022**, *37*, 9290–9308. [CrossRef]
15. Elaziz, M.A.; Xiong, S.; Jayasena, K.P.N.; Li, L. Task scheduling in cloud computing based on hybrid moth search algorithm and differential evolution. *Knowl.-Based Syst.* **2019**, *169*, 39–52. [CrossRef]
16. Yang, Y.; Chen, H.; Heidari, A.A.; Gandomi, A.H. Hunger games search: Visions, conception, implementation, deep analysis, perspectives, and towards performance shifts. *Expert Syst. Appl.* **2021**, *177*, 114864. [CrossRef]
17. Panniem, A.; Puphasuk, P. A Modified Artificial Bee Colony Algorithm with Firefly Algorithm Strategy for Continuous Optimization Problems. *J. Appl. Math.* **2018**, *2018*, 1237823. [CrossRef]
18. Chen, H.L.; Yang, B.; Liu, J.; Liu, D.-Y. A support vector machine classifier with rough set-based feature selection for breast cancer diagnosis. *Expert Syst. Appl.* **2011**, *38*, 9014–9022. [CrossRef]
19. Yadav, D.P.; Saini, P.; Mittal, P. Feature Optimization Based Heart Disease Prediction using Machine Learning. In Proceedings of the 2021 5th IEEE International Conference on Information Systems and Computer Networks (ISCON), Mathura, India, 22–23 October 2021; pp. 1–5.
20. Abdulrahman, S.; Tout, H.; Ould-Slimane, H.; Mourad, A.; Talhi, C.; Guizani, M. A Survey on Federated Learning: The Journey From Centralized to Distributed On-Site Learning and Beyond. *IEEE Internet Things J.* **2021**, *8*, 5476–5497. [CrossRef]

21. Brisimi, T.S.; Chen, R.; Mela, T.; Olshevsky, A.; Paschalidis, I.C.; Shi, W. Federated learning of predictive models from federated electronic health records. *Int. J. Med. Inform.* **2018**, *112*, 59–67. [CrossRef]
22. Nilsson, A.; Smith, S.; Ulm, G.; Gustavsson, E.; Jirstrand, M. A performance evaluation of federated learning algorithms. In Proceedings of the Second Workshop on Distributed Infrastructures for Deep Learning (DIDL), Rennes France, 10–11 December 2018; ACM: New York, NY, USA, 2018; pp. 1–8.
23. Wang, H.; Yurochkin, M.; Sun, Y.; Papailiopoulos, D.; Khazaeni, Y. Federated learning with matched averaging. *arXiv* **2020**, arXiv:2002.06440.
24. Aledhari, M.; Razzak, R.; Parizi, R.; Saeed, F. Federated Learning: A Survey on Enabling Technologies, Protocols, and Applications. *IEEE Access* **2020**, *8*, 140699–140725. [CrossRef]
25. Ma, Z.; Mengying, Z.; Cai, X.; Jia, Z. Fast-convergent federated learning with class-weighted aggregation. *J. Syst. Archit.* **2021**, *117*, 102125. [CrossRef]
26. Salam, M.A.; Taha, S.; Ramadan, M. COVID-19 detection using federated machine learning. *PLoS ONE* **2021**, *16*, e0252573.
27. Cheng, W.; Ou, W.; Yin, X.; Yan, W.; Liu, D.; Liu, C. A Privacy-Protection Model for Patients. *Secur. Commun. Netw.* **2020**, *2020*, 6647562. [CrossRef]
28. Fang, L.; Liu, X.; Su, X.; Ye, J.; Dobson, S.; Hui, P.; Tarkoma, S. Bayesian Inference Federated Learning for Heart Rate Prediction. In Proceedings of the International Conference on Wireless Mobile Communication and Healthcare, Virtual Event, 19 November 2020; Springer: Cham, Switzerland, 2020; pp. 116–130.
29. Babar, M.; Khan, M.; Din, A.; Ali, F.; Habib, U.; Kwak, K.S. Intelligent Computation Offloading for IoT Applications in Scalable Edge Computing Using Artificial Bee Colony Optimization. *Complexity* **2021**, *2021*, 5563531. [CrossRef]
30. Nguyen, H.T.; Sehwag, V.; Hosseinalipour, S.; Brinton, C.; Chiang, M.; Poor, H.V. Fast-Convergent Federated Learning. *IEEE J. Sel. Areas Commun.* **2021**, *39*, 201–218. [CrossRef]
31. Manimurugan, S.; Almutairi, S.; Aborokbah, M.; Narmatha, C.; Ganesan, S.; Chilamkurti, N.; Alzaheb, R.A.; Almoamari, H. Two-Stage Classification Model for the Prediction of Heart Disease Using IoMT and Artificial Intelligence. *Sensors* **2022**, *22*, 476. [CrossRef] [PubMed]
32. Yuan, X.; Chen, J.; Zhang, K.; Wu, Y.; Yang, T. A Stable AI-Based Binary and Multiple Class Heart Disease Prediction Model for IoMT. *IEEE Trans. Ind. Inform.* **2022**, *18*, 2032–2040. [CrossRef]
33. Khan, M.A.; Algarni, F. A Healthcare Monitoring System for the Diagnosis of Heart Disease in the IoMT Cloud Environment Using MSSO-ANFIS. *IEEE Access* **2020**, *8*, 122259–122269. [CrossRef]
34. Chhabra, A.; Singh, G.; Kahlon, K.S. Multi-criteria HPC task scheduling on IaaS cloud infrastructures using meta-heuristics. *Clust. Comput.* **2021**, *24*, 885–918. [CrossRef]
35. Li, C.; Hu, X.; Zhang, L. The IoT-based heart disease monitoring system for pervasive healthcare service. *Procedia Comput. Sci.* **2017**, *112*, 2328–2334. [CrossRef]
36. Khan, M.A. An IoT Framework for Heart Disease Prediction Based on MDCNN Classifier. *IEEE Access* **2020**, *8*, 34717–34727. [CrossRef]
37. Sarmah, S.S. An Efficient IoT-Based Patient Monitoring and Heart Disease Prediction System Using Deep Learning Modified Neural Network. *IEEE Access* **2020**, *8*, 135784–135797. [CrossRef]
38. Makhadmeh, Z.A.; Tolba, A. Utilizing IoT wearable medical device for heart disease prediction using higher order Boltzmann model: A classification approach. *Measurement* **2019**, *147*, 106815. [CrossRef]
39. Ganesan, M.; Sivakumar, N. IoT based heart disease prediction and diagnosis model for healthcare using machine learning models. In Proceedings of the 2019 IEEE International Conference on System, Computation, Automation and Networking (ICSCAN), Pondicherry, India, 29–30 March 2019; pp. 1–5.
40. Albahri, A.S.; Zaidan, A.A.; Albahri, O.S.; Zaidan, B.B.; Alamoodi, A.H.; Shareef, A.H.; Alwan, J.K.; Hamid, R.A.; Aljbory, M.T.; Jasim, A.N.; et al. Development of IoT-based mhealth framework for various cases of heart disease patients. *Health Technol.* **2021**, *11*, 1013–1033. [CrossRef]
41. Gupta, A.; Yadav, S.; Shahid, S.; Venkanna, U. HeartCare: IoT Based Heart Disease Prediction System. In Proceedings of the 2019 International Conference on Information Technology (ICIT), Bhubaneswar, India, 19–21 December 2019; pp. 88–93.
42. Jabeen, F.; Maqsood, M.; Ghanzafar, M.A.; Adil, F.; Khan, S.; Khan, M.F.; Mehmood, I. An IoT based efficient hybrid recommender system for cardiovascular disease. *Peer-to-Peer Netw. Appl.* **2019**, *12*, 1263–1276. [CrossRef]
43. Fedesoriano. Heart Failure Prediction Dataset. Available online: https://www.kaggle.com/fedesoriano/heart-failure-prediction (accessed on 28 November 2022).

Disclaimer/Publisher's Note: The statements, opinions and data contained in all publications are solely those of the individual author(s) and contributor(s) and not of MDPI and/or the editor(s). MDPI and/or the editor(s) disclaim responsibility for any injury to people or property resulting from any ideas, methods, instructions or products referred to in the content.

Article

End-to-End Deep Learning CT Image Reconstruction for Metal Artifact Reduction

Dominik F. Bauer *, Constantin Ulrich †, Tom Russ, Alena-Kathrin Golla, Lothar R. Schad and Frank G. Zöllner

Computer Assisted Clinical Medicine, Mannheim Institute for Intelligent Systems in Medicine, Medical Faculty Mannheim, Heidelberg University, 68167 Mannheim, Germany; constantin.ulrich@dkfz-heidelberg.de (C.U.); tom.russ@medma.uni-heidelberg.de (T.R.); alena-kathrin.golla@medma.uni-heidelberg.de (A.-K.G.); lothar.schad@medma.uni-heidelberg.de (L.R.S.); frank.zoellner@medma.uni-heidelberg.de (F.G.Z.)
* Correspondence: dominik.bauer@medma.uni-heidelberg.de
† Current address: Division of Medical Image Computing, German Cancer Research Center, 69120 Heidelberg, Germany.

Abstract: Metal artifacts are common in CT-guided interventions due to the presence of metallic instruments. These artifacts often obscure clinically relevant structures, which can complicate the intervention. In this work, we present a deep learning CT reconstruction called iCTU-Net for the reduction of metal artifacts. The network emulates the filtering and back projection steps of the classical filtered back projection (FBP). A U-Net is used as post-processing to refine the back projected image. The reconstruction is trained end-to-end, i.e., the inputs of the iCTU-Net are sinograms and the outputs are reconstructed images. The network does not require a predefined back projection operator or the exact X-ray beam geometry. Supervised training is performed on simulated interventional data of the abdomen. For projection data exhibiting severe artifacts, the iCTU-Net achieved reconstructions with SSIM = 0.970 ± 0.009 and PSNR = 40.7 ± 1.6. The best reference method, an image based post-processing network, only achieved SSIM = 0.944 ± 0.024 and PSNR = 39.8 ± 1.9. Since the whole reconstruction process is learned, the network was able to fully utilize the raw data, which benefited from the removal of metal artifacts. The proposed method was the only studied method that could eliminate the metal streak artifacts.

Keywords: image reconstruction; deep learning; metal artifacts; computed tomography

Citation: Bauer, D.F.; Ulrich, C.; Russ, T.; Golla, A.-K.; Schad, L.R.; Zöllner, F.G. End-to-End Deep Learning CT Image Reconstruction for Metal Artifact Reduction. *Appl. Sci.* **2022**, *12*, 404. https://doi.org/10.3390/app12010404

Academic Editors: Lucian Mihai Itu, Constantin Suciu and Anamaria Vizitiu

Received: 9 December 2021
Accepted: 30 December 2021
Published: 31 December 2021

Publisher's Note: MDPI stays neutral with regard to jurisdictional claims in published maps and institutional affiliations.

Copyright: © 2021 by the authors. Licensee MDPI, Basel, Switzerland. This article is an open access article distributed under the terms and conditions of the Creative Commons Attribution (CC BY) license (https://creativecommons.org/licenses/by/4.0/).

1. Introduction

The presence of high attenuation objects in the scanning field leads to artifacts in computed tomography (CT) imaging, which substantially decrease the image quality. The generic term for these kinds of artifacts is metal artifacts, which are a combination of beam hardening, scattering, photon starvation, and edge effects [1]. Metal artifacts are common in CT-guided interventions due to the presence of metallic instruments such as biopsy needles [2–4] or catheters [5]. In many interventions, iodine contrast agent is used, leading to additional beam hardening [6]. These artifacts often obscure clinically relevant structures, which can complicate the intervention. For example, the visibility of liver lesions is significantly reduced during liver biopsy [2] or during transarterial chemoembolization (TACE) [7,8], where catheters are used in combination with contrast agents.

Several CT reconstruction methods have been developed to improve image quality in the presence of metal objects. Statistical iterative reconstruction techniques can be used to correct beam hardening and thus mitigate metal artifacts [9]. Furthermore, dual-energy CT allows one to reconstruct virtual monoenergetic images at high kiloelectron volt levels, which substantially reduces metal artifacts [5,10]. The most common type of metal artifact reduction (MAR) method is based on inpainting projection data that has been affected by metal. In these approaches, the metal objects are first automatically detected (e.g., via thresholding) in the uncorrected CT image. The metal objects are then forward-projected into the sinogram domain to obtain a metal trace. The projection data in this metal trace are treated as missing data and are interpolated, e.g., via linear interpolation (LIMAR) [11]. Meyer et al. proposed a modification of the LIMAR approach called normalized MAR (NMAR) [12]. NMAR uses a forward projection of an image prior to flatten the uncorrected sinogram before interpolation. This additional step smoothes the sinogram, which reduces the streak artifacts caused by interpolation. In NMAR, the image prior is obtained by identifying air, soft tissue, and bone, in either the uncorrected CT or pre-corrected LIMAR image.

With the rapidly increasing popularity of deep learning in medical imaging in recent years [13], a plethora of novel MAR methods have emerged. Deep learning networks are mostly trained in a supervised manner and thus require a metal-free and a corresponding metal-affected dataset. These metal-affected data are commonly synthesized by inserting metallic objects into the metal-free data. Zhang et al. presented a convolutional neural network (CNN) called CNN-MAR, which outputs an improved image prior [14]. This image prior is forward-projected, and the resulting sinogram data are used to fill in the metal trace in the original sinogram. Several CNN approaches that operate in the sinogram domain have been introduced [15–17]. Lossau et al. developed a sophisticated sinogram inpainting approach that works in the presence of motion. A segmentation network identifies the metal trace in the projection domain; a second network fills in the missing sinogram data; and, after reconstruction, a third network reinserts the metal objects in the corrected image [18]. A popular class of deep learning MAR techniques are image-based CNNs. They take the uncorrected images as input and either learn a direct mapping to the artifact-free images [14,19,20] or to the artifact residuals [21]. These image-based methods often rely on input data that has already been pre-corrected to produce reasonable results [14,19]. Another option for MAR in the image domain is unsupervised image-to-image translation, which has the advantage that no synthesized metal artifacts are necessary and thus training can be conducted with unaltered clinical data [22–24]. Compared to supervised models, unsupervised models can achieve similar performance on synthetic data [22]. Lin et al. recently proposed an end-to-end trainable network called Dual Domain Network (DuDoNet) [25]. It consists of a sinogram enhancement network and an image enhancement network, which are connected by a Radon Inversion Layer (RIL). The RIL reconstructs the CT images using the filtered back projection (FBP) and allows gradient propagation during training.

In this work, we present an end-to-end deep learning CT reconstruction called iCTU-Net, for the correction of metal artifacts. The network learns the mapping from the metal-affected sinograms to the artifact-free images. It consists of three parts, which are trained simultaneously: sinogram refinement, back projection, and image refinement. To our knowledge, we are the first to train a single end-to-end deep learning network for the task of reducing metal artifacts with a learnable backprojection operation. Since the whole reconstruction process, including the back projection, is learned, the network is able to freely adapt the reconstruction to the imperfections of the sinogram data. The reconstruction is trained in a supervised manner with simulated interventional training data. We focus on liver interventions; thus, we generate abdominal liver data, including metal objects. We compare our iCTU-Net to the classical NMAR algorithm and to a sinogram refinement and an image refinement deep learning network. Both of these networks employ the same U-Net architecture that is used in our network, which allows for a fair comparison. These reference networks were selected to investigate the performance of deep learning MAR approaches in three different domains: sinogram pre-processing, image post-processing, and reconstruction.

2. Materials and Methods

2.1. iCTU-Net

The design of our iCTU-Net displayed in Figure 1a is based on the iCT-Net by Li et al. [26], which in turn is inspired by the classical FBP. The reconstruction is trained end-to-end, i.e., the inputs of the iCTU-Net are sinograms and the outputs are reconstructed images. The network includes pre-processing layers and aims to emulate the filtration of the sinograms and the back projection into the image domain. Post-processing layers were used to further refine the reconstruction. The network performs the complete CT image reconstruction and does not require a predefined back projection operator or the exact X-ray beam geometry.

In a first step, disturbances in the raw measurement data, such as excessive noise, are supposed to be suppressed using 3×3 convolutions (refining layers). The corrected sinogram is then filtered via 10×1 convolutions (filtering layers). By using 1×1 convolutions after the refining and filtering layers and by applying padding in all convolutions, the refined and filtered sinogram maintains the same size of the input sinogram. The convolutions in the refining layers employ a shrinkage activation function with a threshold of 0.0001 [26]. For the filtering layers, a tanh activation function is used. Afterwards, the refined and filtered sinogram is projected into the image space in a back projection step. This is realized by a $d \times 1$ convolution with N^2 output channels without padding, where d is the number of detector elements and N is the output image size. This convolution connects every detector element with every pixel in the image space. Since the back projection is learned, sinograms acquired with different beam geometries can be used to train the network, such as parallel beam and fan beam. Then, the results for each view angle v are reshaped to images of size $N \times N$ and rotated according to the acquisition angle. The acquisition angle of the projections is the only geometrical information provided to the network. The rotated images are linearly interpolated and cropped to maintain an image size of $N \times N$. The back projected image is then obtained by combining all views with a 1×1 convolution using a leaky Rectified Linear Unit (ReLU) activation function [27]. Finally, the image output is further refined by a U-Net. The U-Net is a popular choice for post-processing to reduce artifact in CT imaging [28].

2.2. Reference MAR Networks

To compare our iCTU-Net to other methods, we implement two deep learning MAR algorithms similar to those of Gjesteby et al. Both networks use pre-corrected NMAR inputs [17,19]. One is based in the projection domain (U-Net Sino), and the other one in the image domain (U-Net Image). To ensure comparability, we use the same U-Net architecture in the iCTU-Net, U-Net Sino, and U-Net Image. In the U-Net Sino, the sinograms are first refined by a U-Net, and the result is then reconstructed using the FBP [17]. In the U-Net Image, the sinograms are first reconstructed with the conventional FBP and then refined with a U-Net [19]. These reference networks were chosen to allow a comparison of sinogram pre-processing, image post-processing, and reconstruction deep learning MAR techniques.

The U-Net architecture is shown in Figure 1d and is similar to the original U-Net by Ronneberger et al. [29]. It has four en- and decoding blocks consisting of 3×3 convolutions, which are connected via skip connections. Zero-padding is used in the convolutions to ensure that the network output is the same size as the network input. The blocks of the top level have 32 channels, which are doubled with each encoding block until the lowest block has 512 channels. Downsampling in the contracting path is performed via 2×2 max-pooling with stride 2, while upsampling in the expansive path is accomplished using 3×3 transposed convolutions with stride 2. All convolutional layers are followed by a ReLU activation function.

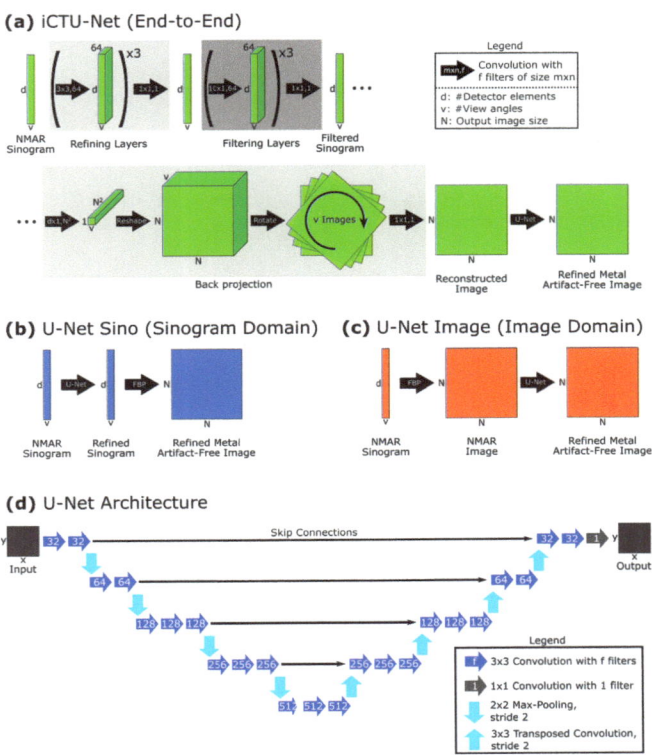

Figure 1. Deep learning architectures for metal artifact reduction. (**a**) iCTU-Net end-to-end CT reconstruction network architecture. (**b**) Sinogram domain U-Net. (**c**) Image domain U-Net. (**d**) U-Net network architecture used in the three networks in (**a**–**c**).

2.3. Data Generation

To simulate the training data, we use the XCAT phantom, which provides highly detailed whole-body anatomies [30]. The phantom includes female and male models of different ages, providing a wide variety of patient geometries. Further customization of anatomies by changing organ sizes is possible. We create 40 different XCAT models for training and 10 additional models for testing, resulting in 3964 and 991 slices of size 512×512 pixel with an in-plane resolution of $1 \times 1 \, mm^2$ and a slice thickness of 2 mm, respectively. Because we choose liver interventions as a use case, we generate abdominal XCATs that include the whole liver.

Organ masks can be easily obtained within the XCAT framework. Utilizing these organ masks, we insert metal structures inside the veins of the XCAT phantoms, emulating contrast agents or interventional instruments, such as catheters. Metal objects are only placed inside thicker blood vessels and have a uniform size, independent of the blood vessel size. This is realized by first eroding the blood vessel masks of the XCAT phantom, using a disk with a radius of 3 pixels as a structuring element. The erosion is performed to exclude the smallest blood vessels. To obtain the final metal mask, we skeletonize the mask and then increase the thickness via dilation using a disk with a radius of 3 pixels. An example is shown in Figure 2, with the metal mask in red, the initial blood vessels in white, and the liver in green. Most of the metal structures are placed inside the liver or in the portal vein beneath the liver.

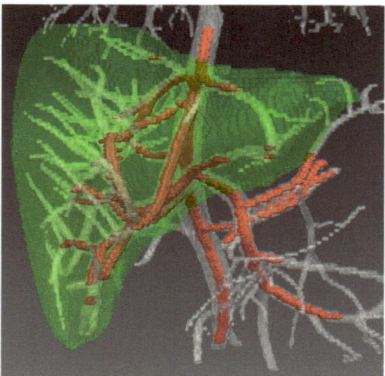

Figure 2. Generation of the metal masks inside blood vessels. The metal masks are shown in red, the initial blood vessel segmentation in white, and the liver in green.

Our data generation pipeline is shown in Figure 3, which starts with the generation of the ground truth data in the first row. First, we create sinograms by forward projecting the XCAT image data using a parallel beam geometry with 736 projection beams and 360 projection angles. A polychromatic X-ray spectrum and the energy-dependence of the absorption coefficients are considered in the forward projection:

$$I = \sum_{i=1}^{N} I_0 \cdot \eta(E_i) \cdot e^{-\int_L \mu(x, E_i) dx}, \qquad (1)$$

with weights of the energy spectrum $\eta(E_i)$. An incident flux of $I_0 = 4 \cdot 10^6$ photons is used, which is slightly increased compared to clinical levels [31], to combat photon starvation due to the presence of the metal objects. The X-ray energy spectrum is generated using the SpekCalc software with a tube peak voltage of 100 kVp and 1 mm aluminium filter [32]. We use 91 energy bins from 10 keV to 100 keV with a uniform size of 1 keV. The organ masks provided by the XCAT framework make it possible to assign an energy-dependent attenuation coefficient $\mu(x, E_i)$ to each organ. The sinograms are then reconstructed via a FBP. Since the energy dependence of the attenuation coefficients is accounted in the forward projection, beam hardening is present in the ground truth data.

To simulate data affected by metal, we utilize the previously mentioned metal mask to insert the attenuation coefficient of iron. Afterwards, metal sinograms are created via forward projection using Equation (1). Noise is then added, the projection data is normalized, and the negative logarithm is applied:

$$p_n = -\ln\left(\frac{Poisson(I) + \mathcal{N}(0, \sigma^2)}{\sum_{i=1}^{N} I_0 \cdot \eta(E_i)}\right). \qquad (2)$$

The photon production, attenuation, and detection is described by a Poisson distribution. Electronic noise of the detector is simulated with a Gaussian distribution $\mathcal{N}$ with a mean value of zero and $\sigma^2 = 40$ [33,34]. A subsequent FBP results in an image containing metal artifacts. As input for the training of our networks we do not use this artifact image; instead, we use data pre-corrected with the NMAR algorithm as shown in the third row of Figure 3. The prior image used for the normalization in NMAR is obtained by segmentation of soft tissue and bone in a LIMAR image [11,12].

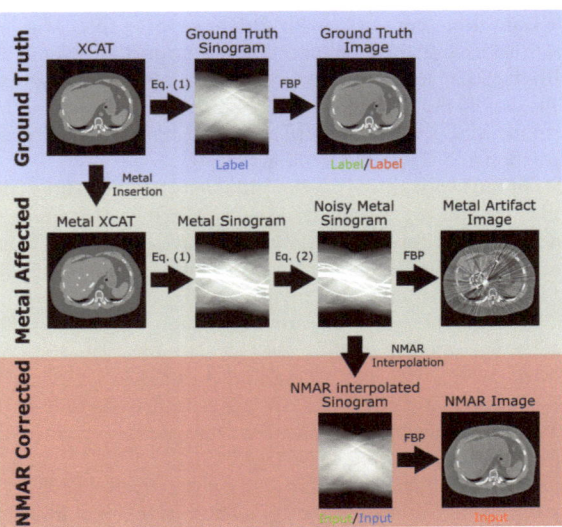

Figure 3. Generation of the metal training dataset. The input and labels for the three used deep learning networks are indicated by the colored words "Input" and "Label" beneath the corresponding images. The color coding corresponds to the colors used for the network architectures in Figure 1 (green: iCTU-Net, blue: U-Net Sino, and red: U-Net Image).

2.4. Training

The networks are trained with the SSIM loss function using the Adam optimizer with a learning rate of 0.001 [35]. We apply L2 regularization to the network weights, with a weighting factor of 10^{-6}. Each network is trained for 25 epochs. The training data in image domain is windowed to $[-1000, 1000]$ HU and then mapped to the interval $[-1, 1]$. The whole image slices are used for training, and no patches are extracted. The sinogram training data is neither windowed nor normalized. The input and label images for the iCTU-Net (green), U-Net Sino (blue), and U-Net Image (red) are noted in Figure 3.

2.5. Evaluation

The reconstructions are evaluated by calculating the peak signal-to-noise ratio (PSNR) and structural similarity (SSIM) for the test data. We set the background values of ground truth and reconstructions to -1000 HU to focus the analysis on the body region where the relevant anatomy is located. For the evaluation, the slices of the test dataset are divided into three categories: no metal, moderate metal artifacts, and severe metal artifacts, with 106, 748, and 137 slices, respectively. This separation allows one to evaluate the reconstructions when no metal is present. A slice is assigned to the severe metal artifact category if the FBP yields an SSIM value of less than 0.7, and to the moderate metal artifact category if the SSIM value of the FBP is greater than or equal to 0.7. The SSIM threshold is chosen, such that the number of slices with severe metal artifacts is similar to the number of slices without metal.

2.6. Experiments

We conduct three experiments. First, we configure our iCTU-Net in an ablation study. Then, we investigate the impact of different sinogram input data for training in an input study. Finally, we compare our best network configuration with state-of-the-art MAR algorithms.

In the ablation study, we investigate different post-processing layers and loss functions. The purpose of the ablation study is to find settings for the iCTU-Net that yield the best reconstructions. The resulting network configuration will be used in following studies.

We train three networks with different post-processing layers after backprojection: no post-processing, three convolution layers, and a U-Net. All of these networks are trained with the SSIM loss and with pre-corrected NMAR sinograms as input. To investigate the influence of the loss function, we additionally train the U-Net post-processing network with the MSE loss. Both SSIM and MSE are commonly used loss functions in CT artifact reduction and CT reconstruction [28].

In the input study, we train the network with different sets of training input data in addition to the previously used pre-corrected NMAR sinograms. The idea behind the input study is to find out how the network behaves for different kinds of sinogram input data. We use sinograms without metal (ground truth sinogram in Figure 3 but with additional noise added via Equation (2)) to investigate the network's performance if no metal is present. In this way, the reconstruction performance and the ability to mitigate metal artifacts can be evaluated separately. We calculate the evaluation metrics for different categories of artifact severity, even though none of the test data contain any metal. Nevertheless, the categories are used to allow fair comparisons to the other networks. We also train a network with uncorrected metal sinograms (noisy metal sinogram in Figure 3), to see if an NMAR pre-correction is necessary.

Finally, in the comparison study, we compare our iCTU-Net with the NMAR sinogram inpainting algorithm and the U-Net Sino and U-Net Image networks described earlier.

3. Results
3.1. Ablation Study

The results of the evaluation metrics for the ablation study are shown in Table 1, and reconstructed images are shown in Figure 4. We first investigate the impact of the different post-processing layers. Compared to using the U-Net for post-processing, using no post-processing and three convolutional layers performs generally worse, especially for severe artifacts. When using no post-processing, clear streak and extinction artifacts are present. Using the three convolutional layers for post-processing improves SSIM and PSNR and the streak and extinction artifacts disappear. However, the geometry of some soft tissue organs such as the liver is not reconstructed correctly, which is particularly evident for severe metal artifacts. Using the U-Net as the final layers of the network substantially improves the evaluation metrics, completely eliminates artifacts, and reconstructs organs more accurately. For no artifacts, the iCTU-Net underperforms compared to the FBP, especially in terms of PSNR. As shown by the arrows in the zoomed regions in Figure 4, the iCTU-Net is not capable of resolving small structures of only a few millimeters in size. From now on, we will only use the U-Net for post-processing as it yields the best results.

Finally, we train the iCTU-Net with the MSE loss. For no artifacts and moderate artifacts, the SSIM and PSNR evaluation metrics for the SSIM and MSE losses are similar. However, the SSIM metric for the MSE loss is considerably worse for severe artifacts and the reconstructions of the MSE iCTU-Net in Figure 4 look grainy. Thus, the network with U-net post-processing layers combined with SSIM loss performs best. Only this network configuration is referred to as iCTU-Net in this work.

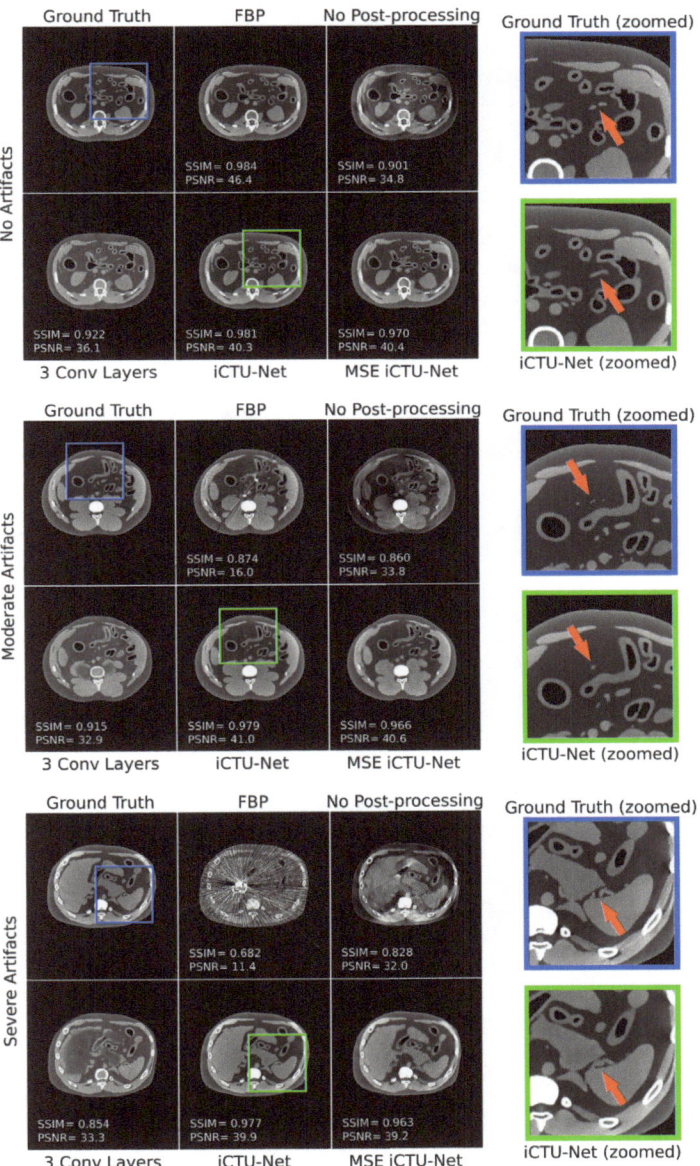

Figure 4. Results of the ablation study, where the ground truth and FBP are compared to different iCTU-Net settings. All networks are trained with pre-corrected NMAR sinograms and the SSIM loss, except for the MSE iCTU-Net, which is trained with the MSE loss. A slice without metal artifacts, with moderate metal artifacts, and with severe metal artifacts is shown. The scans are windowed to [−300 HU, 300 HU] to increase the visibility of the artifacts. The arrows in the zoomed regions indicate small structures that the iCTU-Net cannot resolve accurately.

Table 1. SSIM and PSNR evaluation metrics for the ablation study. All networks are trained with pre-corrected NMAR sinograms and the SSIM loss, except for the MSE iCTU-Net, which is trained with the MSE loss. The best result for each metric is marked **bold**.

		FBP	No Post-Processing	3 Conv Layers	iCTU-Net	MSE iCTU-Net
No Artifacts	SSIM	**0.988 ± 0.020**	0.865 ± 0.038	0.903 ± 0.028	0.969 ± 0.008	0.962 ± 0.010
No Artifacts	PSNR	**50.3 ± 4.9**	31.0 ± 2.9	30.5 ± 4.3	37.7 ± 1.8	37.3 ± 1.3
Moderate Artifacts	SSIM	0.869 ± 0.087	0.859 ± 0.036	0.897 ± 0.038	**0.976 ± 0.007**	0.967 ± 0.009
Moderate Artifacts	PSNR	15.2 ± 3.1	33.1 ± 2.1	34.3 ± 2.8	**39.5 ± 1.8**	39.3 ± 1.5
Severe Artifacts	SSIM	0.625 ± 0.056	0.715 ± 0.06	0.832 ± 0.024	**0.970 ± 0.009**	0.946 ± 0.013
Severe Artifacts	PSNR	10.6 ± 1.4	24.9 ± 5.6	34.2 ± 1.3	**40.7 ± 1.6**	39.6 ± 1.3
All Images	SSIM	0.848 ± 0.125	0.840 ± 0.06	0.889 ± 0.04	**0.975 ± 0.008**	0.964 ± 0.012
All Images	PSNR	18.3 ± 11.6	31.7 ± 4.0	33.9 ± 3.1	**39.5 ± 1.9**	39.1 ± 1.6

3.2. Input Study

In the input study, we investigate different sinogram inputs for the iCTU-Net. The results of the evaluation metrics for the input study are shown in Table 2, and reconstructed images are shown in Figure 5. The SSIM and PSNR in Table 2 show that the network performs similarly independent of the input. The network trained without metal in the input sinogram achieves the best PSNR, and the network trained with the pre-corrected NMAR sinograms achieved the best SSIM. However, these differences in SSIM and PSNR are not significant. For the metal input, some reconstruction inaccuracies close to metal objects can be observed, as indicated by the arrows in the zoomed images in Figure 5. Apart from this, the reconstructions in Figure 5 show no noticeable differences in image quality. Therefore, we continue to use the pre-corrected NMAR sinograms for the iCTU-Net. This allows for a fairer comparison with the deep learning reference methods, since they also use NMAR inputs.

Table 2. SSIM and PSNR evaluation metrics for the input study. The differentiation of artifact severity is not meaningful for No Metal Input because none of the test data contain metal. Since this network is not trained with any metal data, it is not suitable for artifact reduction. However, to allow a reasonable comparison to the other methods, we keep the categories, meaning the same slices are used for evaluation. The best result for each metric is marked **bold**.

		FBP	No Metal Input	Metal Input	iCTU-Net
No Artifacts	SSIM	**0.988 ± 0.020**	0.967 ± 0.006	0.968 ± 0.011	0.969 ± 0.008
No Artifacts	PSNR	**50.3 ± 4.9**	37.8 ± 1.5	37.4 ± 1.9	37.7 ± 1.8
Moderate Artifacts	SSIM	0.869 ± 0.087	0.974 ± 0.004	0.975 ± 0.009	**0.976 ± 0.007**
Moderate Artifacts	PSNR	15.2 ± 3.1	**40.2 ± 1.5**	39.6 ± 1.9	39.5 ± 1.8
Severe Artifacts	SSIM	0.625 ± 0.056	0.968 ± 0.005	0.962 ± 0.012	**0.970 ± 0.009**
Severe Artifacts	PSNR	10.6 ± 1.4	**40.7 ± 0.8**	40.4 ± 1.7	40.7 ± 1.6
All Images	SSIM	0.848 ± 0.125	0.972 ± 0.005	0.972 ± 0.011	**0.975 ± 0.008**
All Images	PSNR	18.3 ± 11.6	**40.0 ± 1.6**	39.5 ± 2.0	39.5 ± 1.9

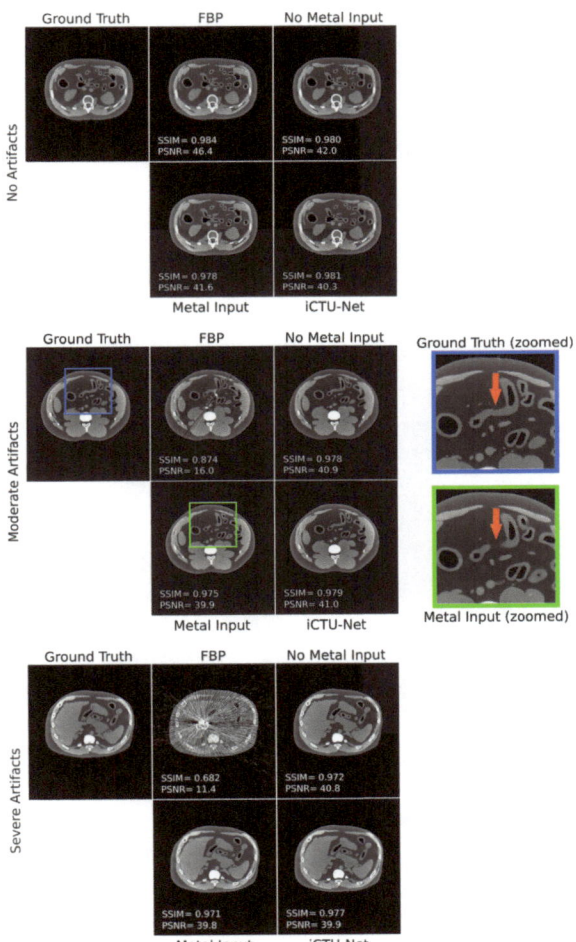

Figure 5. Results of the input study, where the ground truth and FBP are compared to iCTU-Nets trained with different input sinograms. All networks are trained with the U-Net post-processing layers and the SSIM loss, which yield the best results in the ablation study. No Metal Input, Metal Input, and iCTU-Net are, respectively, trained with metal-free, metal, and NMAR pre-corrected sinograms. A slice without metal artifacts, with moderate metal artifacts, and with severe metal artifacts is shown. The scans are windowed to [−300 HU, 300 HU] to increase the visibility of the artifacts. The arrows in the zoomed images indicate an anatomy that the Metal Input network cannot resolve accurately.

3.3. Comparison Study

The results of the evaluation metrics for the comparison study are shown in Table 3, and reconstructed images are shown in Figure 6. The deep learning reference methods U-Net Sino and U-Net Image both perform better than NMAR in terms of SSIM, especially for severe artifacts. In terms of PSNR, they perform worse when artifacts are not present and similarly when artifacts are present. The U-Net Image achieves a slightly higher SSIM than the U-Net Sino, but the performance of both methods is very similar. In Figure 6, no substantial removal of metal artifacts can be observed for the U-Net Sino and U-Net Image, only a smoothing of the streak artifacts is observed for the U-Net Image.

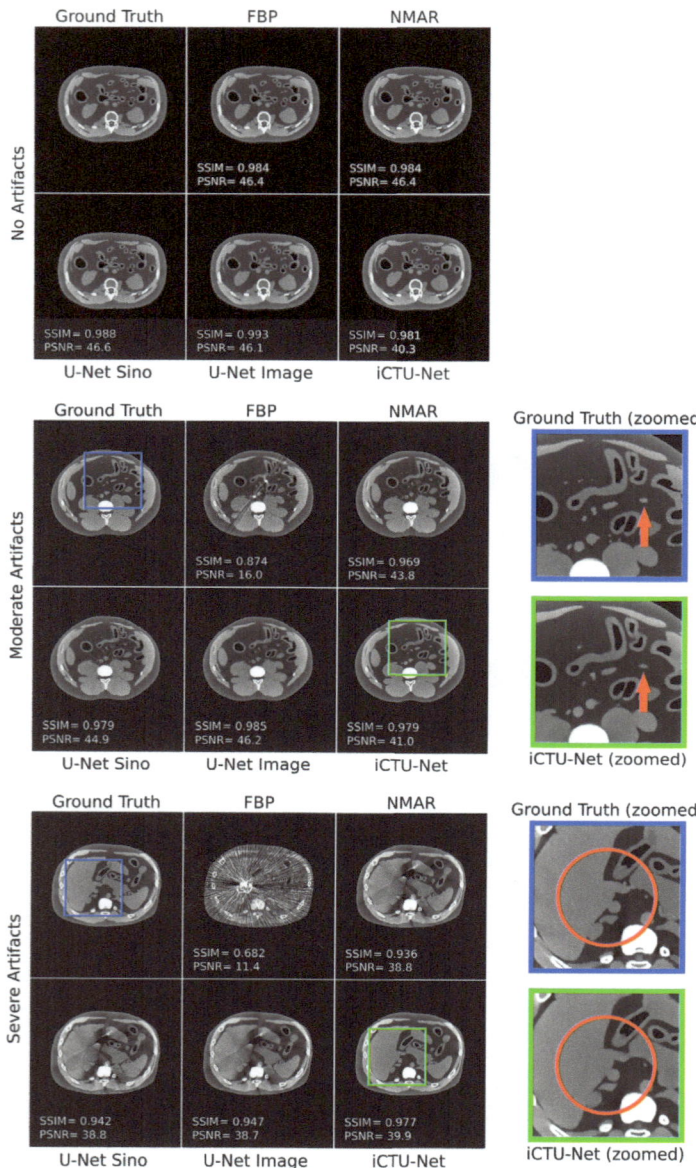

Figure 6. Results of the comparison study, where the ground truth and FBP are compared to NMAR, U-Net Sino, U-Net Image, and iCTU-Net. A slice without metal artifacts, with moderate metal artifacts, and with severe metal artifacts is shown. The scans are windowed to [−300 HU, 300 HU] to increase the visibility of the artifacts. The arrows and circles in the zoomed images indicate anatomies that could only be recovered by the iCTU-Net.

Table 3. SSIM and PSNR evaluation metrics for the comparison study. The best result for each metric is marked **bold**.

		FBP	NMAR	U-Net Sino	U-Net Image	iCTU-Net
No Artifacts	SSIM	0.988 ± 0.020	0.988 ± 0.020	0.990 ± 0.011	**0.993 ± 0.010**	0.969 ± 0.008
	PSNR	**50.3 ± 4.9**	**50.3 ± 4.9**	46.5 ± 2.1	45.3 ± 3.2	37.7 ± 1.8
Moderate Artifacts	SSIM	0.869 ± 0.087	0.976 ± 0.016	0.979 ± 0.012	**0.983 ± 0.011**	0.976 ± 0.007
	PSNR	15.2 ± 3.1	**44.3 ± 3.1**	43.9 ± 2.6	42.8 ± 2.1	39.5 ± 1.8
Severe Artifacts	SSIM	0.625 ± 0.056	0.911 ± 0.042	0.936 ± 0.025	0.944 ± 0.024	**0.970 ± 0.009**
	PSNR	10.6 ± 1.4	38.2 ± 2.4	39.3 ± 1.9	39.8 ± 1.9	**40.7 ± 1.6**
All Images	SSIM	0.848 ± 0.125	0.968 ± 0.032	0.974 ± 0.021	**0.978 ± 0.019**	0.975 ± 0.008
	PSNR	18.3 ± 11.6	**44.1 ± 4.4**	43.5 ± 3.1	42.7 ± 2.6	39.5 ± 1.9

Without artifacts, the iCTU-Net is outperformed by all methods in terms of PSNR and SSIM as they are all FBP-based and already outperformed the iCTU-Net in the ablation study. For moderate artifacts, the iCTU-Net achieves competitive SSIM values compared to the reference methods but performs worse in terms of PSNR. Nevertheless, the iCTU-Net is the only method capable of completely removing moderate metal artifacts, as shown in Figure 6. As indicated by the arrows in the zoomed images in Figure 6, the iCTU-net is also the only method that can restore a blood vessel into which a metal object has been inserted. For severe artifacts, the iCTU-Net performs better than all reference methods with SSIM = 0.970 ± 0.009 and PSNR = 40.7 ± 1.6. The second best method, the U-Net Image, only achieved SSIM = 0.944 ± 0.024 and PSNR = 39.8 ± 1.9. Averaged over all images, the SSIM of the iCTU-Net is competitive with the U-Net Image, but a worse PSNR is achieved. The iCTU-Net is able to remove severe metal artifacts completely, whereas for the other methods strong streak artifacts are still present over the whole image. The iCTU-Net can not only efficiently remove severe artifacts but also reliably restore the anatomy that is obstructed by these artifacts. This is especially evident inside the circles shown in the zoomed images in Figure 6. All other methods fail to restore the anatomy in this region.

4. Discussion

We trained the iCTU-Net with metal-affected data, to investigate its ability to mitigate metal artifacts. The iCTU-Net outperformed the reference methods for reconstructions with severe metal artifacts. Similar results were found for the application of the iCTU-Net to sparse-angle CT reconstruction, where the iCTU-Net showed good performance for a small number of projections [28]. However, the iCTU-Net was not able to resolve small structures of only a few millimeters in size. The reconstructions were slightly blurred, which is probably the reason why the iCTU-Net could not match the quality of the FBP when no metal was present. In the ablation study, it was found that the loss function and the post-processing layers have a major impact on the quality of the reconstruction. We had attempted to sharpen the reconstructed image by combining the SSIM loss with an additional gradient difference loss [36], but no substantial improvements were observed. In the future, we will investigate alternatives to the U-Net as post-processing layers to further optimize the network. The iCTU-Net was trained with a dataset of 3964 slices, of which only 310 contained no metal. Due to this small fraction of metal-free training data, the network might not be able to learn how to properly reconstruct metal-free sinograms. In the input study, we trained the reconstruction network exclusively with metal-free data to test this hypothesis. We found that the network trained with metal-free raw data did not perform better than the iCTU-Net for the no artifact category. Therefore, we can conclude that training the network with mainly metal-affected data does not degrade the quality of the reconstructions. Interestingly, the evaluation metrics for the moderate and severe artifact categories also did not differ substantially. Thus, the network trained with metal-affected input data reconstructs images with metal-affected test data just as well as the network trained without metal reconstructing images that do not include metal. This shows that the iCTU-Net reliably reduces metal artifacts. This is confirmed by the fact that all networks in the input study performed very similarly for all severities of artifacts. The network seems to handle metal objects in the raw data very well.

The input study showed that the iCTU-Net performs similarly regardless of the sinogram input data used. Training the network with uncorrected metal sinograms revealed similar performances compared to the network trained with pre-corrected NMAR sinograms. This means that reconstruction without pre-correction is feasible, which reduces the complexity of the algorithm.

In the comparison study, a sinogram pre-processing and an image post-processing approach were investigated. We have found that the image-based post-processing deep learning approach provides better results than the sinogram pre-processing approach. This is consistent with the findings of Arabi et al. [37]. Since the reference methods are all FBP-based, they are superior to the iCTU-Net in the absence of artifacts due to the aforementioned blurring. However, the artifacts introduced by the FBP cannot be completely mitigated by the reference methods. The iCTU-Net is the only method that removes all metal artifacts and yields the best results of all methods for severe metal artifacts. Since the iCTU-Net is trained end-to-end, the network can fully utilize the raw data and learn to reconstruct an artifact-free image. The U-Net Sino learns to mitigate disturbances in the sinogram with the raw data as input. However, small errors in the sinogram can lead to significant deviations in the reconstruction [28], which the U-Net Sino cannot correct. The U-Net Image only mitigates the artifacts in the image domain introduced by the FBP. In doing so, the network no longer has the original raw data to learn from.

The usage of digital XCAT phantom data for metal data simulation instead of real patient data has several advantages. First of all, with the organ masks provided by the XCAT, metal objects can automatically be inserted in specific body regions. In this work, we inserted iron into the blood vessels. For future studies it would be better to insert attenuation coefficients of materials that are commonly used for contrast agents and catheters. Moreover, for the simulation of polychromatic projections, it is not necessary to segment the images into soft tissue, bone, and metal to assign the corresponding attenuation coefficients, as is done in several other works [14,21,37]. Instead, the organ masks of the XCAT allow for the insertion of energy-dependent attenuation coefficients for every organ. In the future, it will be desirable to test the iCTU-Net on experimental raw data instead of simulated data. However, this requires the iCTU-Net to be adapted to work with the raw data of multirow detector CT scanners. The two-dimensional projection data might lead to restrictions due to GPU memory limitations. Since dual-energy CT has been shown to help reduce metal artifacts [5,10], the iCTU-Net should benefit from the additional spectral information. Photon-counting CT is another spectral technology that can be used to reduce metal artifacts [38]. The energy of individual photons can be measured by energy-resolving detectors [39]. The iCTU-Net is readily applicable to energy-resolved raw data by including the energy information in separate input channels. The additional spectral information in the raw data is expected to mitigate beam hardening artifacts.

We will also investigate the ability of the iCTU-Net to simultaneously mitigate different kinds of artifacts. This is achievable by using a training dataset that contains a combination of artifacts. Promising results for the isolated mitigation of artifacts with the iCTU-Net in low-dose CT and sparse-angle CT have already been shown [28].

5. Conclusions

The presented end-to-end deep learning CT reconstruction algorithm was trained with simulated interventional data to mitigate metal artifacts during reconstruction. We showed that the iCTU-Net reconstruction MAR approach is better suited to mitigate metal artifacts than commonly used sinogram pre-processing and image post-processing deep learning approaches. The iCTU-Net is the only studied method that can eliminate the metal streak artifacts. However, the end-to-end reconstruction approach performs worse than the other approaches when no artifacts are present. Reconstructions without any metal showed that the iCTU-Net is prone to blurring. Because the whole reconstruction is learned, the network is able to fully utilize the raw data, which benefits the removal of metal artifacts. In the future, we will try to improve the network architecture by investigating alternative loss functions and post-processing layers to avoid blurring. We will also train networks with data including different kinds of artifacts to investigate simultaneous mitigation of several types of artifacts.

Author Contributions: Conceptualization, D.F.B.; methodology, D.F.B. and C.U.; software, D.F.B., A.-K.G. and C.U.; validation, D.F.B. and T.R.; formal analysis, D.F.B.; investigation, D.F.B.; resources, L.R.S. and F.G.Z.; data curation, D.F.B.; writing—original draft preparation, D.F.B.; writing—review and editing, D.F.B., C.U., T.R., A.-K.G., L.R.S., and F.G.Z.; visualization, D.F.B.; supervision, L.R.S. and F.G.Z.; project administration, F.G.Z.; funding acquisition, F.G.Z. All authors have read and agreed to the published version of the manuscript.

Funding: This research project is part of the Research Campus M^2OLIE and funded by the German Federal Ministry of Education and Research (BMBF) within the Framework "Forschungscampus: public-private partnership for Innovations" under the funding code 13GW0388A.

Institutional Review Board Statement: Not applicable.

Informed Consent Statement: Not applicable.

Data Availability Statement: The data presented in this study are available on request from the corresponding author. The data are not publicly available.

Conflicts of Interest: The authors declare no conflict of interest.

References

1. Boas, F.E.; Fleischmann, D. CT artifacts: Causes and reduction techniques. *Imaging Med.* **2012**, *4*, 229–240. [CrossRef]
2. Do, T.D.; Heim, J.; Skornitzke, S.; Melzig, C.; Vollherbst, D.F.; Faerber, M.; Pereira, P.L.; Kauczor, H.U.; Sommer, C.M. Single-energy versus dual-energy imaging during CT-guided biopsy using dedicated metal artifact reduction algorithm in an in vivo pig model. *PLoS ONE* **2021**, *16*, e0249921. [CrossRef]
3. McWilliams, S.R.; Murphy, K.P.; Golestaneh, S.; O'Regan, K.N.; Arellano, R.S.; Maher, M.M.; O'Connor, O.J. Reduction of guide needle streak artifact in CT-guided biopsy. *J. Vasc. Interv. Radiol.* **2014**, *25*, 1929–1935. [CrossRef]
4. Stattaus, J.; Kuehl, H.; Ladd, S.; Schroeder, T.; Antoch, G.; Baba, H.A.; Barkhausen, J.; Forsting, M. CT-guided biopsy of small liver lesions: Visibility, artifacts, and corresponding diagnostic accuracy. *Cardiovasc. Interv. Radiol.* **2007**, *30*, 928–935. [CrossRef]
5. Laukamp, K.R.; Zopfs, D.; Wagner, A.; Lennartz, S.; Pennig, L.; Borggrefe, J.; Ramaiya, N.; Hokamp, N.G. CT artifacts from port systems: Virtual monoenergetic reconstructions from spectral-detector CT reduce artifacts and improve depiction of surrounding tissue. *Eur. J. Radiol.* **2019**, *121*, 108733. [CrossRef]
6. Kim, C.; Kim, D.; Lee, K.Y.; Kim, H.; Cha, J.; Choo, J.Y.; Cho, P.K. The optimal energy level of virtual monochromatic images from spectral CT for reducing beam-hardening artifacts due to contrast media in the thorax. *Am. J. Roentgenol.* **2018**, *211*, 557–563. [CrossRef]
7. Lin, M.; Loffroy, R.; Noordhoek, N.; Taguchi, K.; Radaelli, A.; Blijd, J.; Balguid, A.; Geschwind, J.F. Evaluating tumors in transcatheter arterial chemoembolization (TACE) using dual-phase cone-beam CT. *Minim. Invasive Ther. Allied Technol.* **2011**, *20*, 276–281. [CrossRef]
8. Jeong, S.; Kim, S.H.; Hwang, E.J.; Shin, C.i.; Han, J.K.; Choi, B.I. Usefulness of a metal artifact reduction algorithm for orthopedic implants in abdominal CT: Phantom and clinical study results. *Am. J. Roentgenol.* **2015**, *204*, 307–317. [CrossRef]
9. Bismark, R.N.; Frysch, R.; Abdurahman, S.; Beuing, O.; Blessing, M.; Rose, G. Reduction of beam hardening artifacts on real C-arm CT data using polychromatic statistical image reconstruction. *Z. Med. Phys.* **2020**, *30*, 40–50. [CrossRef]
10. Bamberg, F.; Dierks, A.; Nikolaou, K.; Reiser, M.F.; Becker, C.R.; Johnson, T.R. Metal artifact reduction by dual energy computed tomography using monoenergetic extrapolation. *Eur. Radiol.* **2011**, *21*, 1424–1429. [CrossRef]
11. Kalender, W.A.; Hebel, R.; Ebersberger, J. Reduction of CT artifacts caused by metallic implants. *Radiology* **1987**, *164*, 576–577. [CrossRef]
12. Meyer, E.; Raupach, R.; Lell, M.; Schmidt, B.; Kachelrieß, M. Normalized metal artifact reduction (NMAR) in computed tomography. *Med. Phys.* **2010**, *37*, 5482–5493. [CrossRef]
13. Lundervold, A.S.; Lundervold, A. An overview of deep learning in medical imaging focusing on MRI. *Z. Med. Phys.* **2019**, *29*, 102–127. [CrossRef]

14. Zhang, Y.; Yu, H. Convolutional neural network based metal artifact reduction in X-ray computed tomography. *IEEE Trans. Med Imaging* **2018**, *37*, 1370–1381. [CrossRef]
15. Ghani, M.U.; Karl, W.C. Deep learning based sinogram correction for metal artifact reduction. *Electron. Imaging* **2018**, *2018*, 4721–4728. [CrossRef]
16. Ghani, M.U.; Karl, W.C. Fast enhanced CT metal artifact reduction using data domain deep learning. *IEEE Trans. Comput. Imaging* **2019**, *6*, 181–193. [CrossRef]
17. Gjesteby, L.; Yang, Q.; Xi, Y.; Zhou, Y.; Zhang, J.; Wang, G. Deep learning methods to guide CT image reconstruction and reduce metal artifacts. In *Medical Imaging 2017: Physics of Medical Imaging*; International Society for Optics and Photonics: Bellingham, WA, USA, 2017; Volume 10132, p. 101322W.
18. Lossau, T.; Nickisch, H.; Wissel, T.; Morlock, M.; Grass, M. Learning metal artifact reduction in cardiac CT images with moving pacemakers. *Med Image Anal.* **2020**, *61*, 101655. [CrossRef]
19. Gjesteby, L.; Yang, Q.; Xi, Y.; Claus, B.; Jin, Y.; De Man, B.; Wang, G. Reducing metal streak artifacts in CT images via deep learning: Pilot results. In Proceedings of the 14th International Meeting on Fully Three-Dimensional Image Reconstruction in Radiology and Nuclear Medicine, Xi'an, China, 18–23 June 2017; Volume 14, pp. 611–614.
20. Wang, J.; Zhao, Y.; Noble, J.H.; Dawant, B.M. Conditional generative adversarial networks for metal artifact reduction in CT images of the ear. In Proceedings of the International Conference on Medical Image Computing and Computer-Assisted Intervention, Granada, Spain, 16–20 September 2018; pp. 3–11.
21. Huang, X.; Wang, J.; Tang, F.; Zhong, T.; Zhang, Y. Metal artifact reduction on cervical CT images by deep residual learning. *Biomed. Eng. Online* **2018**, *17*, 1–15. [CrossRef]
22. Liao, H.; Lin, W.A.; Zhou, S.K.; Luo, J. ADN: Artifact disentanglement network for unsupervised metal artifact reduction. *IEEE Trans. Med. Imaging* **2019**, *39*, 634–643. [CrossRef]
23. Nakao, M.; Imanishi, K.; Ueda, N.; Imai, Y.; Kirita, T.; Matsuda, T. Regularized three-dimensional generative adversarial nets for unsupervised metal artifact reduction in head and neck CT images. *IEEE Access* **2020**, *8*, 109453–109465. [CrossRef]
24. Lee, J.; Gu, J.; Ye, J.C. Unsupervised CT Metal Artifact Learning using Attention-guided β-CycleGAN. *IEEE Trans. Med Imaging* **2021**, *40*, 3932–3944. [CrossRef]
25. Lin, W.A.; Liao, H.; Peng, C.; Sun, X.; Zhang, J.; Luo, J.; Chellappa, R.; Zhou, S.K. Dudonet: Dual domain network for ct metal artifact reduction. In Proceedings of the IEEE/CVF Conference on Computer Vision and Pattern Recognition, Long Beach, CA, USA, 15–19 June 2019; pp. 10512–10521.
26. Li, Y.; Li, K.; Zhang, C.; Montoya, J.; Chen, G.H. Learning to reconstruct computed tomography images directly from sinogram data under a variety of data acquisition conditions. *IEEE Trans. Med Imaging* **2019**, *38*, 2469–2481. [CrossRef]
27. Maas, A.L.; Hannun, A.Y.; Ng, A.Y. Rectifier nonlinearities improve neural network acoustic models. *Proc. icml* **2013**, *30*, 3.
28. Leuschner, J.; Schmidt, M.; Ganguly, P.S.; Andriiashen, V.; Coban, S.B.; Denker, A.; Bauer, D.; Hadjifaradji, A.; Batenburg, K.J.; Maass, P.; et al. Quantitative Comparison of Deep Learning-Based Image Reconstruction Methods for Low-Dose and Sparse-Angle CT Applications. *J. Imaging* **2021**, *7*, 44. [CrossRef] [PubMed]
29. Ronneberger, O.; Fischer, P.; Brox, T. U-net: Convolutional networks for biomedical image segmentation. In Proceedings of the International Conference on Medical Image Computing and Computer-Assisted Intervention, Munich, Germany, 5–9 October 2015; pp. 234–241.
30. Segars, W.P.; Sturgeon, G.; Mendonca, S.; Grimes, J.; Tsui, B.M.W. 4D XCAT phantom for multimodality imaging research. *Med. Phys.* **2010**, *37*, 4902–4915. [CrossRef]
31. Williamson, J.F.; Whiting, B.R.; Benac, J.; Murphy, R.J.; Blaine, G.J.; O'Sullivan, J.A.; Politte, D.G.; Snyder, D.L. Prospects for quantitative computed tomography imaging in the presence of foreign metal bodies using statistical image reconstruction. *Med. Phys.* **2002**, *29*, 2404–2418. [CrossRef] [PubMed]
32. Poludniowski, G.; Landry, G.; Deblois, F.; Evans, P.; Verhaegen, F. SpekCalc: A program to calculate photon spectra from tungsten anode X-ray tubes. *Phys. Med. Biol.* **2009**, *54*, N433. [CrossRef] [PubMed]
33. Leuschner, J.; Schmidt, M.; Baguer, D.O.; Maaß, P. The lodopab-ct dataset: A benchmark dataset for low-dose ct reconstruction methods. *arXiv* **2019**, arXiv:1910.01113.
34. La Rivière, P.J.; Bian, J.; Vargas, P.A. Penalized-likelihood sinogram restoration for computed tomography. *IEEE Trans. Med Imaging* **2006**, *25*, 1022–1036. [CrossRef]
35. Kingma, D.P.; Ba, J. Adam: A method for stochastic optimization. *arXiv* **2014**, arXiv:1412.6980.
36. Bauer, D.F.; Russ, T.; Waldkirch, B.I.; Tönnes, C.; Segars, W.P.; Schad, L.R.; Zöllner, F.G.; Golla, A.K. Generation of annotated multimodal ground truth datasets for abdominal medical image registration. *Int. J. Comput. Assist. Radiol. Surg.* **2021**, *16*, 1277–1285. [CrossRef] [PubMed]
37. Arabi, H.; Zaidi, H. Deep learning–based metal artefact reduction in PET/CT imaging. *Eur. Radiol.* **2021**, *31*, 6384–6396. [CrossRef] [PubMed]
38. Nasirudin, R.A.; Mei, K.; Panchev, P.; Fehringer, A.; Pfeiffer, F.; Rummeny, E.J.; Fiebich, M.; Noël, P.B. Reduction of metal artifact in single photon-counting computed tomography by spectral-driven iterative reconstruction technique. *PLoS ONE* **2015**, *10*, e0124831.
39. Willemink, M.J.; Persson, M.; Pourmorteza, A.; Pelc, N.J.; Fleischmann, D. Photon-counting CT: Technical principles and clinical prospects. *Radiology* **2018**, *289*, 293–312. [CrossRef] [PubMed]

Article

Generative Adversarial CT Volume Extrapolation for Robust Small-to-Large Field of View Registration

Andrei Puiu [1,2,*], Sureerat Reaungamornrat [3], Thomas Pheiffer [3], Lucian Mihai Itu [1,2], Constantin Suciu [1,2], Florin Cristian Ghesu [3] and Tommaso Mansi [3]

1. Advanta, Siemens SRL, 500097 Brasov, Romania; lucian.itu@siemens.com (L.M.I.); constantin.suciu@siemens.com (C.S.)
2. Department of Automation and Information Technology, Transilvania University of Brasov, 500174 Brasov, Romania
3. Siemens Healthineers, Digital Technology and Innovation, Princeton, NJ 08540, USA; sureerat.reaungamornrat@siemens-healthineers.com (S.R.); thomas.pheiffer@gmail.com (T.P.); florin.ghesu@siemens-healthineers.com (F.C.G.); thomas.mansi@gmail.com (T.M.)
* Correspondence: andrei.puiu@siemens.com

Abstract: Intraoperative Computer Tomographs (iCT) provide near real time visualizations which can be registered with high-quality preoperative images to improve the confidence of surgical instrument navigation. However, intraoperative images have a small field of view making the registration process error prone due to the reduced amount of mutual information. We herein propose a method to extrapolate thin acquisitions as a prior step to registration, to increase the field of view of the intraoperative images, and hence also the robustness of the guiding system. The method is based on a deep neural network which is trained adversarially using self-supervision to extrapolate slices from the existing ones. Median landmark detection errors are reduced by approximately 40%, yielding a better initial alignment. Furthermore, the intensity-based registration is improved; the surface distance errors are reduced by an order of magnitude, from 5.66 mm to 0.57 mm (p-value = 4.18×10^{-6}). The proposed extrapolation method increases the registration robustness, which plays a key role in guiding the surgical intervention confidently.

Keywords: generative adversarial networks; volume extrapolation; self-supervision; volume registration

Citation: Puiu, A.; Reaungamornrat, S.; Pheiffer, T.; Itu, L.M.; Suciu, C.; Ghesu, F.C.; Mansi, T. Generative Adversarial CT Volume Extrapolation for Robust Small-to-Large Field of View Registration. *Appl. Sci.* **2022**, *12*, 2944. https://doi.org/10.3390/app12062944

Academic Editor: Vladislav Toronov

Received: 10 January 2022
Accepted: 10 March 2022
Published: 14 March 2022

Publisher's Note: MDPI stays neutral with regard to jurisdictional claims in published maps and institutional affiliations.

Copyright: © 2022 by the authors. Licensee MDPI, Basel, Switzerland. This article is an open access article distributed under the terms and conditions of the Creative Commons Attribution (CC BY) license (https://creativecommons.org/licenses/by/4.0/).

1. Introduction

Over the past years, the use of medical imaging in computer aided interventions has become more and more popular, supporting clinicians in their workflow and thus reducing the procedural associated risks [1].

This paper is focused on increasing the trustworthiness of liver needle therapies such as Radiofrequency Ablation (RFA) or biopsy, where real time imaging plays a main role in guiding the intervention confidently. Although it is well known that there is a trade-off between radiation dose, acquisition time and image quality, during such surgical interventions all procedures must be carried out as quickly and accurately as possible. A possible solution to this problem is to intraoperatively acquire thin images-that provide low—resolution visualizations of a small liver region—and register them with complete high resolution preoperative images [2].

Registration is a technique used to align two images with respect to the patient's internal structures. Formally, having a reference and a template image $R, T : \mathbb{R}^d \to \mathbb{R}$, registration objective is to find a transformation $\varphi : \mathbb{R}^d \to \mathbb{R}$ such that $R \approx T \circ \varphi$ [3]. Therefore, registration techniques are employed to retrieve high resolution preoperative information such as lesion location and appearance and aggregate it with the thin intra-operative images revealing the real-time needle localization, thus increasing navigation confidence. Based on the operands, there are multiple types of registration including slice-to-volume, projection-to-volume, volume-to-volume, etc. [4]. Herein we focus on the latter, aiming to boost the performance of two Computer Tomograph (CT) volumes rigid registration. Two volumes can be registered using a feature-based approach, an intensity-based approach or a combination of the two techniques. In feature-based registration, a set of corresponding features (e.g., landmarks, center of mass, etc.) are used to compute the transformation φ to register a volume (called the moving or template volume, T) to the space of the other volume (fixed or reference volume, R) [5]. The intensity-based approach can be formulated as an optimization problem, seeking the best set of parameters for the transformation φ to minimize a predefined distance measure: $argmin_\varphi[D(R, T \circ \varphi)]$ [3,6]. However, this approach is not robust due to the potential presence of local minimums caused by image artifacts and sub-optimal distance metrics. Combinations of the two approaches might be used to improve registration accuracy and robustness (e.g., using intensity-based registration as a refinement step for the feature-based registration).

To the best of our knowledge, registration of thin images has been overlooked so far. Since all the registration techniques are highly dependent on the amount of mutual information (common data presented by both images from different perspectives), analysis of thin images is very challenging due to their reduced field of view (FOV). However, during surgeries low-resolution thin CT slabs are acquired to mitigate the patient's exposure risk. In this context, despite performing an initial alignment based on center of mass or geometric center, intensity-based registration is prone to failure given to the distinct fields of view of the operands. To reliably retrieve the corresponding high resolution preoperative data, a feature-based approach must be considered. However landmark detection algorithms might also be affected by the thin volume quality thus yielding a poor registration performance.

We therefore propose a method to extrapolate thin CT slabs, generating additional slices from the few existing ones, hence providing enhanced context information required by registration algorithms to work robustly.

Artificial intelligence in medical applications has received significant attention from the scientific community over the past few years. Due to their potential to model complex problems based on large datasets of examples, deep learning algorithms are nowadays employed for solving a wide variety of tasks such as regression, classification, segmentation and image generation [7,8]. Generative adversarial networks (GANs) [9] are a state of the art method for solving tasks such as synthetic image generation [10–12], segmentation [13], super-resolution [14], denoising [15,16], style-transfer [17,18] and inpainting [19,20].

Image interpolation, also known as image completion or inpainting [21], aims at filling missing regions within an image with coherent and realistic content based on the surrounding information. Thus, in image interpolation, the field of view is well defined. In contrast, image extrapolation [22–24] is a more challenging task since the field of view has to be extended by hallucinating coherent and realistic content outside the boundaries of the existing information.

In this paper, we introduce an extrapolation methodology based on a generator network which increases the field of view of thin intraoperative CT volumes, and improves the accuracy and robustness of a subsequent registration process. To prove the efficiency of the proposed method we focus on the liver area and assume that a thin acquisition would have a thickness of approximately 5 cm. However, this can be easily adjusted for other thicknesses or use-cases.

The paper is organized as follows: Section 2 provides an overview of the proposed methods, including details regarding data, network architecture, optimization and quantification strategies. Section 3 presents the results from a task oriented perspective. Strengths and limitations of this study are discussed in Section 4, including the next steps towards the adoption of our method in real world applications and Section 5 presents some overall conclusions.

2. Materials and Methods

In this section we introduce a self-supervised approach for extrapolating axial slices, thus enhancing the context information required by the registration algorithms to obtain a good alignment. Due to the lack of real intraoperative data, we synthesize thin images by extracting approximately 5 cm thick sub-regions (see Section 2.1) from full CT field of views (Figure 1a).

(a) Thin intraoperative image.　　(b) Full preoperative image.

Figure 1. Examples of CT images: The thin image (a) displays a reduced field of view (thickness) with respect to the z axis, as compared to the full preoperative image (b).

As depicted in Figure 2, given a CT volume $f : \mathbb{R}^d \to \mathbb{R}$ we first use an uniform distribution to build a binary mask $m : \mathbb{R}^d \to \{0,1\}$ to randomly remove 75% of the information through a voxel-wise multiplication, thus yielding an image $g : \mathbb{R}^d \to \mathbb{R}$. We further refer to this image as the grid image, which is defining the extrapolation extent. Next, we simulate a thin acquisition $t : \mathbb{R}^d \to \mathbb{R}$ by extracting a region of interest (ROI) out of the grid image and then employ a deep neural network to restore the missing information, thus extrapolating the thin slab across z direction.

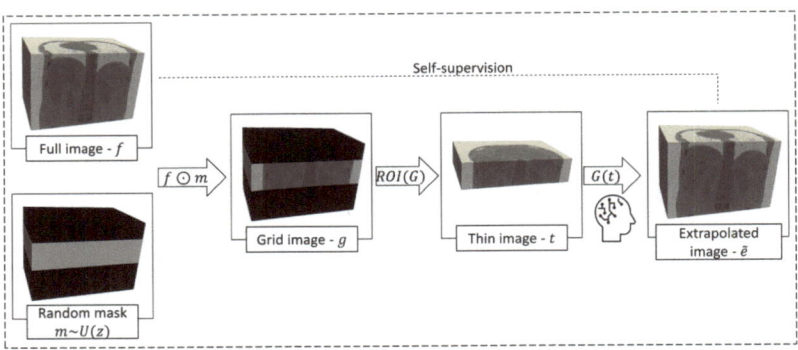

Figure 2. Schematic overview of the workflow. The data are stochastically processed at training time to create a self-supervised learning framework.

2.1. Dataset

The dataset consisted of 1400 high resolution CT images, each of which provided a complete visualization of the liver. Furthermore, from each of these images we only considered an ROI determined by the liver bounding box with respect to the z axis (further, we refer to this as the full image). To generalize the model, we stochastically set the thickness of the full image to the height of the liver bounding box, adding ±25 mm in each direction. All these images have a constant resolution of 512 × 512 in the x-y plane, with a voxel spacing of 0.8 mm, while the mean resolution for the z-axis is of 179.2 voxels (ranging from 24 to 796, with a mean voxel spacing of 1.49 mm). All the images were resampled to a spacing of [3,3,1.5] mm. Further, to create an isotropic grid of size 128 × 128 × 128 voxels, either padding or cropping was performed. To avoid numerical instability and arithmetic overflow when computing the variance, we normalized our data using the Welford's online algorithm [25].

The data were employed to develop a self-supervised learning framework, automatically creating input-output pairs from the ground-truth images: at training time, a quarter of the full volume FOV was randomly extracted simulating an intraoperative volume of varying thickness, as depicted in Figure 3. Further, a deep neural network was employed when reconstructing the original volume, thus extrapolating the thin slab across the z-axis.

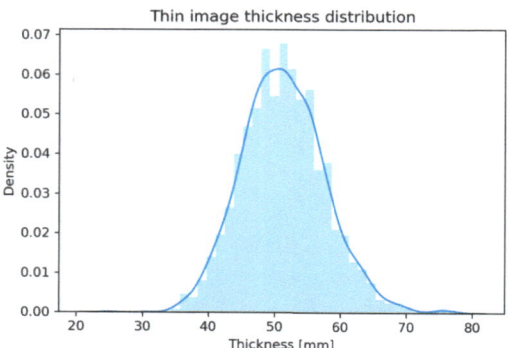

Figure 3. Thickness distribution of the training input data.

We randomly split the data into a training set representing 80% of the data and a testing set representing the remaining 20% of the data. Additionally, we used 100 CT pairs for quantifying the registration performance.

2.2. Proposed Method

We trained our extrapolation network (also referred to as generator) within an adversarial framework, optimizing it to "fool" another neural network (called critic or discriminator) regarding the authenticity of generated samples.

As depicted in Figure 4 the generator network first performed a repetition of the thin slab across the z-axis, increasing the thickness of the input with a factor of four, thus defining the target FOV of the extrapolated image. This repetition adapts the encoder's feature maps to the decoder's dimensions such that we can take advantage of the long term skip connections propagating the information through the network. Moreover, this strategy is beneficial in terms of expanding the receptive field of view at the bottleneck, thus using the limited amount of real information efficiently.

The rest of the generator is a variation of U-net, where each block consists of a sequence of convolution, activation function and instance normalization layers [26]. In the encoder part, downsampling was performed using 2-strided convolutions, until a receptive field of view of 255 × 255 × 255 voxels was obtained at the bottleneck. Nonlinearities are provided by LeakyReLU activations, while the decoder employs ReLUs. Upsampling was performed through interpolation layers followed by 1-strided convolutions.

We used similar blocks as in the generator to create a patch-discriminator [27] conditioned on the grid image (Figure 2—g), which, besides the image to be discriminated, was provided as an input. This image helped the critic to penalize the generator in regards to finding the right extrapolation extent. Instead of outputting a single value, the critic outputs a 8 × 8 × 8 feature-map on which each element discriminates 31 × 31 × 31 voxels patches in the input.

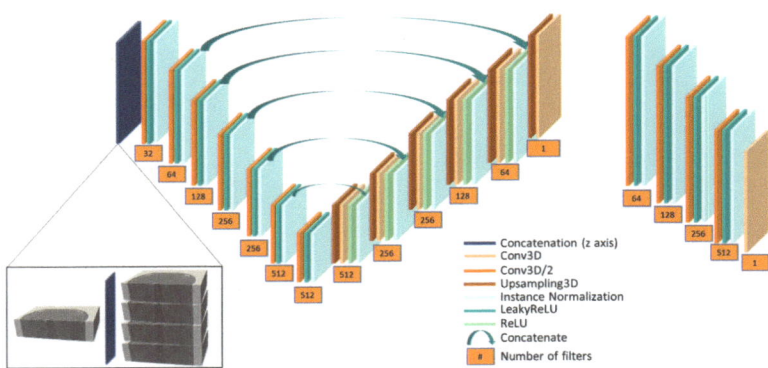

Figure 4. Network architectures; **left**—U-net like generator; **right**—patch discriminator. The first layer of the generator performs a 4× repetition of the thin input to define the extrapolation extent.

2.2.1. Optimization Strategy

We trained the critic to distinguish between fake ($\tilde{e}$) and real samples (f), thus maximizing the Wasserstein distance between the real (P_r) and fake (P_g) data distribution [28]:

$$L_{critic} = E_{\tilde{e} \sim P_g}[D(\tilde{e}, g)] - E_{f \sim P_r}[D(f, g)] + \lambda E_{\hat{e} \sim P_{\hat{e}}}[(||\nabla_{\hat{e}} (D(\hat{e}, g))||_2 - 1)^2] \quad (1)$$

Equation (1) displays the objective function used to train the critic, where the third term is a gradient penalty term used to improve the training stability [29].

Secondly, we trained the generator to produce images which are indistinguishable from the real ones, thus minimizing L_{critic} by optimizing:

$$L_{adv} = -E_{\tilde{e} \sim P_g}[D(\tilde{e}, g)] \quad (2)$$

To further stimulate the generation of image details and consistent internal structures, in addition to the adversarial component, we also used a feature loss [30] penalty. This component aims at minimizing the L_1 distance between features F extracted from real and fake samples, respectively. The feature maps are provided by the third convolution layer of a 3D network trained in brain tumor segmentation [31].

$$L_{feat} = E_{\tilde{e}, f}[||F(\tilde{e}) - F(f)||_1] \quad (3)$$

As depicted in Figure 5, the grid information (volume g) was only used at training time by the critic to constrain the generator to find the right position of the thin slab within the target field of view.

The objective of the generator represents a weighted combination of the two terms of Equations (2) and (3). The weights have been empirically chosen such that the components take values in the same range: $\lambda_{adv} = 1$ and $\lambda_{feat} = 1$, which has been shown to lead to a better performance of the model. When using a larger weight for the supervision signal, as suggested in [27], the adversarial loss became unstable in the early stages of the training, hindering an improvement of the generated images over time.

$$L_{gen} = \lambda_{adv} L_{adv} + \lambda_{feat} L_{feat} \quad (4)$$

Since the cost function used to train GANs stems from another neural network trained jointly, the loss alone can be misleading when trying to identify the best performing model. Therefore, for the current experiment, model selection was performed through a visual inspection of the samples produced by the generator over time.

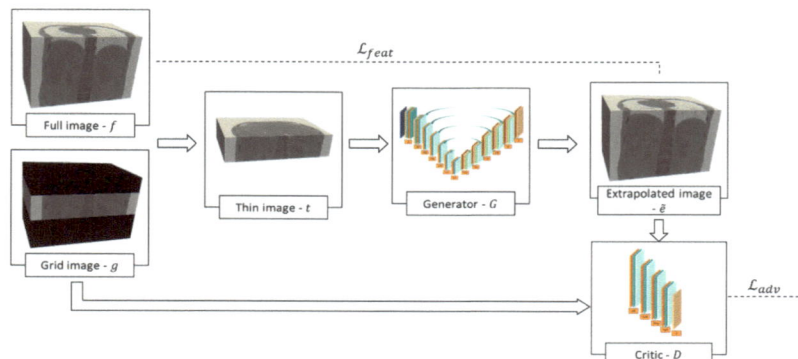

Figure 5. Generator optimization workflow. A conditional GAN was employed in extrapolating thin input volumes, expanding their FOV with a factor of 4.

2.2.2. Image Metadata Retrieval

Since our convolutional neural network (CNN) generator operates on voxel intensity information only, we needed to perform an extra-step to retrieve the metadata of the extrapolated images.

Intuitively, the extrapolated image will have the same spacing and orientation as the thin one. However, the origin and dimension of the image changes due to the addition of synthetic information. Determining the grid dimension of the expanded volume is straight forward since we always quadruple the input field of view on the z-axis:

$$(d\tilde{e}_x, d\tilde{e}_y, d\tilde{e}_z) = (dt_x, dt_y, dt_z \times 4) \tag{5}$$

To compute the origin of the extrapolated volume, we first needed to determine thin slab's location within the extrapolation grid. In the current work, we addressed this issue in the post-processing phase, performing an extra-registration step to determine the extent of extrapolation as further described:

We overlapped the thin slab (sliding it across z direction) at each possible location of the extrapolated volume, calculating the voxel-wise mean squared error (Figure 6—$d_{1..k}$). Next, we determined the extent extrapolation by picking the index which minimized this penalty.

Figure 6. Position regression; **left**—thin slab t; **right**—extrapolated volume $\tilde{e}$.

Further, the origin of the extrapolated image was calculated using the following expression:

$$(o\tilde{e}_x, o\tilde{e}_y, o\tilde{e}_z) = (ot_x, ot_y, ot_z - argmin_{i=1..k}(d_i) \times st_z) \tag{6}$$

where st_z is the spacing of thin volume across z direction.

In our tests, this simple registration step was always accurate because the extrapolation network only had to copy the thin slab's intensities into the output volume without modifying them at all, hence generating relatively few errors.

2.3. Performance Quantification

One of the major challenges in image generation tasks is the lack of a goal standard method to quantify the performance of the generative models. Hence, we herein propose a goal oriented quantification method consisting in two tests: landmark detection [32,33] and registration errors [34].

As we want to perform a feature-based registration of two volumes based on a set of corresponding landmarks, we must encourage accurate detection on the synthetic images. Hence, we first evaluate our extrapolation models based on the euclidean distance between the manual annotations and the landmarks detected on the thin, extrapolated and ground-truth volumes, respectively.

For the registration test, the 100 additional CT pairs mentioned in Section 2.1 were used as follows: we randomly extracted thin slabs from the fixed images and then employed our models for extrapolation. Further, we compared the performance between the registration of ground-truth fixed and full moving images, thin-fixed and full moving images and extrapolated-fixed and full-moving images. We used two metrics for this evaluation: surface distance and DICE, both computed on the liver masks, obtained by using the same segmentation model employed for data preprocessing.

3. Results

Table 1 displays the structural similarity index (SSIM) and the peak signal to noise ratio (PSNR) metrics for the train and test set, respectively. No significant differences were observed between the performance of the two sets, indicating the good generalization power of the generator.

Table 1. Image similarity results.

	Mean ($\pm$std) [mm]	
Metric	**Train Set**	**Test Set**
SSIM	0.726($\pm$0.04)	0.719($\pm$0.05)
PSNR	24.13($\pm$2.19)	23.76($\pm$2.26)

3.1. Landmark Detection Test

We ran a pretrained landmark detection model [33] on three variants of each test image: full, thin and extrapolated. Next, we calculated the Euclidean distance between each detected landmark and the corresponding manual annotation. The results are depicted in Figure 7: the proposed method reduces the median detection error by approximately 40% (from 19.51 mm to 12.08 mm, p-value = 7.38×10^{-37}) while the interquartile range (IQR) is reduced by more than a half, which means that our method increases landmark detection robustness significantly (Table 2).

Since a quarter of the full volume thickness is always used as an input, each extrapolated image should contain (1) that quarter of the FOV (we will refer to it as actual information region) and (2) three quarters of extrapolated (hallucinated) information. All detected landmarks were considered for the blue boxplots, including the ones detected in the extrapolated region. On the other hand, the red boxplots display the detection error on the actual information only, which is more relevant, since we only employed extrapolation to provide more context for detection algorithms, rather than generating synthetic points to be used for registration.

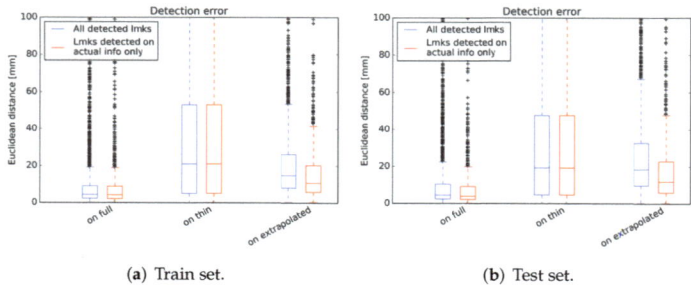

(a) Train set. (b) Test set.

Figure 7. Detection errors. For the blue boxplots all detected landmarks were considered, while the red boxplots only take into account landmarks detected on the region containing the actual information.

Table 2. Landmark detection results on the test set.

	Mean (±std) [mm]	
Image	All Detected Landmarks	Landmarks Detected on Actual Info
Full volume	4.64(±8.02)	4.04(±7.05)
Thin volume	19.51(±43.0)	19.51(±43.0)
Extrapolated volume	18.62(±22.96)	12.08(±16.86)

3.2. Registration Test

Figure 8 displays the registration results of the full moving images with all three variants of the fixed images-full, thin and extrapolated. The blue boxplots display the results of landmark-based registration which is then used as an initialization for the intensity-based registration, depicted in red.

As expected, the best performance was obtained when the full-moving images are registered with full-fixed images (having a median SD of 0.20(±0.08) mm after intensity-based registration), and the worst results were obtained when the full-moving images were registered with thin-fixed images (5.66(±20.56) mm). However, we obtained a registration performance comparable to the one corresponding to full-fixed images (0.57(±2.05) mm) by using the proposed extrapolation method as a prior step, thus reducing the thin slab registration error with a factor of 10 (p-value = 4.18×10^{-6}). The same holds true when considering the DICE score (Figure 8b), which increased due to the extrapolation from 0.67 to 0.88 (median).

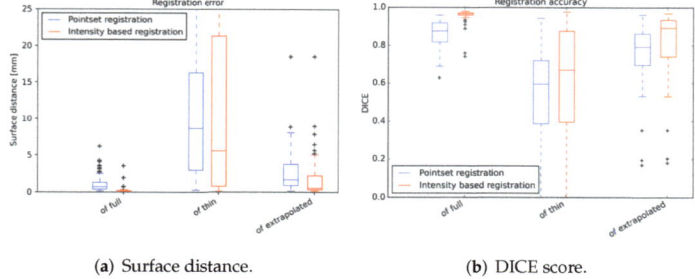

(a) Surface distance. (b) DICE score.

Figure 8. Registration results: landmark-based registration in blue, intensity-based registration in red. Each figure has three groups: left—registration of full-fixed with full-moving images; middle—registration of thin-fixed with full-moving images; right—registration of extrapolated-fixed with full-moving images.

Figure 9 displays a comparison between the registration of thin, full and extrapolated images. While the first row presents a poor registration of the thin image, the third row demonstrates the benefits of our approach, showing a much better overlap between the extrapolated-fixed and full-moving images. However, the best performance was obtained when aligning the full-fixed with the full-moving images (second row).

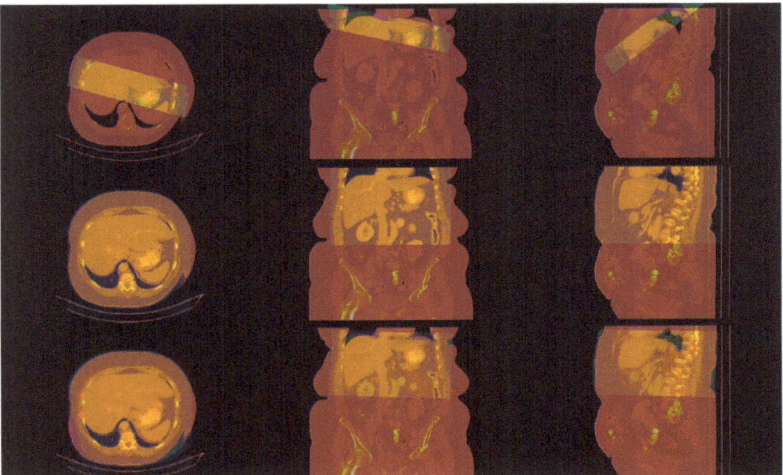

Figure 9. Registration example. The first row displays a poor alignment of the thin image, while the third row displays an improved registration determined by the extrapolation. The middle row presents a very accurate registration of the ground truth full image for comparison.

4. Discussion
4.1. Algorithm Selection

Before defining the method described in this paper, a number of experiments were performed. At first, a 2D Wasserstein GAN with gradient penalty was employed in extrapolating individual coronal slices, which were stacked together afterwards to create a 3D volume. Although the generated 2D samples looked relatively realistic per se (see Figure A1), the resulting 3D volumes had a significant inconsistency across the y direction, leading to a poor detection and registration performance. We have tried to mitigate this inconsistency by using a sequence of five consecutive frames as input to predict the middle one, and by adding to the cost function of total variation loss across the y direction. No substantial improvement was observed.

On 3D data we trained various settings of Least Squared GANs (LSGAN) [35] with no success. The discriminator learned to distinguish fake samples within a few iterations, providing very strong gradients, hence placing the generator into a mode collapse. To mitigate this behavior we tried different strategies such as decreasing the size of the network, Dropout regularization, Gaussian noise addition to layers and/or labels, occasionally flipping the targets, addition of voxel-wise supervision loss component for training the generator, etc. The 3D Wasserstein GAN with gradient penalty demonstrated more stable behavior during the training, when used in conjunction with an extra-supervision loss, such as voxel-wise mean squared error or a feature loss (Equation (3)). Although the perceptual quality of the generated samples is much lower when compared to the 2D counterpart (see Figure A2), the goal oriented metrics (landmark detection error and registration performance) demonstrated a substantial improvement.

4.2. Overall Discussion

We found that our method reduced the median landmark detection error by a very large margin, thus leading to a superior feature-based registration. A better detection yields a better initialization for the intensity-based registration, rendering the alignment of extrapolated images comparable to the alignment of ground-truth images. Due to the enhanced context available in synthetic images, the overlapping error denoted by the median surface distance improved from 5.66 mm to 0.57 mm (p-value = 4.18×10^{-6}).

We note that the paper represents only a proof of concept that recent advances in generative networks can improve the robustness of computer aided interventions significantly using augmented data.

The major challenge of image out-painting remains the determination of the expansion size and direction. To expand the field of view of an image, Ref. [23] propose an extra input for the network, which is a vector representing the padding to be applied in each direction (top, left, bottom and right), hence defining an extrapolation grid. In [22], a binary mask is used to remove 32 pixels from each side of an image, thus solving a symmetrical extrapolation problem. However, our self-supervised learning approach does not require such prior information since it wouldn't be available in real world applications, but it is limited to always quadrupling the thickness of the input. Therefore, to address other use-cases, where the thickness of the intraoperative images is out of distribution with respect to the training data (see Figure 3), the model must be retrained accordingly.

Another important aspect of our proposed method is the need for an extra-registration step to determine the origin of extrapolated volumes. We are aware of the existence of other (maybe more elegant) approaches, but we have chosen this simple strategy to preserve the non-rigid properties offered by the fully convolutional networks, as well as to maintain that the entire workflow is self explainable.

5. Conclusions

We proposed a method for improving the performance of intraoperative image registration by expanding the field of view of thin slabs, thus enhancing the context information required for the matching process. We showed that our approach increased the detection performance by a large margin. Therefore, the feature-based registration provided a much better initialization for the intensity-based refinement step, which produced results comparable to the ones obtained after aligning two high-resolution images having the same field of view.

Author Contributions: Conceptualization, T.M. and S.R.; methodology, A.P., S.R. and T.P.; software, A.P.; validation, A.P., S.R. and T.P.; formal analysis, A.P.; investigation, A.P.; resources, S.R. and F.C.G.; data curation, A.P.; writing—original draft preparation, A.P.; writing—review and editing, S.R., L.M.I., C.S., F.C.G. and T.M.; supervision, L.M.I., C.S. and T.M. All authors have read and agreed to the published version of the manuscript.

Funding: This work was partially supported by a grant of the Romanian Ministry of Education and Research, CNCS—UEFISCDI, project number PN-III-P1-1.1-TE-2019-1804, within PNCDI III.

Institutional Review Board Statement: Not applicable.

Informed Consent Statement: Not applicable.

Data Availability Statement: Due to the nature of this research, participants of this study did not agree for their data to be shared publicly, so supporting data is not available.

Acknowledgments: The concepts and information presented in this paper are based on research results that are not commercially available. Future commercial availability cannot be guaranteed.

Conflicts of Interest: Andrei Puiu, Lucian Mihai Itu and Constantin Suciu are employees of Siemens SRL, Advanta, Brasov, Romania. Sureerat Reaungamornrat and Florin Cristian Ghesu are employees of Siemens Healthineers, Digital Technology and Innovation, Princeton, NJ, USA.

Abbreviations

The following abbreviations are used in this manuscript:

RFA	Radiofrequency ablation
CT	Computer Tomograph
FOV	Field of view
GAN	Generative adversarial network
ROI	Region of interest
CNN	Convolutional neural network
IQR	Interquartile range
SSIM	Structural similarity index
PSNR	Peak signal to noise ratio

Appendix A

Figure A1 displays some examples of extrapolated volumes using a 2D neural network, trained to extrapolate individual coronal frames. Although the network produces relatively plausible images across the extrapolation axis (middle), the resulting 3D volume displays a significant inconsistency across the y direction, which can be seen in the axial (left) and sagittal (right) views.

Figure A2 displays two examples of 3D extrapolation with the proposed approach, where the thin input image is overlayed on the coronal and sagittal views. Extrapolated volumes tend to be blurry and affected by noise, but in terms of structural consistency they are superior to the ones generated with the 2D neural network.

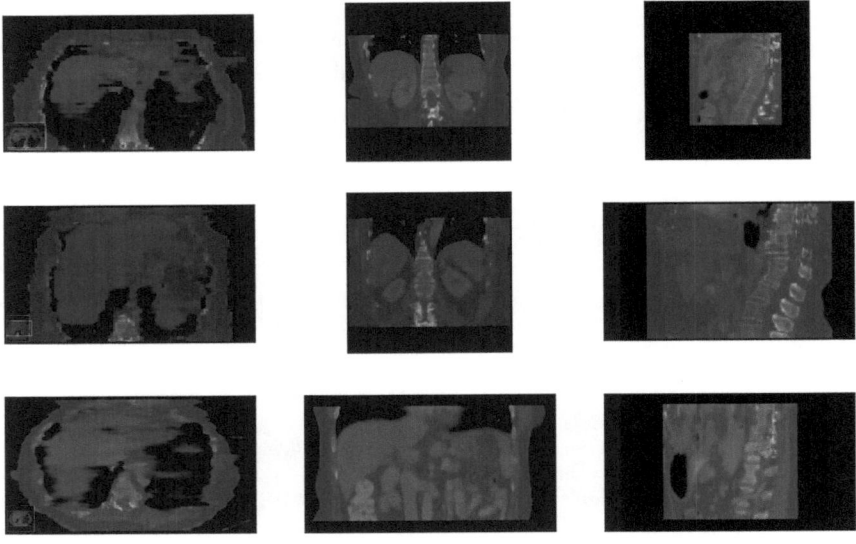

Figure A1. 2D extrapolation examples. **Left**—axial view; **Middle**—coronal frame (extrapolation axis); **Right**—sagittal frame.

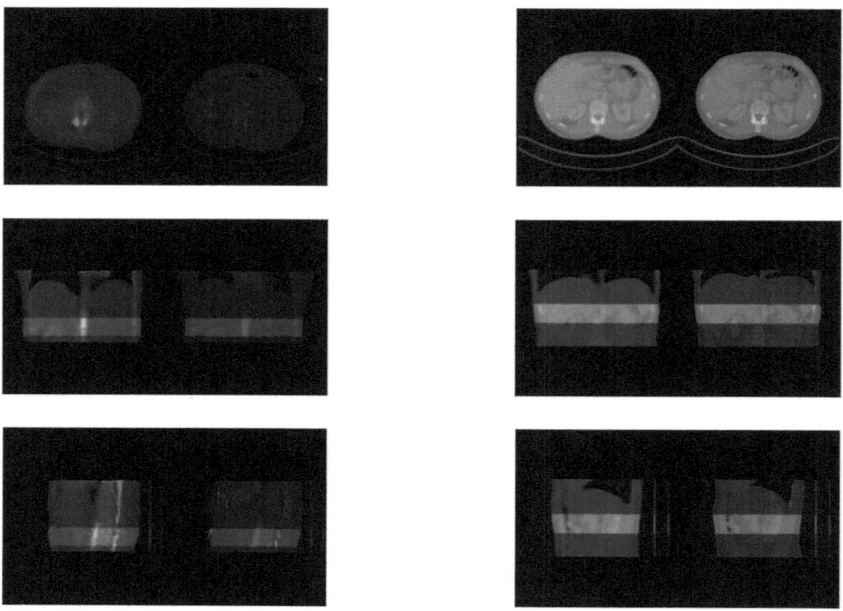

Figure A2. 3D extrapolation examples. Each example provides a comparison of the extrapolated image (**left**) with the ground truth image (**right**). **Top**—axial view; **Middle**—coronal frame (extrapolation axis); **Bottom**—sagittal frame.

References

1. Mauro, M.; Murphy, K.; Thomson, K.; Venbrux, A.; Morgan, R. *Image-Guided Interventions*, 3rd ed.; Elsevier, Inc.: Philadelphia, PA, USA, 2020.
2. Cleary, K.; Peters, T.M. Image-Guided Interventions: Technology Review and Clinical Applications. *Annu. Rev. Biomed. Eng.* **2010**, *12*, 119–142. [CrossRef] [PubMed]
3. Modersitzki, J. *Numerical Methods for Image Registration*; Oxford University Press Inc.: New York, NY, USA, 2004; pp. 27–44.
4. Liao, R.; Zhang, L.; Sun, Y.; Miao, S.; Chefd'Hotel, C. A Review of Recent Advances in Registration Techniques Applied to Minimally Invasive Therapy. *IEEE Trans. Multimed.* **2013**, *15*, 983–1000. [CrossRef]
5. Zitová, B.; Flusser, J. Image Registration Methods: A Survey. *Image Vis. Comput.* **2003**, *21*, 977–1000. [CrossRef]
6. Pluim, J.P.W.; Maintz, J.B.A.; Viergever, M.A. Mutual-information-based registration of medical images: A survey. *IEEE Trans. Med. Imaging* **2003**, *22*, 986–1004. [CrossRef] [PubMed]
7. Mansi, T.; Passerini, T.; Comaniciu, D. (Eds.) *Artificial Intelligence for Computational Modeling of the Heart*, 1st ed.; Academic Press in an Imprint of Elsevier: San Diego, CA, USA, 2019.
8. Fischer, A.; Klein, P.; Radulescu, P.; Gulsun, M.; Mohamed Ali, A.; Schoebinger, M.; Sahbaee, P.; Sharma, P.; Schoepf, U. Deep Learning Based Automated Coronary Labeling for Structured Reporting of Coronary CT Angiography in Accordance with SCCT Guidelines. *J. Cardiovasc. Comput. Tomogr.* **2020**, *14*, S21–S22. [CrossRef]
9. Goodfellow, I.J.; Pouget-Abadie, J.; Mirza, M.; Xu, B.; Warde-Farley, D.; Ozair, S.; Courville, A.; Bengio, Y. Generative Adversarial Networks. *arXiv* **2014**, arXiv:1406.2661.
10. Karras, T.; Aila, T.; Laine, S.; Lehtinen, J. Progressive Growing of GANs for Improved Quality, Stability, and Variation. International Conference on Learning Representations. 2018. Available online: https://openreview.net/forum?id=Hk99zCeAb (accessed on 9 March 2022).
11. Karras, T.; Laine, S.; Aila, T. A Style-Based Generator Architecture for Generative Adversarial Networks. *IEEE Trans. Pattern Anal. Mach. Intell.* **2021**, *43*, 4217–4228. [CrossRef] [PubMed]
12. Park, T.; Liu, M.Y.; Wang, T.C.; Zhu, J.Y. Semantic Image Synthesis with Spatially-Adaptive Normalization. In Proceedings of the 2019 IEEE/CVF Conference on Computer Vision and Pattern Recognition (CVPR), Long Beach, CA, USA, 16–20 June 2019; pp. 2332–2341. [CrossRef]

13. Zhang, Y.; Miao, S.; Mansi, T.; Liao, R. Task Driven Generative Modeling for Unsupervised Domain Adaptation: Application to X-ray Image Segmentation. *arXiv* **2018**, arXiv:1806.07201.
14. You, C.; Li, G.; Zhang, Y.; Zhang, X.; Shan, H.; Li, M.; Ju, S.; Zhao, Z.; Zhang, Z.; Cong, W.; et al. CT Super-resolution GAN Constrained by the Identical, Residual, and Cycle Learning Ensemble (GAN-CIRCLE). *IEEE Trans. Med. Imaging* **2019**, *39*, 188–203. [CrossRef] [PubMed]
15. Yang, Q.; Yan, P.; Zhang, Y.; Yu, H.; Shi, Y.; Mou, X.; Kalra, M.K.; Zhang, Y.; Sun, L.; Wang, G. Low-Dose CT Image Denoising Using a Generative Adversarial Network with Wasserstein Distance and Perceptual Loss. *IEEE Trans. Med. Imaging* **2018**, *37*, 1348–1357. [CrossRef] [PubMed]
16. Vizitiu, A.; Puiu, A.; Reaungamornrat, S.; Itu, L.M. Data-Driven Adversarial Learning for Sinogram-Based Iterative Low-Dose CT Image Reconstruction. In Proceedings of the 2019 23rd International Conference on System Theory, Control and Computing (ICSTCC), Sinaia, Romania, 9–11 October 2019; pp. 668–674. [CrossRef]
17. Zhu, J.-Y.; Park, T.; Isola, P.; Efros, A.A. Unpaired Image-to-Image Translation using Cycle-Consistent Adversarial Networks. In Proceedings of the 2017 IEEE International Conference on Computer Vision (ICCV), Venice, Italy, 22–29 October 2017; pp. 2242–2251. [CrossRef]
18. Karim, A.; Chenming, J.; Marc, F.; Thomas, K.; Tobias, H.; Konstantin, N.; Sergios, G.; Bin, Y. MedGAN: Medical Image Translation using GANs. *Comput. Med. Imaging Graph.* **2019**, *79*, 101684. [CrossRef]
19. Iizuka, S.; Simo-Serra, E.; Ishikawa, H. Globally and Locally Consistent Image Completion. *ACM Trans. Graph.* **2017**, *36*, 107. [CrossRef]
20. Liu, P.; Qi, X.; He, P.; Li, Y.; Lyu, M.R.; King, I. Semantically Consistent Image Completion with Fine-grained Details. *arXiv* **2017**, arXiv:1711.09345.
21. Yang, C.; Lu, X.; Lin, Z.; Shechtman, E.; Wang, O.; Li, H. High-Resolution Image Inpainting using Multi-Scale Neural Patch Synthesis. In Proceedings of the 2017 IEEE Conference on Computer Vision and Pattern Recognition (CVPR), Honolulu, HI, USA, 21–26 July 2017; pp. 4076–4084. [CrossRef]
22. Sabini, M.; Rusak, G. Painting Outside the Box: Image Outpainting with GANs. *arXiv* **2018**, arXiv:1808.08483.
23. Wang, Y.; Tao, X.; Shen, X.; Jia, J. Wide-Context Semantic Image Extrapolation. In Proceedings of the 2019 IEEE/CVF Conference on Computer Vision and Pattern Recognition (CVPR), Long Beach, CA, USA, 15–20 June 2019; pp. 1399–1408. [CrossRef]
24. Sumantri, J.S.; Park, I.K. 360 Panorama Synthesis from a Sparse Set of Images with Unknown Field of View. *arXiv* **2019**, arXiv:1904.03326.
25. Knuth, D.E. *The Art of Computer Programming*, 3rd ed.; Addison-Wesley: Reading, MA, USA, 1997; p. 232.
26. Ulyanov, D.; Vedaldi, A.; Lempitsky, V. Instance Normalization: The Missing Ingredient for Fast Stylization. *arXiv* **2016**, arXiv:1607.08022.
27. Isola, P.; Zhu, J.Y.; Zhou, T.; Efros, A.A. Image-to-Image Translation with Conditional Adversarial Networks. In Proceedings of the 2017 IEEE Conference on Computer Vision and Pattern Recognition (CVPR), Honolulu, HI, USA, 21–26 July 2017; pp. 5967–5976. [CrossRef]
28. Arjovsky, M.; Chintala, S.; Bottou, L. Wasserstein GAN. *arXiv* **2017**, arXiv:1701.07875.
29. Gulrajani, I.; Ahmed, F.; Arjovsky, M.; Dumoulin, V.; Courville, A.C. Improved Training of Wasserstein GANs. *arXiv* **2017**, arXiv:1704.00028.
30. Johnson, J.; Alahi, A.; Fei-Fei, L. Perceptual Losses for Real-Time Style Transfer and Super-Resolution. In *European Conference on Computer Vision*; ECCV 2016: Computer Vision—ECCV 2016, Lecture Notes in Computer Science; Leibe, B., Matas, J., Sebe, N., Welling, M., Eds.; Springer: Cham, Switzerland, 2016; Volume 9906. [CrossRef]
31. Isensee, F.; Kickingereder, P.; Wick, W.; Bendszus, M.; Maier-Hein, K. Brain Tumor Segmentation and Radiomics Survival Prediction: Contribution to the BRATS 2017 Challenge. *arXiv* **2017**, arXiv:1802.10508.
32. Ghesu, F.C.; Georgescu, B.; Mansi, T.; Neumann, D.; Hornegger, J.; Comaniciu, D. An Artificial Agent for Anatomical Landmark Detection in Medical Images. In *Medical Image Computing and Computer-Assisted Intervention—MICCAI 2016*; Ourselin, S., Joskowicz, L., Sabuncu, M.R., Unal, G., Wells, W., Eds.; Springer International Publishing: Cham, Switzerland, 2016; pp. 229–237.
33. Ghesu, F.C.; Georgescu, B.; Zheng, Y.; Grbic, S.; Maier, A.; Hornegger, J.; Comaniciu, D. Multi-Scale Deep Reinforcement Learning for Real-Time 3D-Landmark Detection in CT Scans. *IEEE Trans. Pattern Anal. Mach. Intell.* **2019**, *41*, 176–189. [CrossRef] [PubMed]
34. Chenoune, Y.; Constantinides, C.; Berbari, R.; Roullot, E.; Frouin, F.; Herment, A.; Mousseaux, E. Rigid registration of Delayed-Enhancement and Cine Cardiac MR images using 3D Normalized Mutual Information. In Proceedings of the 2010 Computing in Cardiology, Belfast, UK, 26–29 September 2010; Volume 37, pp. 161–164.
35. Mao, X.; Li, Q.; Xie, H.; Lau, R.Y.K.; Wang, Z.; Smolley, S.P. Least Squares Generative Adversarial Networks. In Proceedings of the 2017 IEEE International Conference on Computer Vision (ICCV), Venice, Italy, 22–29 October 2017; pp. 2813–2821. [CrossRef]

Article

GAN-TL: Generative Adversarial Networks with Transfer Learning for MRI Reconstruction

Muhammad Yaqub [1], Feng Jinchao [1,*], Shahzad Ahmed [1], Kaleem Arshid [1], Muhammad Atif Bilal [2,3], Muhammad Pervez Akhter [2] and Muhammad Sultan Zia [4]

[1] Beijing Key Laboratory of Computational Intelligence and Intelligent System, Faculty of Information Technology, Beijing University of Technology, Beijing 100124, China
[2] Riphah College of Computing, Faisalabad Campus, Riphah International University, Islamabad 38000, Pakistan
[3] College of Geoexploration Science and Technology, Jilin University, Changchun 130061, China
[4] Department of Computer Science, The University of Chenab, Gujranwala 50250, Pakistan
* Correspondence: fengjc@bjut.edu.cn

Citation: Yaqub, M.; Jinchao, F.; Ahmed, S.; Arshid, K.; Bilal, M.A.; Akhter, M.P.; Zia, M.S. GAN-TL: Generative Adversarial Networks with Transfer Learning for MRI Reconstruction. *Appl. Sci.* **2022**, *12*, 8841. https://doi.org/10.3390/app12178841

Academic Editors: Lucian Mihai Itu, Constantin Suciu and Anamaria Vizitiu

Received: 28 July 2022
Accepted: 1 September 2022
Published: 2 September 2022

Publisher's Note: MDPI stays neutral with regard to jurisdictional claims in published maps and institutional affiliations.

Copyright: © 2022 by the authors. Licensee MDPI, Basel, Switzerland. This article is an open access article distributed under the terms and conditions of the Creative Commons Attribution (CC BY) license (https://creativecommons.org/licenses/by/4.0/).

Abstract: Generative adversarial networks (GAN), which are fueled by deep learning, are an efficient technique for image reconstruction using under-sampled MR data. In most cases, the performance of a particular model's reconstruction must be improved by using a substantial proportion of the training data. However, gathering tens of thousands of raw patient data for training the model in actual clinical applications is difficult because retaining k-space data is not customary in the clinical process. Therefore, it is imperative to increase the generalizability of a network that was created using a small number of samples as quickly as possible. This research explored two unique applications based on deep learning-based GAN and transfer learning. Seeing as MRI reconstruction procedures go for brain and knee imaging, the proposed method outperforms current techniques in terms of signal-to-noise ratio (PSNR) and structural similarity index (SSIM). As compared to the results of transfer learning for the brain and knee, using a smaller number of training cases produced superior results, with acceleration factor (AF) 2 (for brain PSNR (39.33); SSIM (0.97), for knee PSNR (35.48); SSIM (0.90)) and AF 4 (for brain PSNR (38.13); SSIM (0.95), for knee PSNR (33.95); SSIM (0.86)). The approach that has been described would make it easier to apply future models for MRI reconstruction without necessitating the acquisition of vast imaging datasets.

Keywords: image reconstruction; MRI; GANs; transfer learning; deep learning

1. Introduction

Magnetic Resonance Imaging (MRI) is a non-ionizing imaging technique used in biomedical research and diagnostic medicine. A strong magnetic field and Radio Frequency (RF) pulses are the foundational elements of MRI. An image is created when antennas placed near the area of the body being examined absorb hydrogen atom radiation, which is present in abundance in all living things. Due to the greater soft-tissue contrast and non-invasive nature of MRI, it is commonly utilized to identify diseases. MRI, on the other hand, has a severe problem in that it takes a long time to acquire sufficient data in k-space. In order to address this issue, k-space imaging approaches with insufficient sampling have been proposed. Compressed sensing [1] and parallel imaging [2] are two commonly used reconstruction approaches for obtaining artifact-free images.

Numerous research organizations and well-known MRI scanner manufacturers are accelerating MRI acquisition. Hardware techniques such as several coils are utilized to sample k-space data in parallel [3]. One of the two main approaches is used in commercial MRI scanners [4] to reconstruct a picture from the coils' under-sampled k-space data. To be more precise, aliased pictures produced by partial k-space conversion are combined into a single coherent image via the Sensitivity Encoder (SENSE) [5]. The inverse Fourier transform (IFT) is calibrated using GRAPPA [6], which uses information from signals in the complex frequency domain. These techniques are examined by [7], along with a hybrid approach that combines the advantages of SENSE with the GRAPPA method's resilience to some flaws. Figure 1, is a summary of data from the PubMed results for "GPU reconstruction" from 2004 to 2020.

Figure 1. Summary of data from the PubMed results for "GPU reconstruction" from 2004 to 2020. The number of papers in respective years is represented in the graph. Number of papers are 5, 9, 13, 24, 36, 29, 48, 79 and 133 in years 2004–2020.

Notable is the fact that deep learning reconstructions have notably shorter reconstruction times while maintaining higher image quality [8,9]. Using a convolutional neural network, the authors of [10] were able to determine the mapping between zero-filled (ZF) images and their corresponding fully-sampled data (CNN). Iterative processes from the ADMM algorithm were used to develop a novel deep architecture for optimizing a CS-based MRI model [11]. In [11], they recreated under-sampled 2D cardiac MR images by use of a convolutional neural network cascade. In terms of both speed and accuracy, our strategy was superior to CS approaches. De-Aliasing Generative Adversary Networks (DAGAN) was proposed by Yang et al. for fast CS-MRI reconstruction in [12]. To keep perceptual image data in the generator network, an adversarial loss was combined with a unique content loss. For MRI de-aliasing [10], created a GAN with a cyclic loss. This network's reconstruction and refinement are carried out using cascaded residual U-Nets. In [13], the authors employed an L1/L2 norm and mixed-cost loss of Least Squares (LS) generator to train their deep residual network with skip connections as a generator for the reconstruction of high-quality MR images. A two-stage GAN technique, according to [14], can estimate missing k-space samples while also removing image artifacts. The self-attention technique was incorporated into a hierarchical deep residual convolutional neural network by [15] in order to improve the under-sampled MRI reconstruction.

Using the self-attention mechanism and the relative average discriminator (SARA-GAN), [16] constructed an artificial neural network in which half of the input data is true, and the other half is false. Research organizations and prominent MRI scanner manufacturers are working hard to speed up the acquisition of MRI scans. For example, numerous coils can be used to sample k-space data simultaneously, as demonstrated by Roemer and colleagues [17]. Under-sampled k-space data generated by the coils are used to reconstruct a picture in commercial MRI scanners [18]. Both approaches are now being applied. If the reader is interested in learning more about how the Sensitivity Encoder (SENSE) works, we suggest [19]. For example, the GeneRalized Autocalibrating Partial Parallel Acquisition (GRAPPA), developed in 2002 by Griswold et al., works on complex frequency domain signals before the IFT. For an overview of these approaches, as well as a hybrid approach that incorporates the benefits of Sense and Grappa, see [20].

The Compressed Sensing (CS) approach [21] provides efficient acquisition and reconstruction of the signal with fewer samples than the Nyquist–Shannon sampling theorem limit when a signal has a sparse representation in a specified transform domain. By selecting a tiny portion of the k-space grid, CS is employed for MRI reconstruction [22]. The IFT of the zero-filled k-space exhibits incoherent artifacts that behave like additive random noise due to the underlying premise that the under-sampling is random. CS, despite being a popular technique today, promotes smooth rebuilding, which could lead to the loss of fine, anatomically significant textural characteristics. Additionally, a sizable amount of runtime is needed. Recently, various machine learning methods for MRI acceleration were suggested. In order to reconstruct MRI from under-sampled k-space data, Ravishankar and Bressler [23] suggested a dictionary-based learning strategy that takes advantage of the sparsity of overlapping image patches highlighting local structure. Using spatio-temporal patches to reconstruct dynamic MRI [24], elaborated on this concept. Both [25,26] used compressed manifold learning based on Laplacian 75 Eigenmaps to reconstruct cardiac MRI and predict respiratory motion.

A variational network (VN) was created in 2018 by [27] to reconstruct intricate multi-channel MR data. In [28], they suggested the MoDL architecture to handle the MRI reconstruction difficulty. Meanwhile, ref. [29] created PI-CNN, which combines parallel imaging with CNN, for high-quality real-time MRI reconstruction. A method for multi-channel image reconstruction based on residual complicated convolutional neural networks was developed by [30] to expedite parallel MR imaging. Reconstructed multi-coil MR data from under-sampled data was successfully produced by [31] using a variable splitting network (VS-Net). Sensitivity encoding and generative adversarial networks (SENSE-GAN) were merged by [32] for rapid multi-channel MRI reconstruction. For the reconstruction of multi-coil MRI, ref. [33] introduced the GrappaNet architecture. The GrappaNet trained the model from beginning to end using neural networks in addition to conventional parallel imaging techniques. Dual-domain cascade U-nets were proposed by [34] for MRI reconstruction. They showed that dual-domain techniques are superior when reconstructing multi-channel data channels simultaneously. A summary of different articles regarding deep learning-based and other models for MRI reconstruction is presented in Table 1.

Training the network parameters and achieving reliable generalization results, all of the aforementioned approaches require a significant size of the dataset. On publicly accessible datasets, the majority of earlier investigations have verified their reconstruction performances. Gathering tens of thousands of multi-channel data points for model training in clinical applications is challenging, though, because retaining raw k-space data is not a common clinical flow. The generalization of learned image reconstruction networks trained on open datasets must therefore be improved. In order to address this issue, numerous transfer learning studies have been carried out lately.

Table 1. A summary of different articles regarding deep learning-based and other models for MRI reconstruction.

Sr. No.	Reference	Methodology	Results	Future Directions
1	[35]	Reconstruction of brain MRI data using a G1M U-Net model	Reconstruction results are derived from practical sampling schemes of accelerated brain MRIs.	Apply to a wide range of datasets with excellent fidelity to fully sample scans.
2	[36]	Use of a GAN with a Cyclic Loss to Reconstruct a CS-MRI	In terms of both running time and image quality, CS-MRI methods performed noticeably better than open-source MRI datasets.	The next step in the study will be to extend Refine GAN to handle dynamic MRI.
3	[37]	The inverse problem was solved using a deep CNN-based optimization model.	Discriminative CNN denoiser creates a versatile, quick, and efficient image restoration framework.	----
4	[38]	Image Reconstruction from Compressively Sensed Random Measurements Using Recon Net	Recon Net offers high-quality reconstructions of simulated and actual data for various measurement rates.	----
5	[39]	To resolve problems with normal-convolutional inverse, direct inversion and a CNN are proposed.	Parallel beam X-ray CT sparse-view network performance is calculated.	It is possible to address strategies for heterogeneous datasets.
6	[40]	CS-based approaches, especially DLMRI, use a coordinate-descent algorithm to optimize.	CNNs were evaluated for their relevance to the MR image reconstruction challenge.	The model will directly address the coil sensitivity maps' redundancy.
7	[41]	The Primal-Dual algorithm for tomographic reconstruction has been learned.	For the Shepp–Logan phantom, they improve peak SNR by 6 dB over competing approaches.	Capable of using complex loss functions with learned reconstruction operators.
8	[42]	Recurrent Neural Networks are used by RIMs to solve inverse problems.	The RIM-3task model is competitive on all noise levels.	----
9	[43]	A pre-trained CNN model was used to augment and classify brain tumor data.	Before and after data augmentation, they outperformed the most sophisticated algorithms with 90.67 accuracies.	Weight-saving CNN fine-grained classification will use differential stochastic classification.
10	[44]	Investigate the overfitting issue using a CapsNet for classifying brain tumors.	Comparative research with CNN found their accuracy rate was 86.56%. Learning rate decreases with iterations.	In the future, look into how adding more layers affects classification accuracy.
11	[45]	A review of medical image classification using deep learning approaches.	They explain deep learning algorithms and how they can be used for medical imaging, noting that the learning rate is proportional to the inverse of iterations.	To apply the strategies to the modalities where they are not employed, more research is needed.
12	[46]	Predicted patient survival using BraTS2017 and U-NET.	With less computational time, 89.6% accuracy was achieved.	----
13	[47]	BRaTS 2013, 2015 used CNN-based two-path architecture to separate brain tumors.	Cascaded input CNN achieved 88.2% accuracy. Analysis of various architectural designs.	Increasing the architecture layers and data set boosted the outcomes even more.

To reconstruct high-quality images from under-sampled k-space data in MRI, ref. [48] created a unique deep learning approach with domain adaptability. The proposed network made use of a pre-trained network, which was then fine-tuned using a sparse set of radial MR datasets or synthetic radial MR datasets. Knoll et al. [49] investigated the effects of image content, sampling pattern, SNR, and image contrast on the generalizability of a pre-trained model in order to show the potential for transfer learning using the VN architecture. To test the ability of networks trained on normal pictures to generalize to T1- and T2-weighted brain images, ref. [50] suggested a transfer-learning approach. Meanwhile, assessed the generalization of the results of a trained U-net for the single-channel MRI reconstruction problem using MRI performed with a variety of scanners, each with a different magnetic field intensity, anatomical variations, and under-sampling masks.

This study aims to investigate the generalizability of a trained GAN model for reconstructing an MRI with insufficient samples in the following circumstances:

- Transfer learning for a private clinical brain test dataset using the proposed GAN model.
- Using datasets from open-source knee and private source brain tests, transfer learning of the proposed GAN model.
- For datasets on the knee and brain with Afs of 2 and 4, transfer learning of the proposed GAN model is conducted.

2. Method and Material

The formulation of the multi-channel image reconstruction problem for parallel imaging is as follows:

$$w = PFRI + m \tag{1}$$

where I is the image we intend to solve, P is the under-sampling mask, w is the collected k-space measurements, m is the noise, F is the Fourier transform and R is the coil sensitivity maps.

By incorporating past knowledge, CS-MRI constricts the solution space in order to solve the inverse problem of Equation (1). Furthermore, the optimization problem can be stated as:

$$I = min_i \frac{1}{2} \|PFR_i - Y\|_2^2 + \lambda S(i) \tag{2}$$

where the prior regularization term is denoted by $S(i)$, and the first term reflects data fidelity in the k-space domain, which ensures that the reconstruction results are consistent with the original under-sampled k-space data. Term λ is a balance parameter that establishes the trade-off between the data fidelity term and the prior knowledge. In a specific sparsity transform domain, $S(i)$ is often an L0 or L1 norm.

Typically, an iterative strategy is necessary to tackle the above optimization issue. The regularization term $S(i)$, which is based on CNN, can now be used to denote, i.e.,

$$I = min_i \frac{1}{2} \|PFR_i - Y\|_2^2 + \lambda \|I - fCNN(I_u|\theta)\|_2^2 \tag{3}$$

Utilizing the training dataset, the model's parameters can be tuned, and the output of CNN is $fCNN(I_u|\theta)$ with the parameters θ. $I_u = F_u^H w$, where H stands for the Hermitian transpose operation, and also refers to the ZF images that were reconstructed from under-sampled k-space data. Recently, MRI reconstruction has also incorporated conditional GAN.

A GAN has a discriminator D and a generator G. Both the discriminator and the generator need to be trained. The generator G can be taught, through training, to predict the distribution of the genuine data that are provided and to produce data that will deceive the discriminator D. Distinguishing between the output of the generator G and the actual data is the discriminator D's goal. Then, after training, the generator can be used independently to generate new samples that are comparable to the original ones.

The conditional GAN loss was therefore applied to the reconstruction of MRI images, which is:

$$min_{\theta_c} max_{\theta_e} \mathcal{L}_b GAN(\theta_c, \theta_e) = E_{I_t - P_{train}(I_t)}[\log D_{\theta_e}(I_t)] \, E_{I_t - p_c(I_t)}[\log D_{\theta_e}(G_{\theta_c}(I_u))] \quad (4)$$

where I_t is the fully sampled ground truth, and I_u is the equivalent reconstructed image produced by the generator. I_u is the ZF image that serves as the generator's input.

2.1. Datasets

The provincial institutional review committee approved the study, and all subjects provided their informed consent for inclusion prior to participating in it. The MRI scanning was authorized by the institutional review board (Miu Hospital Lahore). Private brain tumor MRI datasets were collected from 19 participants utilizing various imaging sequences. We chose 6 participants at random for network testing and 13 for tuning, which corresponded to 218 and 91 images, respectively. The "Stanford Fully Sampled 3D FSE Knees" repository provided the knee datasets used in this inquiry. The raw data were collected using an 8-coil, 3.0T full-body MR system in conjunction with a 3D TSE sequence with proton density weighting and fat saturation comparison. For network tuning, we randomly selected 18 individuals, and for testing, we randomly selected 2 subjects, which corresponded to 1800 and 200 2D images, respectively.

2.2. Model Architecture

Each of the generator networks had the same design and was based on the proposed GAN model residual of CNN. Our proposed GAN architecture, which includes a generator and discriminator, is depicted in Figure 2 in detail. Five convolutional encoding layers and five deconvolutional decoding layers made up the network, with batch normalization and leaky-ReLU activation functions following each layer. The final layer of the k-space network generator G entailed 2 output channels corresponding to real and fake components (see Figure 3).

The discriminator was made up of nine different blocks of convolution layers, followed by leaky-ReLU activation functions and batch normalization, and the final stage is a fully linked layer. The training was conducted using the Adam optimizer (see Figure 4). Using our proposed GAN model and transfer learning, we were able to recover the undersampled MRI data for two different circumstances, as shown in Figure 5. The dataset contains 1800 images from 18 participants and 4500 images from 45 subjects for testing. We divided the training and validation datasets during the training procedure. Eighteen photos were chosen at random for validation during each round. The models in the validation dataset with the best performance, or those with the highest PSNR, were chosen for further independent testing.

Using random sampling trajectories for AF = 2 and AF = 4, retroactively, all fully sampled k-space data were undersampled. We experimented with different filter sizes, changing the filter sizes according to the pool of our data sample. Figures 3 and 4 show the details of the architecture. The networks were trained using the Adam [43] optimizer and the various hyperparameters. The model was trained with an initial learning rate of 10^{-3}, filter size 3×3, Xavier initialization and an 8-batch size with a monotonically decreasing learning rate over 500 epochs.

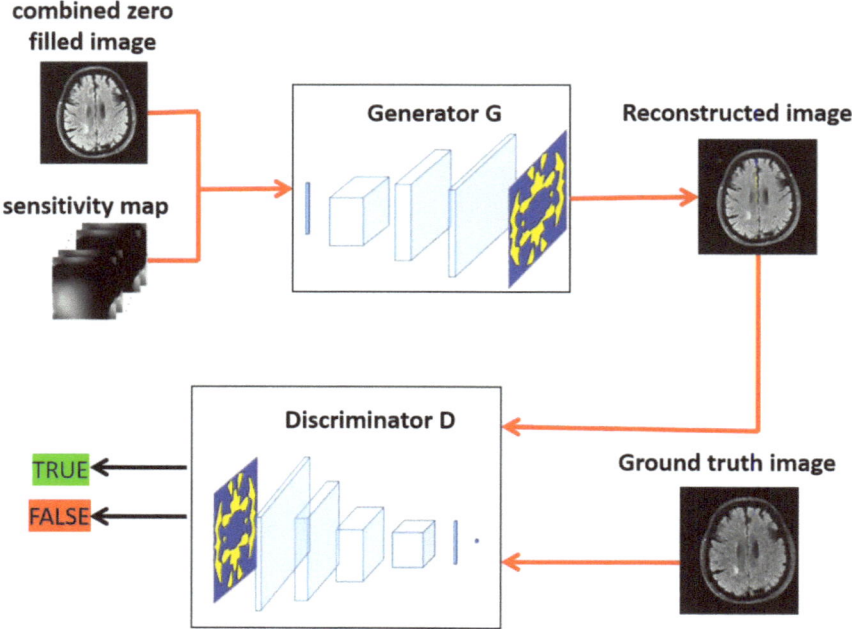

Figure 2. Our proposed GAN architecture, including a generator and discriminator. The generator G takes the input ZF and sensitivity map. The details of both generator G and discriminator D are discussed and shown below in Figures 3 and 4.

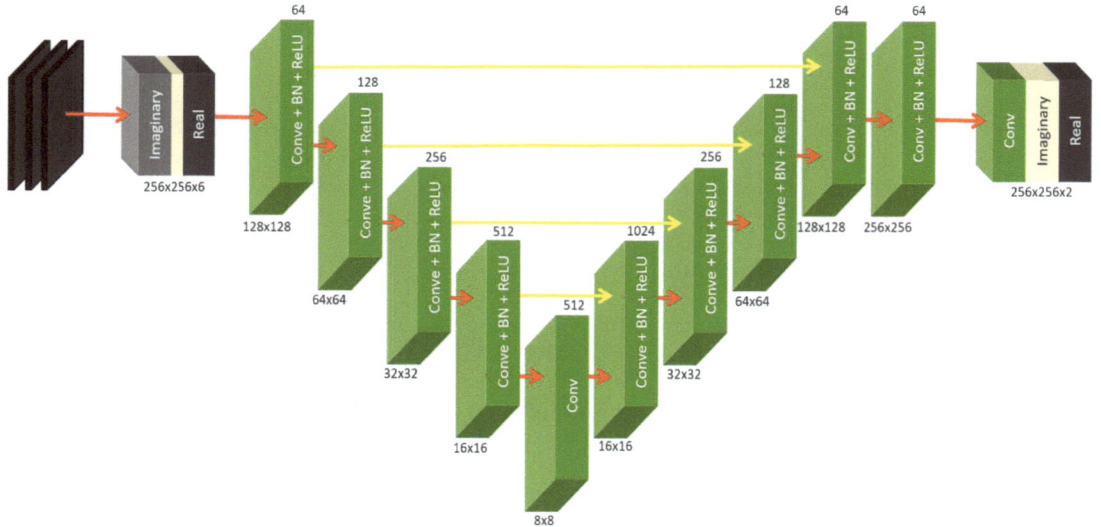

Figure 3. Discriminator architecture.

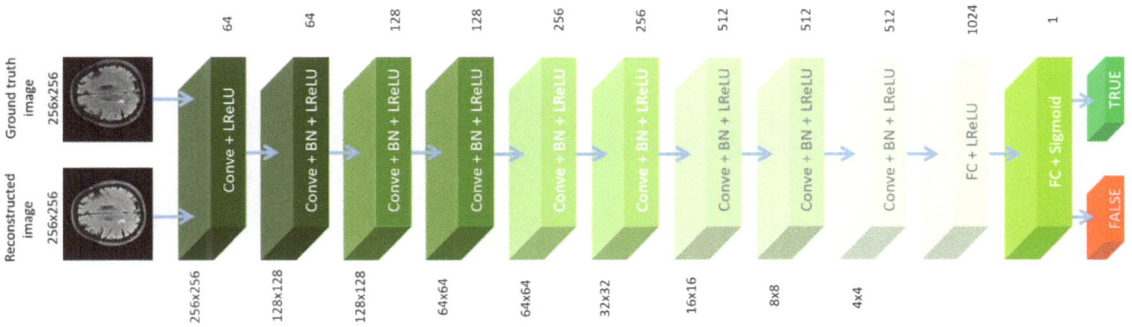

Figure 4. Generator architecture.

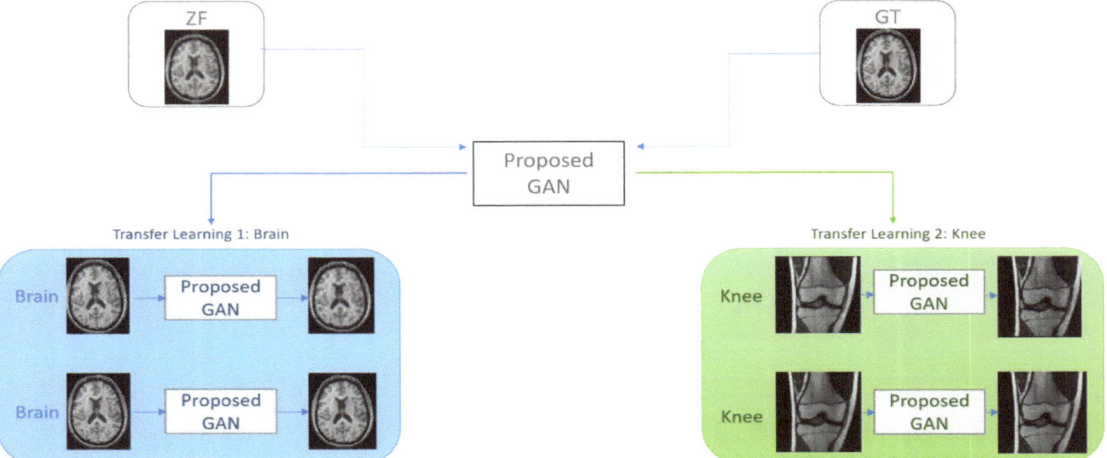

Figure 5. Transfer learning for a GAN reconstruction model proposed for under-sampled MRI reconstruction.

3. Results and Discussion

The experiments were carried out in Python3 with the TensorFlow backend. On a workstation equipped with an NVIDIA GV100GL graphics processor unit (GPU), the reconstruction methods were executed. PSNR and SSIM were used to evaluate the acquired reconstruction outcomes.

We experimented with several filter sizes in order to determine the best filter size, and we ultimately chose the filter size that produced the greatest results on our samples of public and private data. Three distinct filter sizes, including 3×3, 5×5 and 7×7, were used. In comparison to other filter sizes, the 3×3 filters produced greater results. The experimental findings on the private brain and public knee datasets with various filter sizes are shown in Figure 6.

The model produced excellent reconstruction results (PSNR, 37.98; SSIM, 0.97). Then, using a test dataset made up of just a few hundred images from various domains, we applied our model to it. The test results of several brain and knee image reconstruction methods on a private and public dataset are shown in Figures 7–10. As demonstrated by brain images, the findings of Directly Trained (PSNR, 35.78; SSIM, 0.95) were marginally superior to those of the Calgary Model (PSNR, 34.73; SSIM, 0.94) and Image Net (PSNR, 34.25, SSIM, 0.92), which had artifacts.

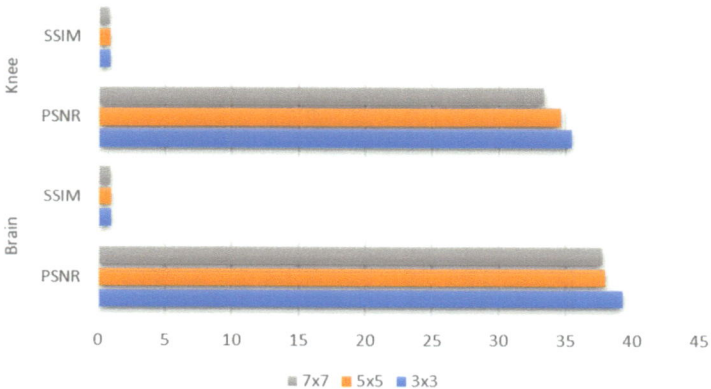

Figure 6. The graph of the PSNR/SSIM values obtained as a result of experiments by using different filter size (3×3, 5×5 and 7×7) on our private brain and public dataset.

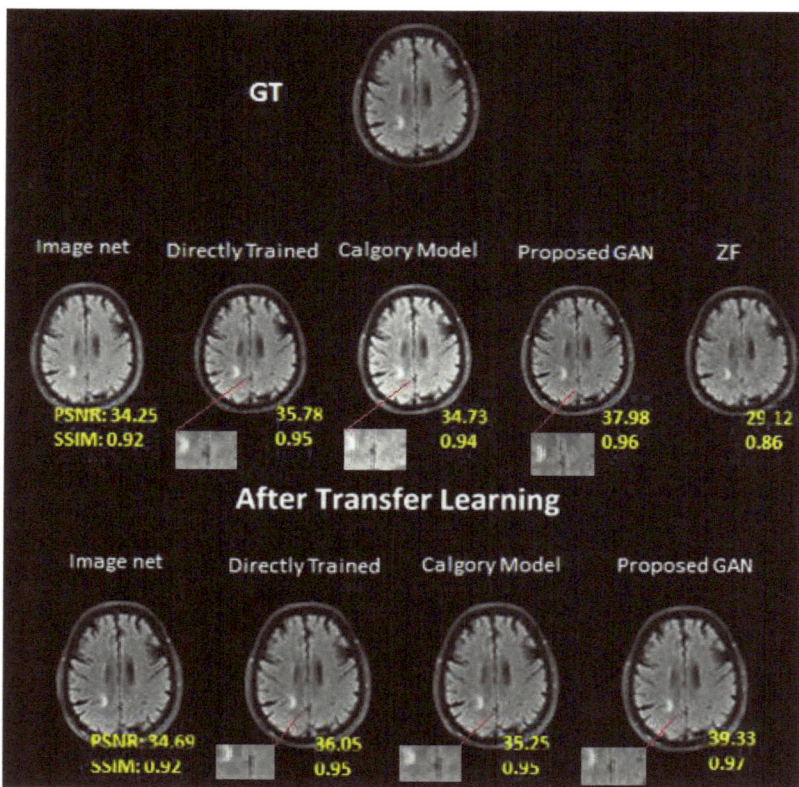

Figure 7. At AF = 2, Reconstruction results on private brain images. From left to right: (i) Image Net (IN); (ii) Directly trained (DT); (iii) Calgary Model (CM); (iv) Proposed GAN; (v) ZF. In the second row: (i) Image net; (ii) Directly Trained; (iii) Calgary Model; (iv) Proposed GAN results of same slices from brain image dataset after applying transfer learning.

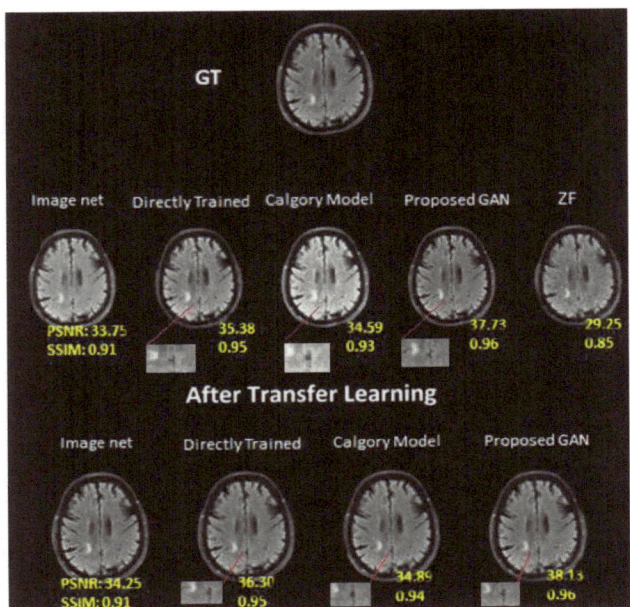

Figure 8. At AF = 4, Reconstruction results on private brain images. From left to right: (i) Image Net; (ii) Directly trained; (iii) Calgary Model; (iv) Proposed GAN; (v) ZF. In the second row: (i) Image net; (ii) Directly Trained; (iii) Calgary Model; (iv) Proposed GAN results of same slices from brain image dataset after applying transfer learning.

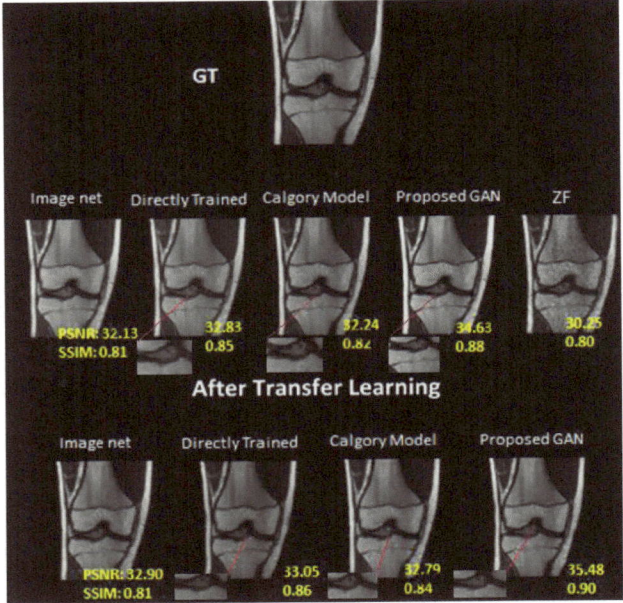

Figure 9. At AF = 2, Reconstruction results on knee dataset. From left to right: (i) Image Net; (ii) Directly trained; (iii) Calgary Model; (iv) Proposed GAN; (v) ZF. In the second row: (i) Image net; (ii) Directly Trained; (iii) Calgary Model; (iv) Proposed GAN results of same slices from knee image dataset after applying transfer learning.

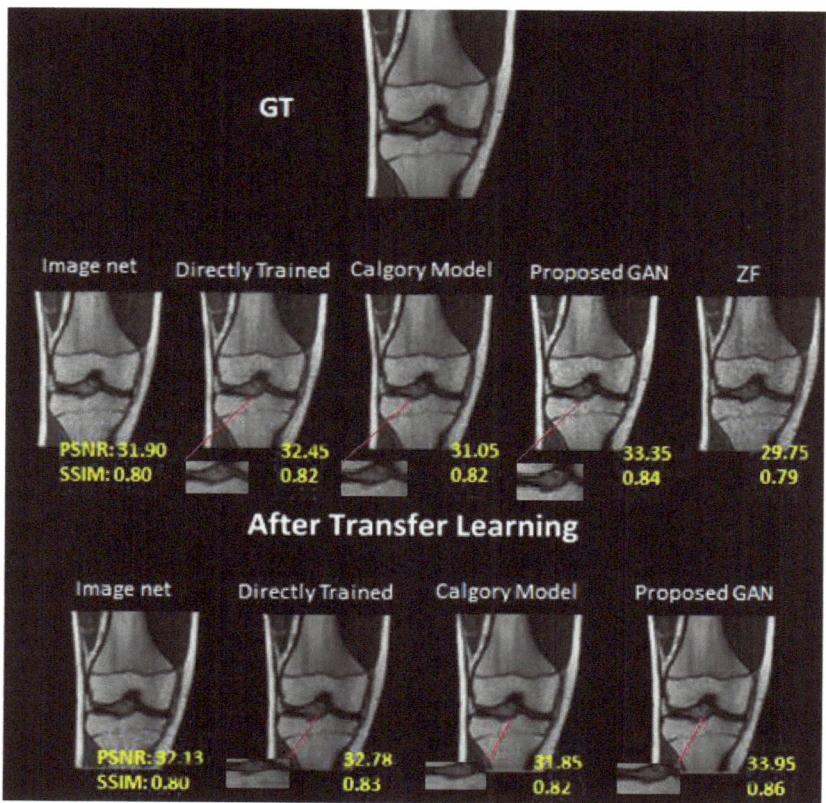

Figure 10. At AF = 4, Reconstruction results on knee dataset. Moving left to right: (i) Image Net; (ii) Directly trained; (iii) Calgary Model; (iv) Proposed GAN; (v) ZF. In the second row: (i) Image net; (ii) Directly Trained; (iii) Calgary Model; (iv) Proposed GAN results of same slices from knee image dataset after applying transfer learning.

The outcomes of our proposed GAN model (PSNR, 37.98; SSIM, 0.97) were superior to those of the other models examined. Our proposed transfer learning model beat other reconstruction techniques (PSNR, 39.33; SSIM, 0.95). Due to a short training dataset, the DT image (PSNR = 36.05; SSIM, 0.95) produced artifacts in the associated brain images. The image net model (PSNR, 34.69; SSIM, 0.92) and CM image (PSNR, 35.25; SSIM, 0.95) had worse results than the other models.

The results of our proposed GAN model at AF 4 were 37.73 PSNR and 0.96 SSIM. Image net (PSNR, 34.25; SSIM, 0.92), CM (PSNR, 34.59; SSIM, 0.93) and DT (PSNR, 35.38; SSIM, 0.95) at AF 4. The image net model performed the worst, the category model performed marginally better and the directly trained model performed better than both. Similarly, the performance was enhanced when transfer learning was applied to the same dataset at AF 4. On the knee dataset, the performance of our proposed GAN was also better than other models. As compared to the brain image dataset, the knee dataset had the lowest accuracy of all mentioned models at AF2 and AF 4. On the knee dataset, our proposed GAN model at AF 2 was 34.63 PSNR and 0.88 SSIM. Image net (PSNR, 32.13; SSIM, 0.81), Calgary Model (PSNR, 32.24; SSIM, 0.82) and Directly Trained (PSNR, 32.83; SSIM, 0.85) at AF 2. The Calgary model outperformed the ImageNet model, whereas the directly trained model outperformed both. The performance was enhanced when transfer learning was applied to the same dataset at AF 2.

There were the following results after applying transfer learning: Directly Trained (PSNR, 33.05; SSIM, 0.86), Image net (PSNR, 32.90; SSIM, 0.81), Calgary Model (PSNR, 32.79; SSIM, 0.84) and proposed GAN model (PSNR, 35.48: SSIM, 0.90).

This study's key contribution is the development of a transfer learning enhanced GAN technique for reconstructing numerous previously unreported multi-channel MR datasets. The findings indicate that TL from our proposed method may be able to lessen variation in image contrast, anatomy and AF between the training and test datasets. With the brain tumor dataset, reconstruction image performance was better.

This demonstrates that the best method might be to generate training and test data with the same contrasts because brain data were initially used to train the proposed model. When the distributions of the training and test datasets were similar, the reconstruction performance was good. The PSNR and SSIM of the images were significantly enhanced after applying transfer learning. This demonstrates that the extra information provided by these reconstructions makes fine-tuning more efficient when data are replicated across domains.

Table 2 shows the comparison of different models' reconstruction results under AF = 2 for brain and knee images. The directly trained model for brain and knee images performed better than the Image Net and Calgary model. Our proposed GAN models beat all other compared methods, having PSNR (37.98) and SSIM (0.96) for brain images and PSNR (34.63) and SSIM (0.88) for knee images. All compared models perform slightly less on knee images than on brain images.

Table 2. Quantitative analysis of PSNR and SSIM values acquired from brain and knee test images using various reconstruction techniques.

	Brain		Knee	
	PSNR	SSIM	PSNR	SSIM
ZF	29.12	0.86	30.25	0.80
ImageNet [51]	34.25	0.92	32.13	0.81
Directly Trained [52]	35.78	0.95	32.83	0.85
Calgary Model [53]	34.73	0.94	32.24	0.82
Proposed GAN	**37.98**	**0.96**	**34.63**	**0.88**

Figure 11 displays the outcomes of reconstruction for knee and brain images at an AF of 2. The x-axis shows various models, and the y-axis shows the value of PSNR. The blue color legend depicts the brain images, and the brown color legend depicts the knee. The proposed GAN model had the highest accuracy (PSNR, 37.98 and 34.63) on brain and knee images, respectively. ZF and image net had the least PSNR compared to other models. Figure 12 displays the outcomes of reconstruction for knee and brain images at an AF of 2. There are multiple model counts on the x-axis, and the SSIM value is displayed on the y-axis. The knee is represented by the brown color legend, and the brain by the blue color legend. In images of the knee and the brain, the proposed GAN model had the highest SSIM (0.96 and 0.88, respectively). When compared to other models, ZF and image net had the lowest SSIM.

Figures 13 and 14 show the reconstruction results at AF 4 for brain and knee images. Our proposed GAN model showed promising results as compared to the other model. Results at AF 4, PSNR and SSIM of all compared models in Figures 13 and 14 were slightly decreased. Furthermore, if we discuss the performance difference at AF 2 and AF 4, the Proposed GAN model improved PSNR by 1.06% and SSIM by 1.01% at AF 2, as compared to AF 4.

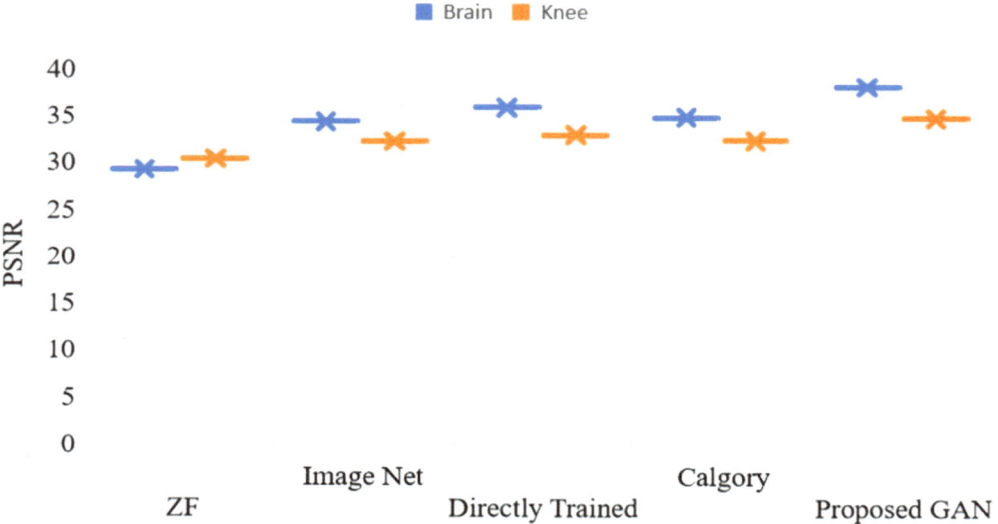

Figure 11. PSNR, values of reconstructed images of the knee and brain using various models, with AF = 2.

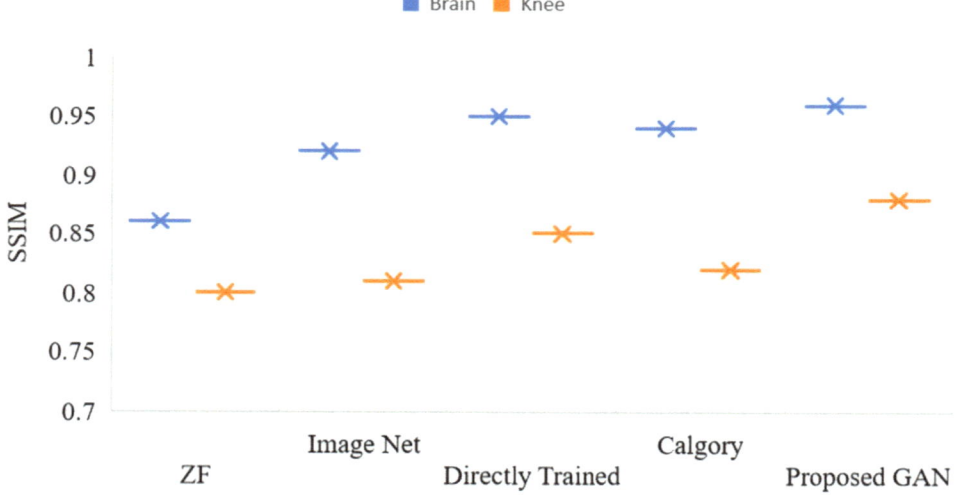

Figure 12. SSIM, values of reconstructed images of the knee and brain using various models, with AF = 2.

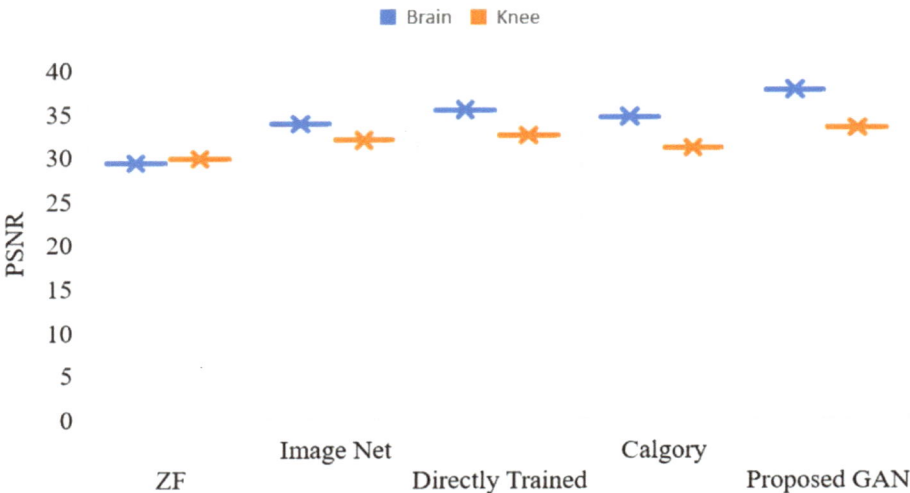

Figure 13. PSNR, values of reconstructed images of the knee and brain using various models, with AF = 4.

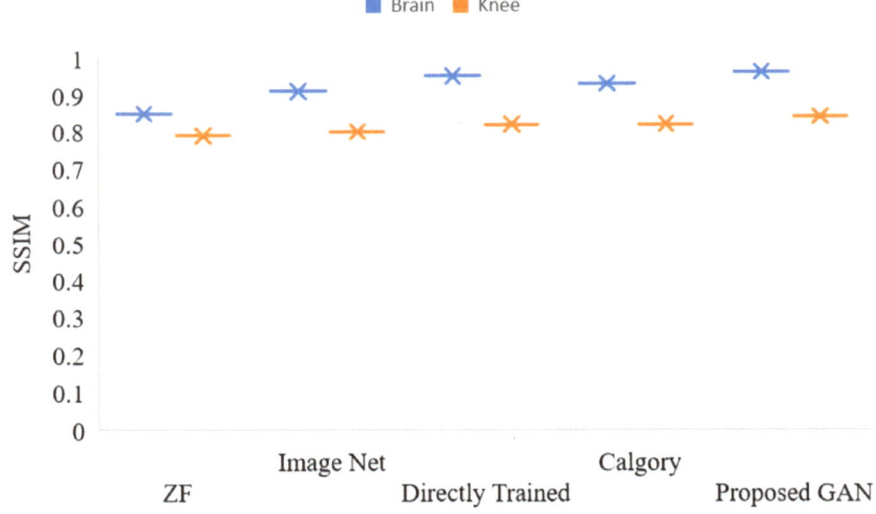

Figure 14. SSIM, values of reconstructed images of the knee and brain using various models, with AF = 4.

After using transfer learning, Table 3 compares the reconstruction outcomes of several models for pictures of the knee and the brain with AF = 2. Our proposed GAN models beat all other compared methods by improved PSNR (39.33) and SSIM (0.97) for brain images and PSNR (35.48) and SSIM (0.90) for knee images. The directly trained model for brain and knee image performance improved by PSNR (36.05), SSIM (0.95) and PSNR (33.05), SSIM (0.86) better than the Image Net and Calgary model. All compared models' performance increased on knee images than brain images after applying transfer learning.

Table 3. PSNR and SSIM quantitatively evaluated values for the brain and knee test images acquired using various reconstruction models after using transfer learning.

	Brain		Knee	
	PSNR	SSIM	PSNR	SSIM
ImageNet [51]	34.69	0.92	32.90	0.81
Directly Trained [52]	36.05	0.95	33.05	0.86
Calgary Model [53]	35.25	0.95	32.79	0.84
Proposed GAN-TL	**39.33**	**0.97**	**35.48**	**0.90**

Reconstruction results for images of the knee and brain at AF = 2 after TL are shown in Figures 15 and 16. Figure 15 displays the PSNR value-based outcomes of reconstruction, while Figure 16 shows the SSIM value-based results of reconstruction after applying TL. The blue color legend depicts the brain images, and the brown color legend depicts the knee. Proposed GAN model PSNR (37.98 and 34.63), SSIM (0.97 and 0.90), on brain and knee images, respectively. Calgary model performance PSNR (35.25 and 32.79) and SSIM (0.95 and 0.84) were slightly better compared to the image net model. Directly trained model PSNR (36.05 and 33.05) and SSIM (0.95 and 0.84) results for brain and knee images, respectively. Performance increased by 3.0 % on brain images compared to knee images.

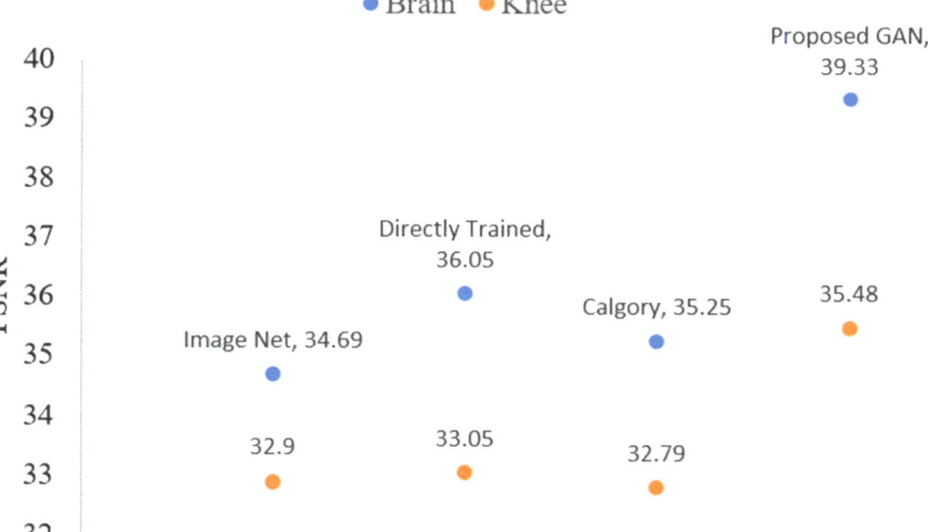

Figure 15. The blue dots represent brain data PSNR values while orange dots represent knee data PSNR values of the reconstructed images when AF = 2 following TL.

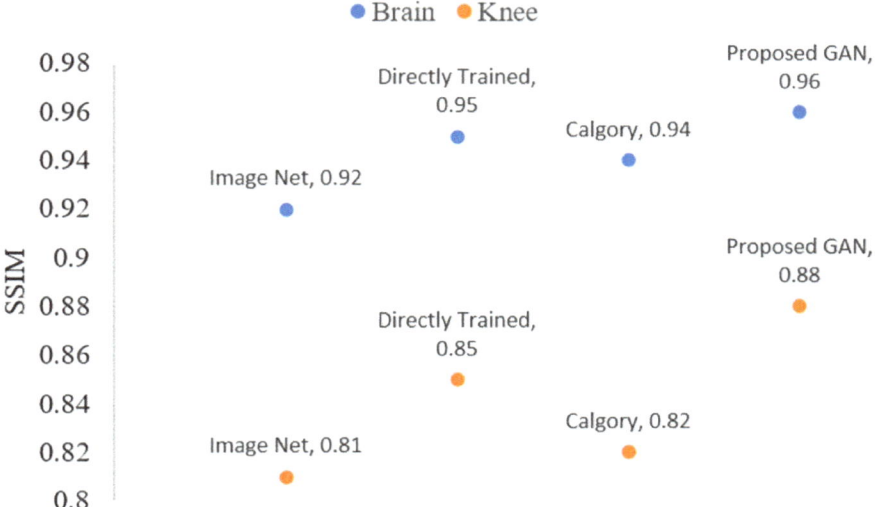

Figure 16. The blue dots represent brain data SSIM values, while orange dots represent knee data SSIM values of the reconstructed images when AF = 2 following TL.

The reconstruction outcomes for images of the knee and brain at AF 4 are shown in Figures 17 and 18. In comparison to the other model, the proposed GAN model demonstrated good results. Additionally, if we compare the performance between AF 2 and AF 4, the proposed GAN model enhanced PSNR and SSIM at AF 2 compared to AF 4 by 1.20 and 2% percent, respectively.

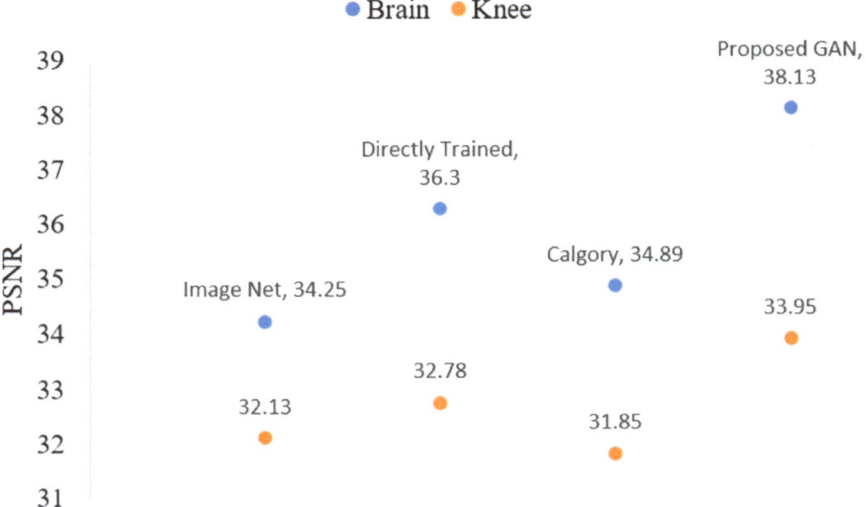

Figure 17. The blue dots represent brain data PSNR values, while orange dots represent knee data PSNR values of the reconstructed images when AF = 4 following TL.

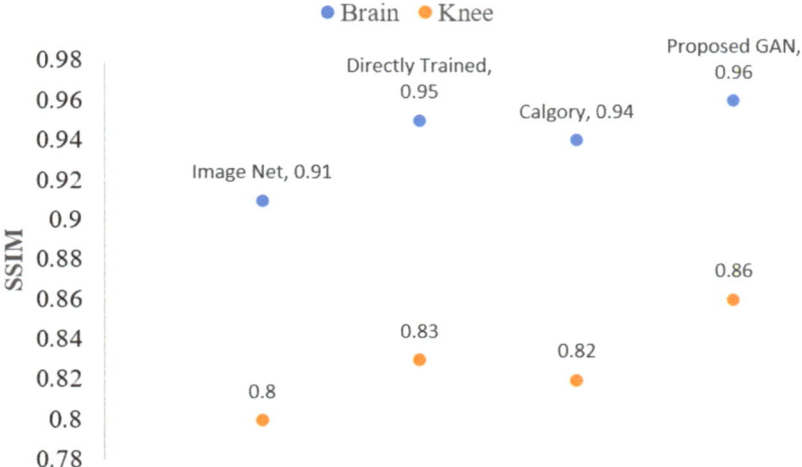

Figure 18. The blue dots represent brain data SSIM values, while orange dots represent knee data SSIM values of the reconstructed images when AF = 4 following TL.

Our research shows that when compared to other techniques, the reconstruction's image's distribution produced by transfer learning is more similar to the distribution of the completely sampled image, which can help with the segmentation and diagnosis of cancerous tumors. We can also successfully use transferred learning across a range of anatomies. We discovered that brain tumor samples converged faster than knee datasets. This might be because only a few transfer learning steps were necessary to achieve the best results because the brain tumor data were located at similar anatomical locations as the training data.

As an alternative, we used a fixed training set and a range of iterations to test model-reconstructed image performance after fine-tuning. It makes sense to draw the conclusion that performance increases with dataset size. Given that there are essentially no data that can be collected, we think the current study is more realistic. The outcomes of reconstruction will be better than using a tiny portion of its own data for training as long as fine-tuning is carried out, regardless of whether the AF is more or less than its own under-sampling AF. A model with a low AF should be chosen for TL because both brain and knee data show that AF = 2 is ideal for fine-tuning. In the future, we will evaluate our transfer learning method's reconstruction performance compared to that of existing unsupervised learning methodologies. The proposed method would facilitate the application of future MRI reconstruction models without requiring the collection of sizable imaging datasets.

4. Conclusions

This work examines the generalization capabilities of a learned proposed GAN model for under-sampled multi-channel MR images in terms of the differences across training and test datasets. Our research demonstrates that the proposed GAN model was used to analyze private brain images, knee images and images with varying AF while utilizing TL and a small tuning dataset. As compared to the results of transfer learning for the brain and knee, fewer training data being used produced superior results, with acceleration factor (AF) 2 (for brain PSNR (39.33) and SSIM (0.97); for knee PSNR (35.48) and SSIM (0.90)) and AF 4 (for brain PSNR (38.13) and SSIM (0.95); for knee PSNR (33.95) and SSIM (0.86)). The proposed method would facilitate the application of future MRI reconstruction models without requiring the collection of sizable imaging datasets.

Author Contributions: Conceptualization, M.Y. and F.J.; methodology, M.Y.; software, M.Y.; validation, F.J., M.Y. and S.A.; formal analysis, M.Y.; investigation, F.J.; resources, M.A.B.; data curation, K.A.; writing—original draft preparation, M.Y.; writing—review and editing, F.J.; visualization, M.P.A.; supervision, F.J. and M.S.Z.; project administration, F.J.; funding acquisition, F.J. All authors have read and agreed to the published version of the manuscript.

Funding: This work was sponsored by the National Science Foundation of China under Grant No. 81871394 and the Beijing Laboratory of Advanced Information Networks.

Informed Consent Statement: Not applicable.

Data Availability Statement: Not applicable.

Acknowledgments: The authors would like to thank sponsored organizations.

Conflicts of Interest: The authors declare no conflict of interest.

References

1. Chen, Y.; Schönlieb, C.-B.; Liò, P.; Leiner, T.; Dragotti, P.L.; Wang, G.; Rueckert, D.; Firmin, D.; Yang, G.J. AI-based reconstruction for fast MRI—A systematic review and meta-analysis. *Proc. IEEE* **2022**, *110*, 224–245. [CrossRef]
2. Feng, C.-M.; Yang, Z.; Chen, G.; Xu, Y.; Shao, L. Dual-octave convolution for accelerated parallel MR image reconstruction. In Proceedings of the AAAI Conference on Artificial Intelligence, Vancouver, BC, Canada, 2–9 February 2021; pp. 116–124.
3. Yang, C.; Liao, X.; Wang, Y.; Zhang, M.; Liu, Q. Virtual Coil Augmentation Technology for MRI via Deep Learning. *arXiv* **2022**, arXiv:2201.07540.
4. Shan, S.; Gao, Y.; Liu, P.Z.; Whelan, B.; Sun, H.; Dong, B.; Liu, F.; Waddington, D.E.J. Distortion-Corrected Image Reconstruction with Deep Learning on an MRI-Linac. *arXiv* **2022**, arXiv:2205.10993.
5. Hollingsworth, K.G. Reducing acquisition time in clinical MRI by data undersampling and compressed sensing reconstruction. *Phys. Med. Biol.* **2015**, *60*, R297. [CrossRef] [PubMed]
6. Lee, J.-H.; Kang, J.; Oh, S.-H.; Ye, D.H. Multi-Domain Neumann Network with Sensitivity Maps for Parallel MRI Reconstruction. *Sensors* **2022**, *22*, 3943. [CrossRef] [PubMed]
7. Scott, A.D.; Wylezinska, M.; Birch, M.J.; Miquel, M.E. Speech MRI: Morphology and function. *Phys. Med.* **2014**, *30*, 604–618. [CrossRef] [PubMed]
8. Oostveen, L.J.; Meijer, F.J.; de Lange, F.; Smit, E.J.; Pegge, S.A.; Steens, S.C.; van Amerongen, M.J.; Prokop, M.; Sechopoulos, I. Deep learning-based reconstruction may improve non-contrast cerebral CT imaging compared to other current reconstruction algorithms. *Eur. Radiol.* **2021**, *31*, 5498–5506. [CrossRef]
9. Lebel, R.M. Performance characterization of a novel deep learning-based MR image reconstruction pipeline. *arXiv* **2020**, arXiv:2008.06559.
10. Lv, J.; Wang, C.; Yang, G.J.D. PIC-GAN: A parallel imaging coupled generative adversarial network for accelerated multi-channel MRI reconstruction. *Diagnostics* **2021**, *11*, 61. [CrossRef]
11. Schlemper, J.; Caballero, J.; Hajnal, J.V.; Price, A.; Rueckert, D. A Deep Cascade of Convolutional Neural Networks for MR image Reconstruction. In *Information Processing in Medical Imaging*; Springer: Berlin/Heidelberg, Germany, 2017.
12. Jiang, M.; Yuan, Z.; Yang, X.; Zhang, J.; Gong, Y.; Xia, L.; Li, T. Accelerating CS-MRI reconstruction with fine-tuning Wasserstein generative adversarial network. *IEEE Access* **2019**, *7*, 152347–152357. [CrossRef]
13. Mardani, M.; Gong, E.; Cheng, J.Y.; Vasanawala, S.S.; Zaharchuk, G.; Xing, L.; Pauly, J.M. Deep generative adversarial neural networks for compressive sensing MRI. *IEEE Trans. Med. Imaging* **2018**, *38*, 167–179. [CrossRef] [PubMed]
14. Sandilya, M.; Nirmala, S.; Saikia, N. Compressed Sensing MRI Reconstruction Using Generative Adversarial Network with Rician De-noising. *Appl. Magn. Reson.* **2021**, *52*, 1635–1656. [CrossRef]
15. Wu, Y.; Ma, Y.; Liu, J.; Du, J.; Xing, L. Self-attention convolutional neural network for improved MR image reconstruction. *Inf. Sci.* **2019**, *490*, 317–328. [CrossRef]
16. Rempe, M.; Mentzel, F.; Pomykala, K.L.; Haubold, J.; Nensa, F.; Kröninger, K.; Egger, J.; Kleesiek, J. k-strip: A novel segmentation algorithm in k-space for the application of skull stripping. *arXiv* **2022**, arXiv:2205.09706.
17. Bydder, M.; Larkman, D.; Hajnal, J. Combination of signals from array coils using image-based estimation of coil sensitivity profiles. *Magn. Reson. Med.* **2002**, *47*, 539–548. [CrossRef] [PubMed]
18. Shitrit, O.; Riklin Raviv, T. Accelerated Magnetic Resonance Imaging by Adversarial Neural Network. In *Deep Learning in Medical Image Analysis and Multimodal Learning for Clinical Decision Support*; Springer: Berlin/Heidelberg, Germany, 2017; pp. 30–38.
19. Pruessmann, K.P.; Weiger, M.; Scheidegger, M.B.; Boesiger, P. SENSE: Sensitivity encoding for fast MRI. *Magn. Reson. Med.* **1999**, *42*, 952–962. [CrossRef]
20. HashemizadehKolowri, S.; Chen, R.-R.; Adluru, G.; Dean, D.C.; Wilde, E.A.; Alexander, A.L.; DiBella, E.V. Simultaneous multi-slice image reconstruction using regularized image domain split slice-GRAPPA for diffusion MRI. *Med. Image Anal.* **2021**, *70*, 102000. [CrossRef]

21. Candès, E.J. Compressive sampling. In Proceedings of the International Congress of Mathematicians, Madrid, Spain, 22–30 August 2006.
22. Liu, B.; Zou, Y.M.; Ying, L. SparseSENSE: Application of compressed sensing in parallel MRI. In Proceedings of the 2008 International Conference on Information Technology and Applications in Biomedicine, Shenzhen, China, 30–31 May 2008.
23. Wen, B.; Ravishankar, S.; Bresler, Y. Structured overcomplete sparsifying transform learning with convergence guarantees and applications. *Int. J. Comput. Vis.* **2015**, *114*, 137–167. [CrossRef]
24. Qin, C.; Schlemper, J.; Caballero, J.; Price, A.N.; Hajnal, J.V.; Rueckert, D. Convolutional recurrent neural networks for dynamic MR image reconstruction. *IEEE Trans. Med. Imaging* **2018**, *38*, 280–290. [CrossRef]
25. Ruijsink, B.; Puyol-Antón, E.; Usman, M.; van Amerom, J.; Duong, P.; Forte, M.N.V.; Pushparajah, K.; Frigiola, A.; Nordsletten, D.A.; King, A.P.; et al. Semi-automatic Cardiac and Respiratory Gated MRI for Cardiac Assessment during Exercise. In *Molecular Imaging, Reconstruction and Analysis of Moving Body Organs, and Stroke Imaging and Treatment*; Springer: Berlin/Heidelberg, Germany, 2017; pp. 86–95.
26. Bhatia, K.K.; Caballero, J.; Price, A.N.; Sun, Y.; Hajnal, J.V.; Rueckert, D. Fast reconstruction of accelerated dynamic MRI using manifold kernel regression. In Proceedings of the International Conference on Medical Image Computing and Computer-Assisted Intervention, Munich, Germany, 5–9 October 2015; pp. 510–518.
27. Hammernik, K.; Klatzer, T.; Kobler, E.; Recht, M.P.; Sodickson, D.K.; Pock, T.; Knoll, F. Learning a variational network for reconstruction of accelerated MRI data. *Magn. Reson. Med.* **2018**, *79*, 3055–3071. [CrossRef]
28. Aggarwal, H.K.; Mani, M.P.; Jacob, M. MoDL: Model-based deep learning architecture for inverse problems. *IEEE Trans. Med. Imaging* **2018**, *38*, 394–405. [CrossRef] [PubMed]
29. Zhou, Z.; Han, F.; Ghodrati, V.; Gao, Y.; Yin, W.; Yang, Y.; Hu, P. Parallel imaging and convolutional neural network combined fast MR image reconstruction: Applications in low-latency accelerated real-time imaging. *Med. Phys.* **2019**, *46*, 3399–3413. [CrossRef] [PubMed]
30. Du, T.; Zhang, H.; Li, Y.; Pickup, S.; Rosen, M.; Zhou, R.; Song, H.K.; Fan, Y. Adaptive convolutional neural networks for accelerating magnetic resonance imaging via k-space data interpolation. *Med. Image Anal.* **2021**, *72*, 102098. [CrossRef] [PubMed]
31. Schlemper, J.; Qin, C.; Duan, J.; Summers, R.M.; Hammernik, K. Σ-net: Ensembled Iterative Deep Neural Networks for Accelerated Parallel MR Image Reconstruction. *arXiv* **2019**, arXiv:1912.05480.
32. Lv, J.; Wang, P.; Tong, X.; Wang, C. Parallel imaging with a combination of sensitivity encoding and generative adversarial networks. *Quant. Imaging Med. Surg.* **2020**, *10*, 2260. [CrossRef]
33. Arvinte, M.; Vishwanath, S.; Tewfik, A.H.; Tamir, J.I. Deep J-Sense: Accelerated MRI Reconstruction via Unrolled Alternating Optimization. In Proceedings of the International Conference on Medical Image Computing and Computer-Assisted Intervention, Strasbourg, France, 27 September–1 October 2021.
34. Souza, R.; Bento, M.; Nogovitsyn, N.; Chung, K.J.; Loos, W.; Lebel, R.M.; Frayne, R. Dual-domain cascade of U-nets for multi-channel magnetic resonance image reconstruction. *Magn. Reson. Imaging* **2020**, *71*, 140–153. [CrossRef]
35. Li, Z.; Sun, N.; Gao, H.; Qin, N.; Li, Z. Adaptive subtraction based on U-Net for removing seismic multiples. *IEEE Trans. Geosci. Remote Sens.* **2021**, *59*, 9796–9812. [CrossRef]
36. Chen, Y.; Firmin, D.; Yang, G. Wavelet improved GAN for MRI reconstruction. In *Medical Imaging 2021: Physics of Medical Imaging*; SPIE: Bellingham, WA, USA, 2021; Volume 11595, pp. 285–295.
37. Zhang, K.; Zuo, W.; Gu, S.; Zhang, L. Learning deep CNN denoiser prior for image restoration. In Proceedings of the IEEE Conference on Computer Vision and Pattern Recognition, Honolulu, HI, USA, 21–26 July 2017.
38. Kulkarni, K.; Lohit, S.; Turaga, P.; Kerviche, R.; Ashok, A. Reconnet: Non-iterative reconstruction of images from compressively sensed measurements. In Proceedings of the IEEE Conference on Computer Vision and Pattern Recognition, Las Vegas, NV, USA, 27–30 June 2016.
39. Jin, K.H.; McCann, M.T.; Froustey, E.; Unser, M. Deep convolutional neural network for inverse problems in imaging. *IEEE Trans. Image Process.* **2017**, *26*, 4509–4522. [CrossRef]
40. Song, P.; Weizman, L.; Mota, J.F.; Eldar, Y.C.; Rodrigues, M.R.D. Coupled dictionary learning for multi-contrast MRI reconstruction. *IEEE Trans. Med. Imaging* **2019**, *39*, 621–633. [CrossRef]
41. Adler, J.; Öktem, O. Learned primal-dual reconstruction. *IEEE Trans. Med. Imaging* **2018**, *37*, 1322–1332. [CrossRef]
42. Putzky, P.; Welling, M. Recurrent inference machines for solving inverse problems. *arXiv* **2017**, arXiv:1706.04008.
43. Sajjad, M.; Khan, S.; Muhammad, K.; Wu, W.; Ullah, A.; Baik, S.W. Multi-grade brain tumor classification using deep CNN with extensive data augmentation. *J. Comput. Sci.* **2019**, *30*, 174–182. [CrossRef]
44. Afshar, P.; Mohammadi, A.; Plataniotis, K.N. Brain tumor type classification via capsule networks. In Proceedings of the 2018 25th IEEE International Conference on Image Processing (ICIP), Athens, Greece, 7–10 October 2018.
45. Zhang, J.; Xie, Y.; Wu, Q.; Xia, Y. Medical image classification using synergic deep learning. *Med. Image Anal.* **2019**, *54*, 10–19. [CrossRef] [PubMed]
46. Isensee, F.; Kickingereder, P.; Wick, W.; Bendszus, M.; Maier-Hein, K.H. Brain tumor segmentation and radiomics survival prediction: Contribution to the brats 2017 challenge. In Proceedings of the International MICCAI Brainlesion Workshop, Quebec City, QC, Canada, 14 September 2017.
47. Khan, H.; Shah, P.M.; Shah, M.A.; ul Islam, S.; Rodrigues, J.J.P.C. Cascading handcrafted features and Convolutional Neural Network for IoT-enabled brain tumor segmentation. *Comput. Commun.* **2020**, *153*, 196–207. [CrossRef]

48. Han, Y.; Yoo, J.; Kim, H.H.; Shin, H.J.; Sung, K.; Ye, J.C. Deep learning with domain adaptation for accelerated projection-reconstruction MR. *Magn. Reson. Med.* **2018**, *80*, 1189–1205. [CrossRef] [PubMed]
49. Healy, J.J.; Curran, K.M.; Serifovic Trbalic, A. Deep Learning for Magnetic Resonance Images of Gliomas. In *Deep Learning for Cancer Diagnosis*; Springer: Berlin/Heidelberg, Germany, 2021; pp. 269–300.
50. Shabbir, A.; Ali, N.; Ahmed, J.; Zafar, B.; Rasheed, A.; Sajid, M.; Ahmed, A.; Dar, S.H. Satellite and scene image classification based on transfer learning and fine tuning of ResNet50. *Math. Probl. Eng.* **2021**, *2021*, 5843816. [CrossRef]
51. Waddington, D.E.; Hindley, N.; Koonjoo, N.; Chiu, C.; Reynolds, T.; Liu, P.Z.; Zhu, B.; Bhutto, D.; Paganelli, C.; Keall, P.J.J.a.p.a. On Real-time Image Reconstruction with Neural Networks for MRI-guided Radiotherapy. *arXiv* **2022**, arXiv:2202.05267.
52. Guo, P.; Wang, P.; Zhou, J.; Jiang, S.; Patel, V.M. Multi-institutional collaborations for improving deep learning-based magnetic resonance image reconstruction using federated learning. In Proceedings of the IEEE/CVF Conference on Computer Vision and Pattern Recognition, Montreal, QC, Canada, 11–17 October 2021.
53. Yiasemis, G.; Sonke, J.-J.; Sánchez, C.; Teuwen, J. Recurrent Variational Network: A Deep Learning Inverse Problem Solver applied to the task of Accelerated MRI Reconstruction. In Proceedings of the IEEE/CVF Conference on Computer Vision and Pattern Recognition, New Orleans, LA, USA, 19–24 June 2022.

Article

One-Step Enhancer: Deblurring and Denoising of OCT Images

Shunlei Li [1,2], Muhammad Adeel Azam [1,2], Ajay Gunalan [1,2] and Leonardo S. Mattos [1,*]

1 Department of Advanced Robotics, Istituto Italiano di Tecnologia, 16163 Genoa, Italy
2 Department of Informatics, Bioengineering, Robotics and Systems Engineering, University of Genoa, 16145 Genoa, Italy
* Correspondence: leonardo.demattos@iit.it

Citation: Li, S.; Azam, M.A.; Gunalan, A.; Mattos, L.S. One-Step Enhancer: Deblurring and Denoising of OCT Images. *Appl. Sci.* **2022**, *12*, 10092. https://doi.org/10.3390/app121910092

Academic Editor: Lucian Mihai Itu

Received: 23 August 2022
Accepted: 5 October 2022
Published: 7 October 2022

Publisher's Note: MDPI stays neutral with regard to jurisdictional claims in published maps and institutional affiliations.

Copyright: © 2022 by the authors. Licensee MDPI, Basel, Switzerland. This article is an open access article distributed under the terms and conditions of the Creative Commons Attribution (CC BY) license (https://creativecommons.org/licenses/by/4.0/).

Abstract: Optical coherence tomography (OCT) is a rapidly evolving imaging technology that combines a broadband and low-coherence light source with interferometry and signal processing to produce high-resolution images of living tissues. However, the speckle noise introduced by the low-coherence interferometry and the blur from device motions significantly degrade the quality of OCT images. Convolutional neural networks (CNNs) are a potential solution to deal with these issues and enhance OCT image quality. However, training such networks based on traditional supervised learning methods is impractical due to the lack of clean ground truth images. Consequently, this research proposes an unsupervised learning method for OCT image enhancement, termed one-step enhancer (OSE). Specifically, OSE performs denoising and deblurring based on a single step process. A generative adversarial network (GAN) is used for this. Encoders disentangle the raw images into a content domain, blur domain and noise domain to extract features. Then, the generator can generate clean images from the extracted features. To regularize the distribution range of retrieved blur characteristics, KL divergence loss is employed. Meanwhile, noise patches are enforced to promote more accurate disentanglement. These strategies considerably increase the effectiveness of GAN training for OCT image enhancement when used jointly. Both quantitative and qualitative visual findings demonstrate that the proposed method is effective for OCT image denoising and deblurring. These results are significant not only to provide an enhanced visual experience for clinicians but also to supply good quality data for OCT-guide operations. The enhanced images are needed, e.g., for the development of robust, reliable and accurate autonomous OCT-guided surgical robotic systems.

Keywords: optical coherence tomography; image enhancement; generative adversarial network; unsupervised learning

1. Introduction

Optical coherence tomography (OCT) is an imaging technology able to produce high-resolution images of living tissues. Most OCT devices used in clinical studies have a resolution of approximately 10 μm and a depth of penetration up to 2 mm in soft tissues [1]. However, OCT image quality is significantly degraded by speckle noise introduced by the low-coherence interferometry used in the imaging process and by blur arising from relative motions between the device and the tissue [2]. This has a strong impact on subsequent analysis and makes clinical application challenging. Therefore, efficient OCT image enhancement methods are urgently needed [3].

By improving the light source, hardware-based approaches reduce the noise of the detector and scanner to some extent, but the speckle in the imaging system cannot be eliminated. Software-based approaches such as non-local means or block-matching and 3D filtering (BM3D) can provide good results, but need laborious efforts of parameter tuning for different noise levels [4]. Block matching and 4D collaborative filtering (BM4D) expands BM3D to three-dimensional picture volumes [5]. Sliding window, adaptive statistical-based filters, and patch correlation–based filters are the three main classes of digital filters used to denoise images [6]. However, these methods have limitations that reduce their potential for clinical applications, such as a long processing time and excessive smoothness [7].

Recently, convolutional neural networks (CNNs) have started to be considered as a potential solution for such image enhancement tasks. For example, Kai Zhang et al. proposed a feed-forward denoising convolutional neural network (DnCNN) able to handle Gaussian denoising with unknown noise levels based on a residual learning strategy [8]. In addition, Rico-Jimenez et al. proposed a self-fusion network that was pre-trained to fuse 3 frames to achieve near-real-time processing frame rates [9]. However, supervised learning methods such as these are laborious in terms of training data acquisition, requiring well-paired training images (images with noise and blur and corresponding clean images). Furthermore, the use of standard CNNs may lead to loss of details due to averaging processes [10]. These characteristics make standard CNNs impractical for OCT image enhancement. To overcome these limitations, Chunhao Tian et al. proposed a generative adversarial network (GAN) for the problem of restoring low-resolution OCT fundus images to their high-resolution counterparts [11]. In addition, several other methods based on GAN have been proposed for unpaired image enhancement, such as CycleGAN [12], SNR-GAN [10], and SiameseGAN [13].

Another interesting unsupervised learning strategy for OCT image enhancement is disentangled representation. This strategy divides each feature into narrowly specified variables and encodes them as distinct dimensions. Recently, it has been used for image-to-image translation, such as in BicycleGAN [14] and cross-cycleGAN [15]. In addition, DRGAN implemented this unsupervised learning method for reducing speckle with disentangled representation [16]. However, even though these GAN-based models provided promising results for OCT image despeckling, the problem of blurriness of OCT images still needs to be solved.

This paper presents a novel solution for simultaneous denoising and deblurring of OCT images without requiring a well-paired training dataset. This is achieved with a deep learning GAN architecture that exploits disentangled representation, as shown in Figure 1. After training, the encoder for content and the generator for a clean image can enhance the original image quality. More specifically, the proposed method learns to disentangle noise, blur and content from raw OCT images and then uses this knowledge to generate enhanced images. In order to accommodate for little content information, Kullback—Leibler (KL) divergence [17] loss is used to regularize the distribution range of extracted blur attributes. As shown in Figure 2, the content encoders learn to extract content features from unpaired clean and raw images, while the blur and noise encoders capture blur and noise information from low-quality raw images.

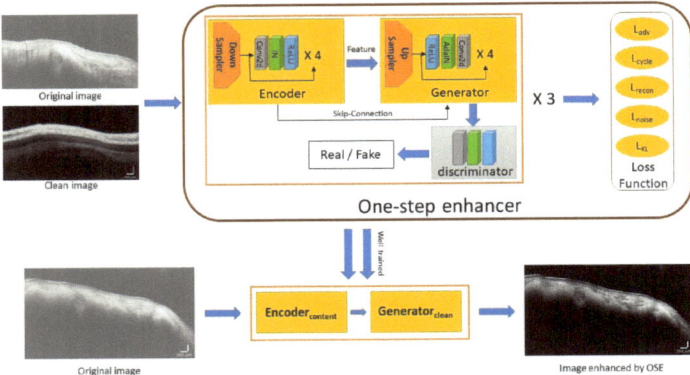

Figure 1. Workflow of proposed OSE image enhancement method: unpaired raw and clean OCT images are used as the input of an unsupervised learning strategy based on disentangled representation and GAN. This process allows the one-step enhancer (OSE) to learn to extract content from low-quality OCT images and generate clean images.

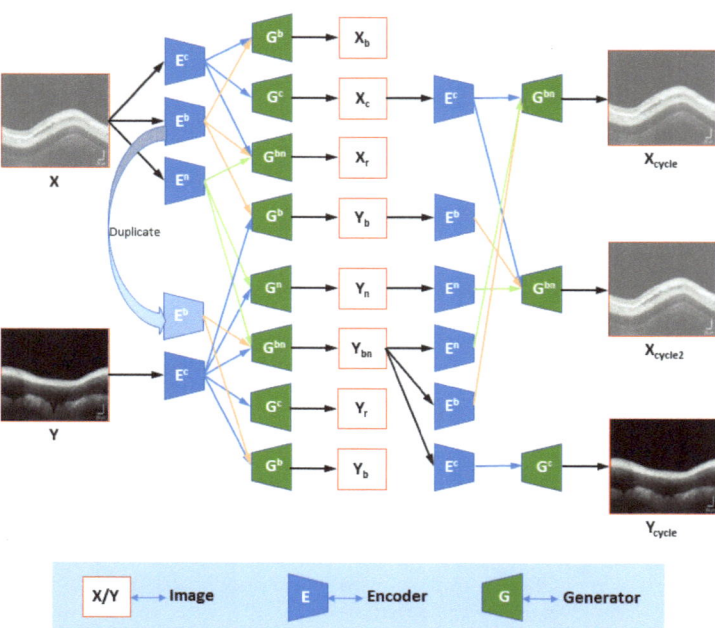

Figure 2. Framework of proposed image enhancement method. X and Y are inputs, where subscripts b, c, r, n, bn and cycle are blurred, clean, reconstructed, noisy, blurred and noisy, and cycled, respectively. The encoder and generator's superscripts c, b, n, and bn is content, blur, noise, and blur-noise, respectively.

The next sections of this paper are organized as follows: in Section 2, we explain related work, including GAN-based deblurring, GAN-based denoising, and disentangled representation. In Section 3, we describe our proposed method, including the problem formulation, definition of loss functions, method implementation and assessment method. In Section 4, we present experiments and results. Finally, conclusions are presented in Section 5.

2. Related Work
2.1. GAN-Based Speckle Removal

OCT images are known to suffer from speckle noise, which are artifacts produced mostly by the coherent nature of the image formation process. Recently, various GAN-based models have been developed to remove such noise from OCT images based on knowledge extracted from unpaired training data. These include SNR-GAN [10], ARM-SRGAN [18], nonlocal-GAN [19], and DRGAN [16].

SNR-GAN was proposed by Yan Guo et al. to establish an end-to-end structure-aware noise reduction GAN that uses cycle GAN to translate data between noisy and clean domains [20]. In order to preserve subtle features during denoising, they used regional structural similarity index (SSIM) loss of image patches instead of the entire image. This method enabled promising improvements in terms of signal-to-noise ratio (SNR), contrast-to-noise ratio (CNR) and SSIM index with a processing speed of 0.3 s per image.

The ARM-SRGAN is a GAN-based method developed for fast and reliable generation of super-resolution (SR) images without relying on a paired training dataset of low- and super-resolution images [18].

The nonlocal-GAN method, unlike cycle-GAN based methods that include two generators and two discriminators, is based on only one generator and one discriminator [19].

The discriminator can learn the features of noise in noisy OCT images and then direct the denoising generator without reference images.

Finally, DRGAN was proposed by Yongqiang Huang et al. as an unsupervised denoising method that disentangles the noisy image into content and noise spaces by using corresponding encoders. It then predicts the denoised OCT image based on the extracted content features [16]. According to the published results, DRGAN outperforms the methods mentioned above in noise reduction and detail preservation.

2.2. GAN-Based Deblurring

Boyu Lu et al. proposed a method for unsupervised deblurring via disentangled representations with a single image [21]. To properly encode blur information into the deblurring framework, the model disentangles the content and blur characteristics from blurred images.

3. Proposed Method

3.1. Problem Formulation

Overall, the learning process for image enhancement based on unpaired data is realized using disentanglement to decode the image features and GAN to generate clean images. For implementing this, the proposed method consists of three parts: (1) encoders for content (E^c) and features (E^b, E^n and E^{bn} for blur, noise and blur-noise); (2) generators of blurred, noisy, blurred-noisy, and clean images (G^b, G^n, G^{bn}, G^c); and (3) discriminators for blurred, noisy, blurred-noisy, and clean image discrimination (D^b, D^n, D^{bn}, D^c).

An overview of the proposed architecture is shown in Figure 2. Given an input blur-noise data X and unpaired clean data Y, the content encoder E^c extracts content information from corresponding samples, and E^b, E^n estimate the feature information in X. Then G^b, G^n, and G^{bn} take features and content information to generate corresponding images, and G^c generates a clean image. Finally, the discriminators distinguish between the real and generated images.

Since clean images should only contain content components, a well-trained content encoder E^c should allow the generation of the desired enhanced images. This is achieved by exploiting information from the blur and noise domains. For the blur domain, considering the content information of E^c and the blur features of E^b, the generated blur images guide the encoder E^c towards extracting content information from blurred images. Similarly, generating and then distinguishing noisy images from clean ones guides E^c towards extracting content from noisy images. In addition, we enforce the last layers of the encoders for content, blur and noise to share weights, which contributes to guiding E^c towards learning how to effectively extract content information from raw images.

Specifically, E^c encodes inputs X and Y as content features F_x^c and F_y^c, respectively. The blur feature F_x^b and noise feature F_x^n are encoded from X by E^b and E^n. Then, as shown in Equations (1) and (2), the reconstructed Y_r is generated from F_y^c using G^c, and the reconstructed X_r is generated from F_x^c, F_x^b, F_x^n using G^{bn}.

$$Y_r = G^c(F_y^c) \qquad (1)$$

$$X_r = G^{bn}(F_x^c, F_x^b, F_x^n) \qquad (2)$$

Generators are used to generate new images based on the features described above according to Equations (3)–(7).

$$X_b = G^b(F_x^c, F_x^b) \qquad (3)$$

$$X_c = G^c(F_x^c) \qquad (4)$$

$$Y_b = G^b(F_y^c, F_x^b) \qquad (5)$$

$$Y_n = G^n(F_y^c, F_x^n) \qquad (6)$$

$$Y_{bn} = G^b(F_y^c, F_x^b, F_x^n) \tag{7}$$

Disentanglement is then used to handle unpaired inputs and generate new images from a corresponding domain. Features are obtained from generated images: $F^c - X_c$, $F^b - Y_b$, $F^n - Y_n$, $F^c - Y_{bn}$, $F^b - Y_{bn}$, $F^n - Y_{bn}$. Finally, cycled inputs are obtained as follows:

$$X_{cycle} = G^{bn}(F^c - X_c, F^b - Y_{bn}, F^n - Y_{bn}) \tag{8}$$

$$X_{cycle2} = G^{bn}(F^c - X_c, F^b - Y_b, F^n - Y_n) \tag{9}$$

$$Y_{cycle} = G^{bn}(F^c - Y_{bn}) \tag{10}$$

After training the model and addressing disentanglement, E^c can extract content features from low-quality images, and then clean images can be obtained using G^c.

3.2. Loss Function

The overall loss function includes five subfunctions: domain adversarial loss (L_{adv}), cycle consistency loss (L_{cycle}), reconstruction loss (L_{recon}), noise patch loss (L_{noise}) and KL divergence loss (L_{KL}). Their interconnections with the processing framework is illustrated in Figure 3.

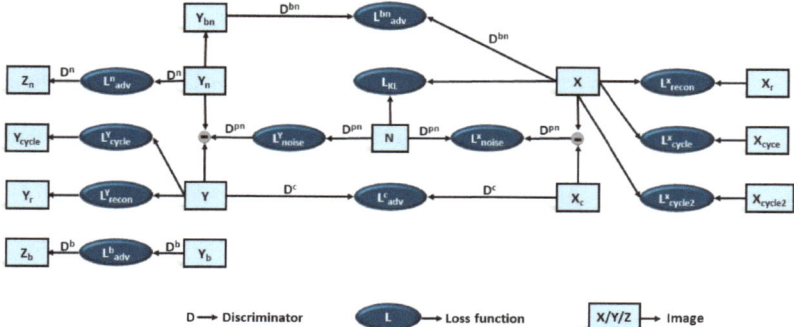

Figure 3. Diagram illustrating the inputs and outputs of loss functions. The inputs include X: original images, Y: clean images, Z: real blurred/noisy images.

(1) Domain adversarial loss: L_{adv} pushes the discriminators to pick the best encoders and generators to minimize the adversarial loss functions, which include content information loss L_{adv}^c, blur feature loss L_{adv}^b, noise feature loss L_{adv}^n, and blur-noise feature loss L_{adv}^{bn}. The domain adversarial loss is defined as:

$$L_{adv} = \underset{E,G}{\operatorname{argmin}}\ \underset{D}{\max}(L_{adv}^c + L_{adv}^b + L_{adv}^n + L_{adv}^{bn}) \tag{11}$$

where E stands for the encoder, G for the generator, and D for the discriminator. The four adversarial loss functions are defined below, where Z_b and Z_n are real blurred and noisy images, and $\mathbb{E}$ is the expectation operator.

$$L_{adv}^c = \mathbb{E}[\log D^c(Y)] + \mathbb{E}[1 - \log D^c(X_c)] \tag{12}$$

$$L_{adv}^b = \mathbb{E}[\log D^b(Z_b)] + \mathbb{E}[1 - \log D^b(Y_b)] \tag{13}$$

$$L_{adv}^n = \mathbb{E}[\log D^n(Z_n)] + \mathbb{E}[1 - \log D^n(Y_n)] \tag{14}$$

$$L_{adv}^{bn} = \mathbb{E}[\log D^{bn}(X)] + \mathbb{E}[1 - \log D^{bn}(Y^{bn})] \tag{15}$$

(2) Cycle consistency loss: inspired by CycleGAN [20], cycle consistency loss is introduced to guarantee that the enhanced image X_c represents a proper reconstruction of the raw sample image, and that Y^{bn} can be translated back to the original clean image domain. The cycle consistency loss further limits the space of the generated samples and preserves the content of original images.

$$\begin{aligned} L_{cycle} =& \mathbb{E}[||X - X_{cycle}||_1] \\ &+ \mathbb{E}[||X - X_{cycle2}||_1] + \mathbb{E}[||Y - Y_{cycle}||_1] \end{aligned} \quad (16)$$

where $||.||_1$ represents the l_1-norm.

(3) Reconstruction loss: the reconstruction loss is applied to facilitate $X = X_r$ and $Y = Y_r$. Consequently, G^c and G^{bn} can reconstruct the inputs to generate a clean counterpart of X and a blur-noise counterpart of Y.

$$L_{recon} = \mathbb{E}[||X - X_r||_1] + \mathbb{E}[||Y - Y_r||_1] \quad (17)$$

(4) Noise patch loss: to overcome the obstacle of multiple types of noise, we leverage noise patches from the background of raw images and use a discriminator D^{pn} to distinguish between real noise and generated noise as follows:

$$L_{noise}^X = \mathbb{E}[\log D^{pn}(N)] + \mathbb{E}[\log D^{pn}(X - X_b)] \quad (18)$$

$$L_{noise}^Y = \mathbb{E}[\log D^{pn}(N)] + \mathbb{E}[\log D^{pn}(Y - Y_b)] \quad (19)$$

According to Equations (18) and (19), the noise patch loss is given by:

$$L_{noise} = \underset{E,G}{\operatorname{argmin}}\ \underset{D}{\max}(L_{noise}^X + L_{noise}^Y) \quad (20)$$

(5) KL divergence loss: to guarantee that the blur encoder $E^b(X)$ only encodes blur components, Y_b is generated from $E^c(Y)$ and $E^b(X)$ in Equation (5). This discourages $E^b(X)$ from encoding content information from X. Furthermore, KL divergence loss is used to regularize the blur feature distribution $F_b = E^b(X)$ to bring it closer to a normal distribution $p(F) \sim N(0,1)$. KL divergence is minimized to obtain the KL loss as described in [22]:

$$L_{KL} = \frac{1}{2}\sum_{i=1}^{N}(\mu_i^2 + \sigma_i^2 - \log(\sigma_i^2) - 1) \quad (21)$$

where μ and σ are the mean and standard deviation of F_b, and N is its dimension. The KL divergence loss can reduce the gap between the prior $p(F)$ and the learned distributions. This further suppresses any content information contained in F_b.

Considering the equations above, the overall loss function can be written as:

$$\begin{aligned} L =& \lambda_{adv}L_{adv} + \lambda_{cycle}L_{cycle} + \lambda_{recon}L_{recon} \\ &+ \lambda_{noise}L_{noise} + \lambda_{KL}L_{KL} \end{aligned} \quad (22)$$

where the subscripted λ are the coefficients of each corresponding loss function.

3.3. Implementation and Data

The proposed network architecture has a structure similar to DRGAN [16]. The content encoder is composed of an input convolutional layer, a down sampler and four residual blocks. The noise encoder consists of an input convolutional layer, a down sampler and an adaptive average pooling layer with a 1×1 convolutional layer. The blur encoder has four strided convolution layers and one fully connected layer. For the generator, the architectures are symmetric to the content encoder, but vary for generating images of different domains. We use skip-connections between E^c and G^c, with SPADE [23] and adaptive instance normalization [24], to fuse features from different levels. Then, the discriminator applies a series of convolutional and pooling layers to give a binary judgement.

This model was implemented in PyTorch with a Ubuntu 20.04 operation system and an NVIDIA Quadro RTX 8000 GPU. During training, the Adam optimizer was used, and the learning rate was set to 0.0002 for 80 epochs. According to information in [16,21], the hyper-parameters in our framework were experimentally set to the following values: $\lambda_{adv} = 1$, $\lambda_{cycle} = 10$, $\lambda_{recon} = 10$, $\lambda_{noise} = 1$, and $\lambda_{KL} = 0.01$.

We acquired two datasets for this study. One consisted of 30 low-quality OCT images and 30 clean OCT images from three different pork larynxes. The second dataset contained the same number of images captured from two ex vivo rabbit eyes. These custom datasets were captured using a commercial OCT device (TEL320C1, Thorlabs, Inc., Newton, NJ, USA). The pixel size was 0.40×2.47 μm (width × height), and the size of each image was set to $10{,}000 \times 1024$ pixels. Therefore, the FOV was 4.00×2.53 mm.

The clean images in our datasets were obtained using the Speckle Averaging function of the ThorImageOCT software (version 5.2), which uses 4 successive A-scans to compute the mean and variance values used in the averaging process.

A test set was formed by randomly selecting 20 images with noise and blur and 20 corresponding clean images from each dataset. The remaining 40 images were randomly combined into pairs to form a training set. Since tissue information is concentrated in the middle part of the OCT images, all images were center-cropped to a pixel size of 900×450 to improve training efficiency.

To obtain noise features for the noise loss function, a window with pixel size of 256×256 was used to extract noisy patches from the background of low-quality images. This window was applied to the images in the training set using a stride of 8 pixels. This process extracted a total of 19,360 patches from low-quality images for the input X, and 19,360 patches from clean images for the input Y.

3.4. Experimental Method and Performance Metrics

An ablation study was performed to evaluate the performance of each component in the proposed image enhancement method. This consisted in evaluating the performances of each module separately: first, the denoise and the deblur modules were independently assessed. Then, the performance of the proposed model, which combines both operations into a single step operation, was evaluated. More specifically, we removed speckle and blur from the original images separately, and then used the proposed model to perform a one-step image enhancement.

In addition, to benchmark the performance of the proposed image enhancer, non-learning (BM3D [25]), supervised learning (DnCNN [8]) and unsupervised learning (DR-GAN [16]) methods were implemented. The BM3D software implements the traditional OCT image enhancement method, while the DnCNN and DRGAN models were trained on the same dataset used to train our new model.

Performance evaluation used the same test set described above. In addition, the processing time of each method was evaluated both on CPU and GPU. Finally, a visual assessment of the four image enhancement methods was performed using sample images from the test set.

Two metrics were used for quantitative performance assessment: peak signal-to-noise ratio (PSNR) and structure similarity index measure (SSIM). PSNR is commonly used to measure the quality of reconstruction of lossy image compression codecs. It offers an approximation to human perception of reconstruction quality based on differences between the reconstructed and the reference image. SSIM, on the other hand, measures the similarity between two images. The overall index of SSIM evaluates luminance, contrast and structural differences.

$$PSNR = 10 log_{10} \left(\frac{(max(I_g))^2}{\frac{1}{MN} \sum_i \sum_j (I_c(i,j) - I_g(i,j))^2} \right) \qquad (23)$$

where I_g and I_c are, respectively, the generated and the averaging clean images. M and N are the size of the image.

$$SSIM = \frac{(2\mu_{I_g}\mu_{I_c} + C_1)(2\sigma_{I_g I_c} + C_2)}{(\mu_{I_g}^2 + \mu_{I_c}^2 + C_1)(\sigma_{I_g}^2 + \sigma_{I_c}^2 + C_2)} \quad (24)$$

where $\mu_{I_g}, \mu_{I_c}, \sigma_{I_g}, \sigma_{I_c}$ and $\sigma_{I_g I_c}$ are the local means, standard deviations and cross-covariance of images I_g and I_c. The constants C_1 and C_2 are set according to the literature [26].

4. Experimental Results

4.1. Ablation Study Results

Visual results from the ablation study are presented in Figure 4, while the quantitative results are shown in Table 1. A visual analysis of Figure 4 shows that the denoise module is effective in removing noise from the original OCT image (raw image). The data in Table 1 demonstrates improvements of 10.59 in PSNR and 0.24 in SSIM, which confirms this module works properly. However, the problem of unclear tissue layers is not addressed.

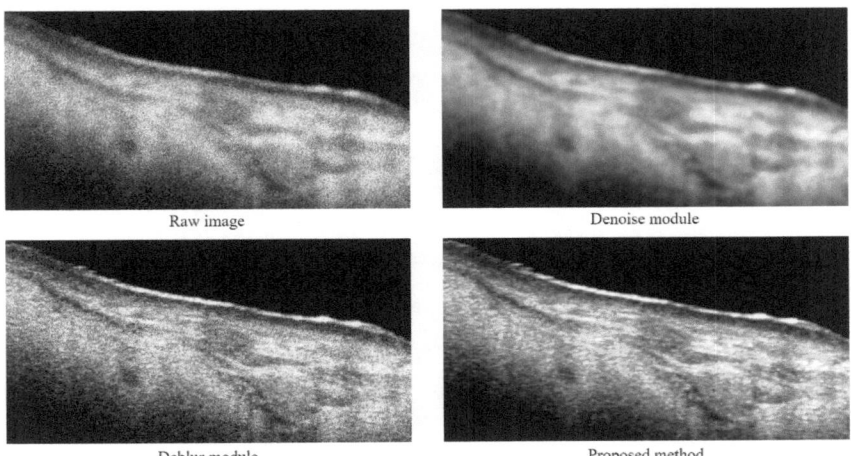

Figure 4. Sample images from the ablation study performed using OCT images from a pork larynx to evaluate the denoise module, the deblur module, and the proposed OSE method.

Table 1. Ablation study results.

Method	Metrics (Mean ± std)	
	PSNR	SSIM
Original images	8.94 ± 2.01	0.34 ± 0.14
Denoise module	19.53 ± 1.87	0.58 ± 0.20
Deblur module	17.55 ± 1.52	0.47 ± 0.12
OSE	**26.71 ± 2.21**	**0.81 ± 0.16**

The deblur module, on the other hand, removes blur and makes the layers more visible. This can be observed in Figure 4 by comparing the result of the deblur module with the raw image. In this case, the PSNR improved from 8.94 to 17.55, and the SSIM improved from 0.34 to 0.47. However, noise is still present in the image, and this limits the image enhancement performance.

The proposed method combining both modules provides better enhancement performance than each single module applied separately. A visual inspection of the result in Figure 4 shows that the proposed method was able to effectively enhance the original raw image. This is corroborated by the data in Table 1, which shows the proposed method achieved top performances in terms of PSNR and SSIM.

4.2. Performance Comparison Results

Figure 5 provides a visual comparison of the image enhancement results achieved with the proposed OSE and with the other state of the art methods: the 3-D block-matching algorithm (BM3D), the supervised learning-based method DnCNN, and the unsupervised learning-based method DRGAN. It can be observed that all methods achieved satisfactory speckle reduction performance, but only OSE was effective in also removing the blurring effects on the image details, as shown in the selected magnified areas.

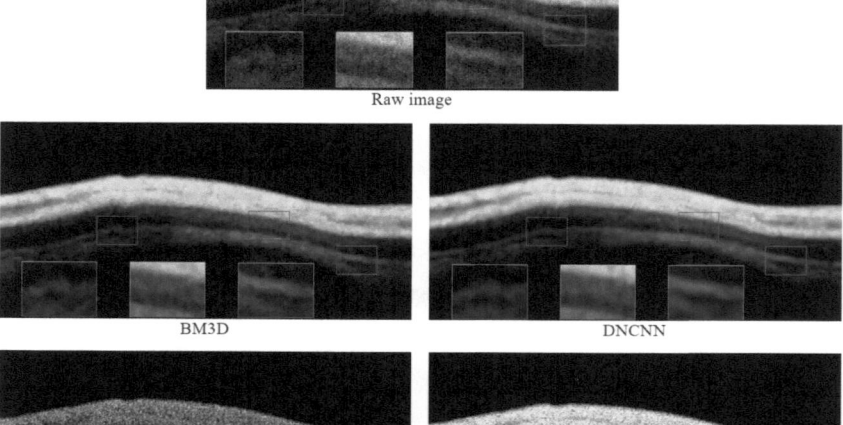

Figure 5. Sample images from the image enhancement performance study comparing the proposed OSE with the state of the art methods BM3D, DnCNN and DRGAN. The image areas marked in red are magnified and shown as inset pictures to facilitate the visual assessment of the different algorithms.

The data in Table 2 summarizes the quantitative performance metrics obtained for the four different enhancement methods. OSE improved PSNR from 8.94 to 26.71 and SSIM from 0.34 to 0.81, outperforming all the other methods in terms of both denoising and deblurring.

Table 3 shows the mean processing time of the methods assessed for 10,000 × 1024 pixel images. Considering this data, we can see that although BM3D provides good image enhancement results, it takes much more time than the other methods to process the OCT images. In addition, we can note that DnCNN performed better than DRGAN but was slower. Furthermore, as explained earlier, DnCNN is a supervised learning method that requires a well-paired set of images for training. On the other hand, OSE provided top image enhancement performance with the best processing speed when the computations were performed on a GPU. It achieved a mean processing rate of 8.3 fps on the GPU.

Table 2. Quantitative performance comparison with state-of-the-art methods for OCT image enhancement.

Method	Metrics (Mean ± std)	
	PSNR	SSIM
Original images	8.94 ± 2.01	0.34 ± 0.14
BM3D	24.11 ± 1.04	0.71 ± 0.08
DnCNN	23.99 ± 2.70	0.78 ± 0.24
DRGAN	16.77 ± 1.04	0.61 ± 0.10
OSE	**26.71 ± 2.21**	**0.81 ± 0.16**

Table 3. Mean processing time for 10,000 × 1024 pixel OCT images

Method	Mean Processing Time (s)	
	CPU	GPU
BM3D	45.69	-
DnCNN	11.14	0.17
DRGAN	**3.77**	0.14
OSE	3.86	**0.12**

5. Conclusions

This paper presented a novel deep learning model for one-step denoising and deblurring of OCT images. This one-step enhancer (OSE) is trained using an unsupervised learning strategy, which allows learning from a mixed dataset of unpaired OCT images. For this, the method uses disentangled representation and generative adversarial network (GAN) to extract content, blur and noise features from raw OCT images, and then learns to generate clean images. The proposed OSE was assessed through both an ablation study and a comparative performance evaluation based on the quantitative metrics PSNR and SSIM. The ablation study demonstrated that the method produced effective denoise and deblur modules, which enabled high performance levels when used in a combined model. The comparative analysis showed the proposed method outperformed state-of-the-art methods for OCT image enhancement, indicating that our one-step enhancer is a valuable alternative for speckle and blur reduction in OCT images.

Author Contributions: Conceptualization, S.L.; Data curation, A.G.; Investigation, S.L. and L.S.M.; Methodology, S.L. and L.S.M.; Software, S.L. and M.A.A.; Supervision, L.S.M.; Writing—original draft, S.L.; Writing—review & editing, M.A.A., A.G. and L.S.M. All authors have read and agreed to the published version of the manuscript.

Funding: This research received no external funding.

Institutional Review Board Statement: Not applicable.

Informed Consent Statement: Not applicable.

Data Availability Statement: The data presented in this study is openly available in IIT's Dataverse repository at https://doi.org/10.48557/RI3EQG, reference "OCT dataset of pork larynx and rabbit eyes".

Conflicts of Interest: The authors declare no conflict of interest.

References

1. Podoleanu, A.G. Optical coherence tomography. *J. Microsc.* **2012**, *247*, 209–219. [CrossRef] [PubMed]
2. Drexler, W.; Morgner, U.; Ghanta, R.K.; Kärtner, F.X.; Schuman, J.S.; Fujimoto, J.G. Ultrahigh-resolution ophthalmic optical coherence tomography. *Nat. Med.* **2001**, *7*, 502–507. [CrossRef] [PubMed]
3. Larin, K.V.; Ghosn, M.G.; Bashkatov, A.N.; Genina, E.A.; Trunina, N.A.; Tuchin, V.V. Optical clearing for OCT image enhancement and in-depth monitoring of molecular diffusion. *IEEE J. Sel. Top. Quantum Electron.* **2011**, *18*, 1244–1259. [CrossRef]
4. Chong, B.; Zhu, Y.K. Speckle reduction in optical coherence tomography images of human finger skin by wavelet modified BM3D filter. *Opt. Commun.* **2013**, *291*, 461–469. [CrossRef]

5. Maggioni, M.; Katkovnik, V.; Egiazarian, K.; Foi, A. Nonlocal transform-domain filter for volumetric data denoising and reconstruction. *IEEE Trans. Image Process.* **2012**, *22*, 119–133. [CrossRef]
6. Adabi, S.; Turani, Z.; Fatemizadeh, E.; Clayton, A.; Nasiriavanaki, M. Optical coherence tomography technology and quality improvement methods for optical coherence tomography images of skin: A short review. *Biomed. Eng. Comput. Biol.* **2017**, *8*, 1179597217713475. [CrossRef] [PubMed]
7. Li, M.; Idoughi, R.; Choudhury, B.; Heidrich, W. Statistical model for OCT image denoising. *Biomed. Opt. Express* **2017**, *8*, 3903–3917. [CrossRef] [PubMed]
8. Zhang, K.; Zuo, W.; Chen, Y.; Meng, D.; Zhang, L. Beyond a Gaussian denoiser: Residual learning of deep CNN for image denoising. *IEEE Trans. Image Process.* **2017**, *26*, 3142–3155. [CrossRef] [PubMed]
9. Rico-Jimenez, J.J.; Hu, D.; Tang, E.M.; Oguz, I.; Tao, Y.K. Real-time OCT image denoising using a self-fusion neural network. *Biomed. Opt. Express* **2022**, *13*, 1398–1409. [CrossRef] [PubMed]
10. Guo, Y.; Wang, K.; Yang, S.; Wang, Y.; Gao, P.; Xie, G.; Lv, C.; Lv, B. Structure-aware noise reduction generative adversarial network for optical coherence tomography image. In Proceedings of the International Workshop on Ophthalmic Medical Image Analysis, Shenzhen, China, 17 October 2019; pp. 9–17.
11. Tian, C.; Yang, J.; Li, P.; Zhang, S.; Mi, S. Retinal fundus image superresolution generated by optical coherence tomography based on a realistic mixed attention GAN. *Med. Phys.* **2022**, *49*, 3185–3198. [CrossRef] [PubMed]
12. Manakov, I.; Rohm, M.; Kern, C.; Schworm, B.; Kortuem, K.; Tresp, V. Noise as domain shift: Denoising medical images by unpaired image translation. In Proceedings of the Domain Adaptation and Representation Transfer and Medical Image Learning with Less Labels and Imperfect Data, International Workshop on Medical Image Learning with Less Labels and Imperfect Data, Shenzhen, China, 17 October 2019; pp. 3–10.
13. Kande, N.A.; Dakhane, R.; Dukkipati, A.; Yalavarthy, P.K. SiameseGAN: A generative model for denoising of spectral domain optical coherence tomography images. *IEEE Trans. Med. Imaging* **2020**, *40*, 180–192. [CrossRef]
14. Zhu, J.Y.; Zhang, R.; Pathak, D.; Darrell, T.; Efros, A.A.; Wang, O.; Shechtman, E. Toward multimodal image-to-image translation. In Proceedings of the Advances in Neural Information Processing Systems, Long Beach, CA, USA, 4–9 December 2017; Volume 30.
15. Lee, H.Y.; Tseng, H.Y.; Huang, J.B.; Singh, M.; Yang, M.H. Diverse image-to-image translation via disentangled representations. In Proceedings of the European Conference on Computer Vision (ECCV), Munich, Germany, 8–14 September 2018; pp. 35–51.
16. Huang, Y.; Xia, W.; Lu, Z.; Liu, Y.; Chen, H.; Zhou, J.; Fang, L.; Zhang, Y. Noise-powered disentangled representation for unsupervised speckle reduction of optical coherence tomography images. *IEEE Trans. Med. Imaging* **2020**, *40*, 2600–2614. [CrossRef] [PubMed]
17. Hershey, J.R.; Olsen, P.A. Approximating the Kullback Leibler Divergence Between Gaussian Mixture Models. In Proceedings of the 2007 IEEE International Conference on Acoustics, Speech and Signal Processing—ICASSP'07, Honolulu, HI, USA, 15–20 April 2007; Volume 4, pp. IV-317–IV-320. [CrossRef]
18. Das, V.; Dandapat, S.; Bora, P.K. Unsupervised super-resolution of OCT images using generative adversarial network for improved age-related macular degeneration diagnosis. *IEEE Sens. J.* **2020**, *20*, 8746–8756. [CrossRef]
19. Guo, A.; Fang, L.; Qi, M.; Li, S. Unsupervised denoising of optical coherence tomography images with nonlocal-generative adversarial network. *IEEE Trans. Instrum. Meas.* **2020**, *70*, 1–12. [CrossRef]
20. Zhu, J.Y.; Park, T.; Isola, P.; Efros, A.A. Unpaired image-to-image translation using cycle-consistent adversarial networks. In Proceedings of the IEEE International Conference on Computer Vision, Venice, Italy, 22–29 October 2017; pp. 2223–2232.
21. Lu, B.; Chen, J.C.; Chellappa, R. Unsupervised domain-specific deblurring via disentangled representations. In Proceedings of the IEEE/CVF Conference on Computer Vision and Pattern Recognition, Long Beach, CA, USA, 15–20 June 2019; pp. 10225–10234.
22. Kingma, D.P.; Welling, M. Auto-encoding variational bayes. *arXiv* **2013**, arXiv:1312.6114.
23. Park, T.; Liu, M.Y.; Wang, T.C.; Zhu, J.Y. Semantic image synthesis with spatially-adaptive normalization. In Proceedings of the IEEE/CVF Conference on Computer Vision and Pattern Recognition, Long Beach, CA, USA, 15–20 June 2019; pp. 2337–2346.
24. Karras, T.; Laine, S.; Aila, T. A style-based generator architecture for generative adversarial networks. In Proceedings of the IEEE/CVF Conference on Computer Vision and Pattern Recognition, Long Beach, CA, USA, 15–20 June 2019; pp. 4401–4410.
25. Mäkinen, Y.; Azzari, L.; Foi, A. Collaborative filtering of correlated noise: Exact transform-domain variance for improved shrinkage and patch matching. *IEEE Trans. Image Process.* **2020**, *29*, 8339–8354. [CrossRef] [PubMed]
26. Wang, Z.; Bovik, A.; Sheikh, H.; Simoncelli, E. Image quality assessment: From error visibility to structural similarity. *IEEE Trans. Image Process.* **2004**, *13*, 600–612. [CrossRef] [PubMed]

Article

Deep Learning of Retinal Imaging: A Useful Tool for Coronary Artery Calcium Score Prediction in Diabetic Patients

Rubén G. Barriada [1,*], Olga Simó-Servat [2], Alejandra Planas [2,3], Cristina Hernández [2,3], Rafael Simó [2,3] and David Masip [1]

1. AIWell Research Group, Faculty of Computer Science, Multimedia and Telecommunications, Universitat Oberta de Catalunya, 08018 Barcelona, Spain; dmasip@uoc.edu
2. Diabetes and Metabolism Research Group, VHIR, Endocrinology Department, Vall d'Hebron University Hospital, Autonomous University Barcelona, 08035 Barcelona, Spain; olga.simo@vhir.org (O.S.-S.); a.planas@vhebron.net (A.P.); cristina.hernandez@vhir.org (C.H.); rafael.simo@vhir.org (R.S.)
3. CIBERDEM (Instituto de Salud Carlos III), 28029 Madrid, Spain
* Correspondence: rgbarriada@uoc.edu

Abstract: Cardiovascular diseases (CVD) are one of the leading causes of death in the developed countries. Previous studies suggest that retina blood vessels provide relevant information on cardiovascular risk. Retina fundus imaging (RFI) is a cheap medical imaging test that is already regularly performed in diabetic population as screening of diabetic retinopathy (DR). Since diabetes is a major cause of CVD, we wanted to explore the use Deep Learning architectures on RFI as a tool for predicting CV risk in this population. Particularly, we use the coronary artery calcium (CAC) score as a marker, and train a convolutional neural network (CNN) to predict whether it surpasses a certain threshold defined by experts. The preliminary experiments on a reduced set of clinically verified patients show promising accuracies. In addition, we observed that elementary clinical data is positively correlated with the risk of suffering from a CV disease. We found that the results from both informational cues are complementary, and we propose two applications that can benefit from the combination of image analysis and clinical data.

Keywords: retina fundus imaging; deep learning; medical imaging; convolutional neural networks

1. Introduction

According to the World Health Organisation, ischaemic heart disease and stroke are nowadays leading causes of death globally [1]. Cardiovascular disease (CVD) progression is dramatically increasing every year and thus many efforts to improve risk predictors are needed. The assessment of cardiovascular risk can be done from a wide variety of information and parameters derived from patients' history such as individual age, gender, smoking status, blood pressure, body mass index (BMI), or metabolic parameters (i.e., glucose and cholesterol levels) [2]. However, this information is not always updated, centralized or available.

Coronary artery calcium (CAC) is a clinically validated strong marker of a CVD [3]. CAC scoring has proved to be a consistent and reproducible method of assessing risk for major cardiovascular outcomes, especially useful in asymptomatic people for planning primary prevention [4]. Nevertheless, CAC score obtention needs the use of CT scans, which are expensive and involve radiation risks.

Retinal Fundus Imaging (RFI) provides relevant information about the vascular system of the eye. There is previous evidence that RFI can be a predictor of CVD [5–9], kidney chronic disease [10] or dementia [11]. Preceding research have shown that capillary vascular and foveal area are the most prevalent regions in predicting the CAC scoring from RFI [12]. In addition, RFI is usually acquired on a regular basis from patients that suffer type 2 diabetes, being CVD one of the most common morbidity [13].

Deep Learning (DL) algorithms have obtained significant improvements in almost all the applications of computer vision [14,15], including medical imaging applications. Particularly, automated retina image classification outperformed human accuracy in diabetic retinopathy diagnosis [16]. This paper proposes to apply DL, more concretely a convolutional neural network (CNN), to the prediction of CAC score from RFI. Using data gathered in a previous project [13] we conjecture that a CNN can predict whether a patient may be diagnosed with a CAC score greater than 400, i.e., the threshold defined by cardiologists to indicate a very high risk of CVD disease. In addition, we explore the use of two clinical variables, age and presence of diabetic retinopathy, which previous studies suggest that are highly correlated with CAC > 400 [13]. The experiments show that both informational cues are complementary, and when properly combined, they can significantly increase the precision or recall of the prediction. We validate the proposal in two different applications, clinical diagnosis and large-scale retrieval. The results show that RFI can be used for CVD risk prediction with low-cost image acquisition, which may have a significant impact in the strategies for selecting those patients in whom the screening of CVD should be prioritized.

2. Materials and Methods

We used two state-of-the art CNN architectures, VGG [17] (16 and 19 layers deep versions) and ResNet [18] to learn a model of the retina images from patients with CAC > 400 and CAC < 400. We consider it as a binary (two class) classification problem. We used a 5-fold crossvalidation approach to validate the neural networks, and we picked the best performing one for the rest of the experiments in the paper. We split retina images into left and right eye data sets, and tested them independently. Table 1 summarizes the results. VGG16 mean accuracies outperform in both cases ResNet, although differences are not statistically significant. Given their equivalence, we used VGG16 as the image classifier for the rest of the paper. Table 2 described the main components of the VGG16 architecture, which is composed of 13 convolutional and 3 fully-connected layers.

Table 1. Comparison of prediction accuracy between DL architectures.

CNN Architecture	VGG16	VGG19	ResNet
Mean	0.68	0.67	0.64
StdDev	0.045	0.088	0.12
CI (95%)	(0.64, 0.72)	(0.59, 0.75)	(0.53, 0.75)

2.1. Transfer Learning

The amount of labelled data is scarce, which poses significant difficulties in training a large neural network (130 M parameters). One of the most common approaches to cope with small sample size problems is the use of transfer learning [19]. The CNN is first trained on a large data set of images (ImageNet [14] in our particular study), and then the model parameters are fine-tuned according to the retina image training set. We removed the last fully-connected layer from VGG16 (FC-1000, adjusted to the ImageNet dataset), adding a two neuron layer (FC-2) to distinguish only two classes: patients labelled CAC > 400 and CAC < 400. These layer parameters where randomly initialized and fined tuned inside the learning process. The model was trained for 60 epochs, batch size of 8, using learning rate 0.0001 and weight decay 4×10^{-2}. Cross-entropy loss and stochastic gradient descent were used for loss function and optimizer respectively. The use of pretrained filters significantly improves the accuracies, being the first CNN layers local filters for detecting specific image compounds, and the top (classification layers) of the network high level representations of the decision boundary. All the experiments were performed using PyTorch [20] Python library.

Table 2. VGG 16-layer model architecture (configuration "D") [17].

CNN Configuration Input 224 × 224 RGB Image
2 × conv3-64 / maxpool / 2 × conv3-128 / maxpool / 3 × conv3-256 / maxpool / 3 × conv3-512 / maxpool / 3 × conv3-512 / maxpool / 2 × FC-4096 / FC-2 / softmax

2.2. Clinical Data Classification

We trained classical machine learning classifiers on the clinical data available, to analyse their prediction capabilities and their complementarity with image based classification. We used the out-of-the-box implementation from Python scikit-learn package of the classifiers: Logistic Regression, K Nearest Neighbours [21], Support Vector Machines [22], Gaussian Process, Decision Tree [23], Random Forest [24], Ada Boost [25], Quadratic Classifier and Naive Bayes.

2.3. RFI Dataset Description

We used a RFI dataset from $N = 76$ patients, i.e., 152 retinal images (left/right eye), 66 of them labelled as CAC > 400 (positive samples) and 86 labelled as CAC < 400 (negative samples). These images were selected from the PRECISED study (ClinicalTrial.gov NCT02248311). In this study, age between 46 to 76 years was an inclusion criteria. It should be noted that the incidence of cardiovascular events before 45 years is very low and, therefore, a screening strategy at younger ages are not cost-effective. Images have 3 channels (RGB). Image dimensions vary from 2576 × 1934 × 3, 2032 × 1934 × 3 to 1960 × 1934 × 3 (width, height and color channels). Figure 1 depicts 2 examples of each eye (one with CAC < 400 and one with CAC > 400) extracted from the database.

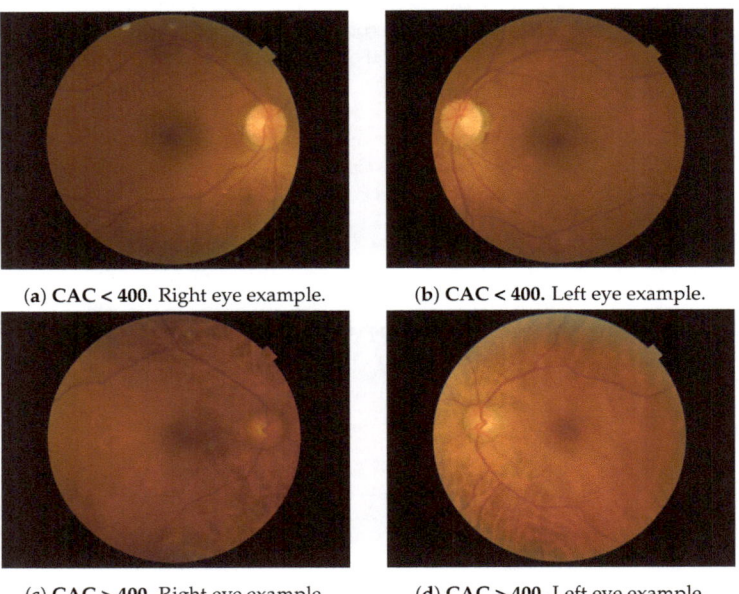

(a) CAC < 400. Right eye example. (b) CAC < 400. Left eye example.

(c) CAC > 400. Right eye example. (d) CAC > 400. Left eye example.

Figure 1. RFI dataset samples (2576 × 1934 × 3).

Prior to the training and inference, we perform the following preprocessing procedure: first, images are color-normalized following the guidelines of the PyTorch library for transfer learning of ImageNet-based learned models. We substract the precomputed $mean = [0.485, 0.456, 0.406]$ and divide each pixel by the standard deviation $std = [0.229, 0.224, 0.225]$ used on the ImageNet pretraining (See for more details: https://pytorch.org/vision/stable/models.html, accessed on 18 January 2022) Therefore, RFI images are loaded to a range of [0, 1] and normalized in that way. Second, all images are rescaled to a 224 × 224 standardized size to fit in the CNN input model. We do not apply postprocessing step to the CNN output, and we just select as the most suitable class the one that has maximum output on the last fully connected layer.

2.4. Metrics

To quantify the model performance, we report the following statistical measures of the model predictions:

- Accuracy: The global percentage of correct predictions.
- Precision: The average of the retrieved items that were relevant, defined as $TP/(TP + FP)$. TP stands for true positives, and FP stands for false positives.
- Recall: The number of relevant items retrieved $TP/(TP + FN)$, where FN stands for false negatives.
- F1-measure: The harmonic mean of the precision and recall, defined as $2 \times (precision \times recall/(precision + recall))$.
- Confusion Matrix: Illustrates the distribution performance ordered as: $\begin{pmatrix} TN & FP \\ FN & TP \end{pmatrix}$

3. Results

3.1. Automated Positive CAC Score Prediction Using DL

In this section, we report the results of applying the DL model over fundus images to automatically predict patients presenting a CAC score > 400 and thus a potential cardiovascular risk. The included subjects were enrolled in the PRECISED study (ClinicalTrial. gov NCT02248311). The study was conducted according to the declaration of Helsinki and was approved by the local ethics committee. All subjects provided written informed consent before study entry.

We performed a 5-Fold stratified Cross Validation, where the image set is split in 5 folds, being 4 folds used for training and 1 for testing. The process is repeated 5 times until each sample has appeared in test once. We report the mean accuracy along the 5 trials and the 95% confidence intervals. We used VGG16 as the DL architecture (see Section 2). Training time was 141 s using NVidia GPU 12 Gb 2080Ti 12 Gb and inference time 0.85 s on the same hardware.

Results suggest that even with a small sample size effect, the VGG16 can model better than change both classes (see Table 3). Notice that the model does not perform equally using images from each eye. Although differences are not statistically significant, the retinas differ from the left to the right eye.

Table 3. Accuracy, Precision, Recall and F1-measure and confusion matrix $\begin{pmatrix} TN & FP \\ FN & TP \end{pmatrix}$ using fundus image and DL.

Model	Accuracy	Precision	Recall	F1	Conf Mat
RFI (Left Eye)	0.67	0.48	0.67	0.56	35 8 17 16
RFI (Right Eye)	0.72	0.52	0.77	0.62	38 5 16 17

3.2. Classification Complementarity: Combination of Clinical and RFI Data

We evaluated the use of clinical data using the classifiers defined in Section 2.2 on the same patients dataset, and following a 5-Fold Cross Validation approach. Table 4 depicts the results for the variables age (ranging from 45 to 76 years old, mean = 65.75) and previous diagnosis of diabetic retinopathy. We also used the combination (concatenation) of both. Results show significantly better than chance accuracies in predicting the CAC > 400 label, especially when the combination of variables is used.

Table 4. Accuracy of classifiers depending on classification variables used.

Classifier	Age	DR	Age + DR
Logistic Regression	0.60	0.70	0.70
KNN	0.56	0.44	0.61
SVM	0.54	0.70	0.69
Gaussian Process	0.54	0.70	0.70
Decision tree	0.62	0.70	0.65
Random Forest	0.57	0.70	0.64
AdaBoost	0.57	0.70	0.70
Quadratic Classifier	0.55	0.70	0.70
Naive Bayes	0.55	0.70	0.70

We further explored the predicted labels for both the image and the clinical data classifiers, and we found that individual predictions differ considerably between modalities. This suggest that an ensemble of both informational cues could take benefit from the complementarity in the predictions. Particularly, we propose two settings that target two specific applications:

- Clinical Diagnosis: we follow a conservative prediction protocol that combines both modalities to reduce the False Negative ratio, i.e., the likelihood of having positive patients diagnosed as negative. This application is conceived for healthcare systems where resources availability is not constrained. After analyzing the retinography, this procedure may suggest further clinical tests (heart scan) for uncertain CAC positive predictions even though the patient probability of having $CAC < 400$ is still considerable.
- Large Scale Retrieval: where we give priority to improve the precision of the ensemble. The goal in this case is to build and application that may search in large databases, and may send to screen the patients with CAC > 400. As the resources are limited, only the patients with high certainty of having CAC > 400 should be further screened. Contrary to the first application, it is designed for efficient resource management and prioritize patients that urgently need further clinical screening.

3.2.1. Application: Clinical Diagnosis

In this application, the ensemble is constructed by first running the model on the clinical data, and then for all the samples that resulted in a $CAC < 400$ classification run the image DL model. We only report $CAC < 400$ if in both cases the result was negative.

This setting, shown in Figure 2, allows the health system to minimize the number of False Negatives (people that urgently needs a treatment but would not be spotted just using one of the methods). Table 5 shows the results. The first row shows the accuracy, recall, precision, F1 measure and confusion matrix for the VGG16 based image prediction (CNN). For each classifier tested on clinical data, we provide its results, and the results using the combination of both the image and clinical classifiers.

There is a consistent trend in the results that show significant improvements in the F1-measure, in all cases. The number of false negatives is reduced about 50%, being the most critical case to be addressed with this setting (patients with CAC > 400 predicted as negative). Besides, we also have increased considerably the true positives (and consequently the recall) by using the image prediction, detecting more patients with CAC > 400. This improvement comes at the cost of increasing the number of false positives (FP). This scenario is less problematic (in terms of health prediction system) since FP are patients diagnosed positive that in fact are CAC < 400, thus not being in danger.

Table 5. Accuracy, Precision, Recall and F1-measure and confusion matrix $\begin{pmatrix} TN & FP \\ FN & TP \end{pmatrix}$ for the clinical diagnosis application.

Classifier	Accuracy	Recall	Precision	F1	Conf Mat	
CNN (RFI)	0.72	0.52	0.77	0.62	38	5
					16	17
Logistic Regression	0.68	0.52	0.68	0.59	35	8
					16	17
Protocol	0.74	0.76	0.68	0.71	31	12
					8	25
KNN	0.61	0.55	0.55	0.55	28	15
					15	18
Protocol	0.64	0.76	0.57	0.65	24	19
					8	25
SVM	0.70	0.45	0.75	0.57	38	5
					18	15
Protocol	0.78	0.76	0.74	0.75	34	9
					8	25
Gaussian Process	0.71	0.48	0.76	0.59	38	5
					17	16
Protocol	0.78	0.76	0.74	0.75	34	9
					8	25
Decision Tree	0.67	0.58	0.63	0.60	32	11
					14	19
Protocol	0.71	0.79	0.63	0.70	28	15
					7	26
Random Forest	0.59	0.42	0.54	0.47	31	12
					19	14
Protocol	0.66	0.73	0.59	0.65	26	17
					9	24
Ada Boost	0.68	0.58	0.66	0.61	33	10
					14	19
Protocol	0.72	0.79	0.65	0.71	29	14
					7	26
Quadratic Classifier	0.70	0.45	0.75	0.57	38	5
					18	15
Protocol	0.78	0.76	0.74	0.75	34	9
					8	25
Naive Bayes	0.70	0.45	0.75	0.57	38	5
					18	15
Protocol	0.78	0.76	0.74	0.75	34	9
					8	25

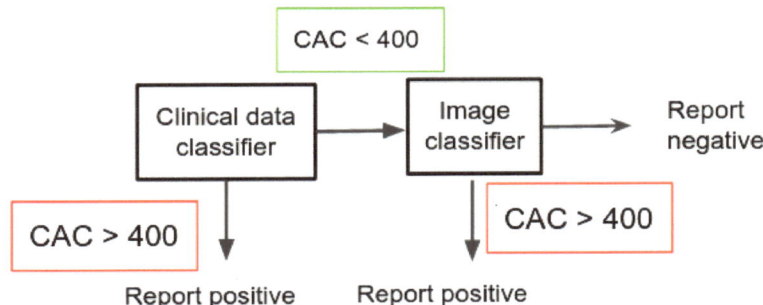

Figure 2. Description of the protocol used in the clinical diagnosis application.

3.2.2. Application: Large Scale Retrieval

In this case, for all the positive samples resulting on the clinical model, we run the VGG16 classifier approach to double check if it is CAC > 400. The ensemble only reports a positive case if in both models the samples result in a positive detection. The model minimizes the number of false positives (saving resources by avoiding the screening of patients that could be diagnosed CAC > 400). This protocol is depicted in Figure 3.

This protocol may find utility in large scale database retrieval, typically hundreds of thousands of patients. In this cases it will not be possible to test with CT scan all the positive patients, and the priority should be to devote the resources to the most certain ones. As can be seen in Table 6, the experiments show a consistent reduction on the FP ratio (up to 90%), and a large improvement of the precision (91%), which shows that when the ensemble predicts a patient with CAC > 400, there is a 90% of chance of being correct. Notice that, globally, this restrictive setting underperforms the individual classifiers, as there is a significant increase in the false negative ratio (positive patients that where discarded).

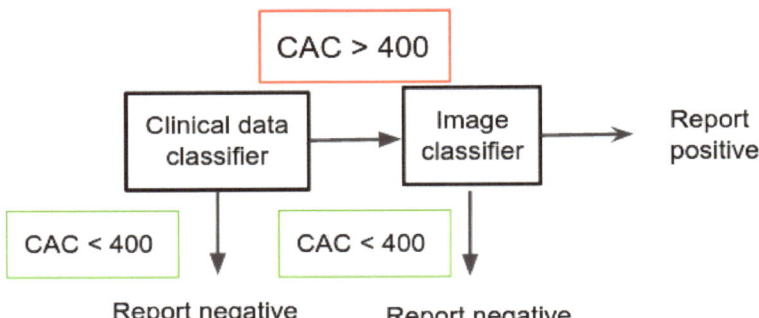

Figure 3. Description of the protocol used in the large scale retrieval application.

Table 6. Accuracy, Precision, Recall and F1-measure and confusion matrix $\begin{pmatrix} TN & FP \\ FN & TP \end{pmatrix}$ for the large scale retrieval application.

Classifier	Accuracy	Recall	Precision	F1	Conf Mat
CNN (RFI)	0.72	0.52	0.77	0.62	38 5 16 17
Logistic Regression	0.68	0.52	0.68	0.59	35 8 16 17
Protocol	0.67	0.27	0.90	0.42	42 1 24 9

Table 6. *Cont.*

Classifier	Accuracy	Recall	Precision	F1	Conf Mat	
KNN	0.61	0.55	0.55	0.55	28	15
					15	18
Protocol	0.68	0.30	0.91	0.45	42	1
					23	10
SVM	0.70	0.45	0.75	0.57	38	5
					18	15
Protocol	0.64	0.21	0.88	0.34	42	1
					26	7
Gaussian Process	0.71	0.48	0.76	0.59	38	5
					17	16
Protocol	0.66	0.24	0.89	0.38	42	1
					25	8
Decision Tree	0.67	0.58	0.63	0.60	32	11
					14	19
Protocol	0.68	0.30	0.91	0.45	42	1
					23	10
Random Forest	0.62	0.48	0.57	0.52	31	12
					17	16
Protocol	0.67	0.24	1.00	0.39	43	0
					25	8
Ada Boost	0.68	0.58	0.66	0.61	33	10
					14	19
Protocol	0.68	0.30	0.91	0.45	42	1
					23	10
Quadratic Classifier	0.70	0.45	0.75	0.57	38	5
					18	15
Protocol	0.64	0.21	0.88	0.34	42	1
					26	7
Naive Bayes	0.70	0.45	0.75	0.57	38	5
					18	15
Protocol	0.64	0.21	0.88	0.34	42	1
					26	7

4. Discussion

Recent research [8] suggests that a Deep CNN (EfficientNet) provides promising accuracies when applied to RFI in predicting CV risk. The study was performed on a large scale cohort of Asian patients, mostly non-diabetic, obtaining scant correlation (0.47) between the RetiCAC CNN score and the CAC assessed by CT scan. In the present pilot study we have focused on type 2 diabetic population, thus using a more homogeneous dataset for making the training. For this reason and due to the use of VGG and ResNet DL architectures the accuracy has been significantly improved.

Results using methods over both clinical and image data show acceptable accuracies in the assessment of a high CAC score. Yet, the two established protocols showed complementary capacities in comparison to independently use both modalities. Experimental studies suggest that it is possible to tailor a specific ensemble to the particular needs of the application, improving either the precision or the recall, although there is a trade off between both performance indicators.

The protocol shows that an application defined specifically for clinical diagnosis, guarantees missing only few patients with CAC > 400, reducing considerably the false negatives and maximizing the recall (75% aprox.). However, this procedure has a significant number of false positives which will impact on resource costs but not in individual diagnosis capabilities.

A second application is conceived for image retrieval of large databases where resources for further cardiological testing are scarce. The ensemble may maximize the precision by reducing the number of false positives, but the application clearly worsens predictions at individual level. Nevertheless, it may find utility when prioritization is urgently needed, as is the case of low income countries.

5. Conclusions

This study provides significant evidences that deep learning methods could be used for evaluating cardiovascular risk by using CAC as unique biomarker. Preliminary results show promising accuracies on a small sample sized database by using classical DL architectures. On the other hand, our findings show that there are clinical variables which also correlate positively with CAC > 400. A very simple preliminary study shows above chance accuracies, and more important, complementarity with the results obtained by image analysis. Based on that, we build two applications that optimize a precision or a recall criteria for a specific application.

This is a preliminary work that proves that there exists discriminative information in the retinal images. Further research challenges are: data acquisition and model improvements. Results can be significantly improved gathering more clinical data (increasing the number of relevant variables) or the number of images (more patients). In addition, more efforts can be put in the specialization of the DL architecture, taking advantage of higher detailed information in the retinal images (thickness, dimension, tortuosity of the vessels, etc.) or increasing the model capacity with current developments in self-supervision (to mitigate the small sample size problem) and curriculum learning (training the machine learning models in a meaningful order [26]).

Author Contributions: Conceptualization, R.G.B., C.H., R.S. and D.M.; methodology, R.G.B., C.H., R.S. and D.M.; software, R.G.B.; validation, R.G.B. and D.M.; formal analysis, R.G.B., C.H., R.S. and D.M.; investigation, R.G.B., O.S.-S., A.P., C.H., R.S. and D.M.; resources, O.S.-S., A.P., C.H., R.S. and D.M.; data curation, O.S.-S., A.P., C.H. and R.S.; writing—original draft preparation, R.G.B. and D.M.; writing—review and editing, R.G.B., R.S., C.H. and D.M.; funding acquisition, R.S., C.H. and D.M. All authors have read and agreed to the published version of the manuscript.

Funding: This research was funded by "RTI2018-095232-B-C22" grant from the Spanish Ministry of Science, Innovation and Universities (FEDER funds).

Institutional Review Board Statement: The study was conducted according to the guidelines of the Declaration of Helsinki, and was approved by the local ethics committee (Ethical Committee for Clinical Research of the Vall d'Hebron University Hospital), with a reference number PR(AG)127/2014.

Informed Consent Statement: Informed consent was obtained from all subjects involved.

Acknowledgments: Authors acknowledge the support from the NVIDIA Hardware grant program.

Conflicts of Interest: The authors declare no conflict of interest.The funders had no role in the design of the study; in the collection, analyses, or interpretation of data; in the writing of the manuscript, or in the decision to publish the results.

Abbreviations

The following abbreviations are used in this manuscript:

BMI	Body Mass Index
CAC	Coronary Artery Calcium
CNN	Convolutional Neural Network
CT	Computed Tomography
CVD	Cardiovascular Diseases
DL	Deep Learning
DR	Diabetic Retinopathy
FN	False Negative
FP	False Positive

	TN	True Negative
	TP	True Positive
	RFI	Retinal Fundus Imaging

References

1. WHO. *The Top 10 Causes of Death*; WHO: Geneva, Switzerland, 2019.
2. Goff, D.C., Jr.; Lloyd-Jones, D.M.; Bennett, G.; Coady, S.; D'Agostino, R.B.; Gibbons, R.; Greenland, P.; Lackland, D.T.; Levy, D.; O'Donnell, C.J.; et al. 2013 ACC/AHA guideline on the assessment of cardiovascular risk: A report of the American College of Cardiology/American Heart Association Task Force on Practice Guidelines. *Circulation* **2013**, *129*, S49–S73. [CrossRef] [PubMed]
3. Yeboah, J.; McClelland, R.L.; Polonsky, T.S.; Burke, G.L.; Sibley, C.T.; O'Leary, D.; Carr, J.J.; Goff, D.C.; Greenland, P.; Herrington, D.M. Comparison of novel risk markers for improvement in cardiovascular risk assessment in intermediate-risk individuals. *JAMA* **2012**, *308*, 788–795. [CrossRef] [PubMed]
4. Greenland, P.; Blaha, M.J.; Budoff, M.J.; Erbel, R.; Watson, K.E. Coronary Calcium Score and Cardiovascular Risk. *J. Am. Coll. Cardiol.* **2018**, *72*, 434–447. [CrossRef] [PubMed]
5. Cheung, C.Y.L.; Ikram, M.K.; Chen, C.; Wong, T.Y. Imaging retina to study dementia and stroke. *Prog. Retin. Eye Res.* **2017**, *57*, 89–107. [CrossRef]
6. Poplin, R.; Varadarajan, A.V.; Blumer, K.; Liu, Y.; McConnell, M.V.; Corrado, G.S.; Peng, L.; Webster, D.R. Prediction of cardiovascular risk factors from retinal fundus photographs via deep learning. *Nat. Biomed. Eng.* **2018**, *2*, 158–164. [CrossRef]
7. Cheung, C.Y.l.; Zheng, Y.; Hsu, W.; Lee, M.L.; Lau, Q.P.; Mitchell, P.; Wang, J.J.; Klein, R.; Wong, T.Y. Retinal vascular tortuosity, blood pressure, and cardiovascular risk factors. *Ophthalmology* **2011**, *118*, 812–818. [CrossRef]
8. Rim, T.H.; Lee, C.J.; Tham, Y.C.; Cheung, N.; Yu, M.; Lee, G.; Kim, Y.; Ting, D.S.; Chong, C.C.Y.; Choi, Y.S.; et al. Deep-learning-based cardiovascular risk stratification using coronary artery calcium scores predicted from retinal photographs. *Lancet Digit. Health* **2021**, *3*, e306–e316. [CrossRef]
9. Cheung, C.Y.; Xu, D.; Cheng, C.Y.; Sabanayagam, C.; Tham, Y.C.; Yu, M.; Rim, T.H.; Chai, C.Y.; Gopinath, B.; Mitchell, P.; et al. A deep-learning system for the assessment of cardiovascular disease risk via the measurement of retinal-vessel calibre. *Nat. Biomed. Eng.* **2021**, *5*, 498–508. [CrossRef]
10. Sabanayagam, C.; Xu, D.; Ting, D.S.W.; Nusinovici, S.; Banu, R.; Hamzah, H.; Lim, C.; Tham, Y.C.; Cheung, C.Y.; Tai, E.S.; et al. A deep learning algorithm to detect chronic kidney disease from retinal photographs in community-based populations. *Lancet Digit. Health* **2020**, *2*, e295–e302. [CrossRef]
11. McGrory, S.; Cameron, J.R.; Pellegrini, E.; Warren, C.; Doubal, F.N.; Deary, I.J.; Dhillon, B.; Wardlaw, J.M.; Trucco, E.; MacGillivray, T.J. The application of retinal fundus camera imaging in dementia: A systematic review. *Alzheimers Dement* **2016**, *6*, 91–107. [CrossRef]
12. Son, J.; Shin, J.Y.; Chun, E.J.; Jung, K.H.; Park, K.H.; Park, S.J. Predicting high coronary artery calcium score from retinal fundus images with deep learning algorithms. *Transl. Vis. Sci. Technol.* **2020**, *9*, 28. [CrossRef] [PubMed]
13. Simó, R.; Bañeras, J.; Hernández, C.; Rodríguez-Palomares, J.; Valente, F.; Gutiérrez, L.; Alujas, T.G.; Ferreira, I.; Aguade-Bruix, S.; Montaner, J.; et al. Diabetic retinopathy as an independent predictor of subclinical cardiovascular disease: baseline results of the PRECISED study. *BMJ Open Diabetes Res. Care* **2019**, *7*, e000845. [CrossRef] [PubMed]
14. Deng, J.; Dong, W.; Socher, R.; Li, L.J.; Li, K.; Fei-Fei, L. ImageNet: A large-scale hierarchical image database. In Proceedings of the 2009 IEEE Conference on Computer Vision and Pattern Recognition, Miami, FL, USA, 20–25 June 2009; pp. 248–255. [CrossRef]
15. Mnih, V.; Kavukcuoglu, K.; Silver, D.; Rusu, A.A.; Veness, J.; Bellemare, M.G.; Graves, A.; Riedmiller, M.; Fidjeland, A.K.; Ostrovski, G.; et al. Human-level control through deep reinforcement learning. *Nature* **2015**, *518*, 529–533. [CrossRef] [PubMed]
16. Ting, D.S.W.; Cheung, C.Y.L.; Lim, G.; Tan, G.S.W.; Quang, N.D.; Gan, A.; Hamzah, H.; Garcia-Franco, R.; San Yeo, I.Y.; Lee, S.Y.; et al. Development and validation of a deep learning system for diabetic retinopathy and related eye diseases using retinal images from multiethnic populations with diabetes. *JAMA* **2017**, *318*, 2211–2223. [CrossRef]
17. Simonyan, K.; Zisserman, A. Very deep convolutional networks for large-scale image recognition. *arXiv* **2014**, arXiv:1409.1556.
18. He, K.; Zhang, X.; Ren, S.; Sun, J. Deep residual learning for image recognition. *arXiv* **2015**, arXiv:1512.03385.
19. Ho, N.; Kim, Y.C. Evaluation of transfer learning in deep convolutional neural network models for cardiac short axis slice classification. *Sci. Rep.* **2021**, *11*, 1839. [CrossRef]
20. Paszke, A.; Gross, S.; Massa, F.; Lerer, A.; Bradbury, J.; Chanan, G.; Killeen, T.; Lin, Z.; Gimelshein, N.; Antiga, L.; et al. Pytorch: An imperative style, high-performance deep learning library. *Adv. Neural Inf. Process. Syst.* **2019**, *32*, 8026–8037.
21. Cunningham, P.; Delany, S. k-Nearest neighbour classifiers. *ACM Comput. Surv.* **2007**, *54*, 1–25. [CrossRef]
22. Cortes, C.; Vapnik, V. Support-vector networks. *Mach. Learn.* **1995**, *20*, 273–297. [CrossRef]
23. Praagman, J. *Classification and Regression Trees*; Breiman, L., Friedman, J.H., Olshen, R.A., Stone, C.J., Eds.; The Wadsworth Statistics/Probability Series; Belmont: Wadsworth, OH, USA, 1985; 358p.
24. Breiman, L. Random forests. *Mach. Learn.* **2001**, *45*, 5–32. [CrossRef]
25. Freund, Y.; Schapire, R.; Abe, N. A short introduction to boosting. *J. Jpn. Soc. Artif. Intell.* **1999**, *14*, 1612.
26. Kim, M.; Yun, J.; Cho, Y.; Shin, K.; Jang, R.; Bae, H.j.; Kim, N. Deep learning in medical imaging. *Neurospine* **2019**, *16*, 657. [CrossRef] [PubMed]

Article

Normalizing Flows for Out-of-Distribution Detection: Application to Coronary Artery Segmentation

Costin Florian Ciușdel [1,2,*], Lucian Mihai Itu [1,2], Serkan Cimen [3], Michael Wels [4], Chris Schwemmer [4], Philipp Fortner [5], Sebastian Seitz [5], Florian Andre [5], Sebastian Johannes Buß [5], Puneet Sharma [3] and Saikiran Rapaka [3]

1. Advanta, Siemens SRL, 15 Noiembrie Bvd, 500097 Brasov, Romania; lucian.itu@siemens.com
2. Automation and Information Technology, Transilvania University of Brasov, Mihai Viteazu nr. 5, 500174 Brasov, Romania
3. Digital Services, Digital Technology & Innovation, Siemens Healthineers, 755 College Road, Princeton, NJ 08540, USA; serkan.cimen@siemens-healthineers.com (S.C.); sharma.puneet@siemens-healthineers.com (P.S.); saikiran.rapaka@siemens-healthineers.com (S.R.)
4. Computed Tomography-Research & Development, Siemens Healthcare GmbH, 91301 Forchheim, Germany; michael.wels@siemens-healthineers.com (M.W.); chris.schwemmer@siemens-healthineers.com (C.S.)
5. Das Radiologische Zentrum-Radiology Center, 74889 Sinsheim, Germany; dr.fortner@das-radiologische-zentrum.de (P.F.); dr.seitz@das-radiologische-zentrum.de (S.S.); florian.andre@med.uni-heidelberg.de (F.A.); prof.buss@das-radiologische-zentrum.de (S.J.B.)
* Correspondence: costin.ciusdel@unitbv.ro

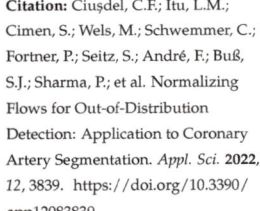

Citation: Ciușdel, C.F.; Itu, L.M.; Cimen, S.; Wels, M.; Schwemmer, C.; Fortner, P.; Seitz, S.; André, F.; Buß, S.J.; Sharma, P.; et al. Normalizing Flows for Out-of-Distribution Detection: Application to Coronary Artery Segmentation. *Appl. Sci.* **2022**, *12*, 3839. https://doi.org/10.3390/app12083839

Academic Editor: Claudio Belvedere

Received: 28 February 2022
Accepted: 6 April 2022
Published: 11 April 2022

Publisher's Note: MDPI stays neutral with regard to jurisdictional claims in published maps and institutional affiliations.

Copyright: © 2022 by the authors. Licensee MDPI, Basel, Switzerland. This article is an open access article distributed under the terms and conditions of the Creative Commons Attribution (CC BY) license (https://creativecommons.org/licenses/by/4.0/).

Abstract: Coronary computed tomography angiography (CCTA) is an effective imaging modality, increasingly accepted as a first-line test to diagnose coronary artery disease (CAD). The accurate segmentation of the coronary artery lumen on CCTA is important for the anatomical, morphological, and non-invasive functional assessment of stenoses. Hence, semi-automated approaches are currently still being employed. The processing time for a semi-automated lumen segmentation can be reduced by pre-selecting vessel locations likely to require manual inspection and by submitting only those for review to the radiologist. Detection of faulty lumen segmentation masks can be formulated as an Out-of-Distribution (OoD) detection problem. Two Normalizing Flows architectures are investigated and benchmarked herein: a Glow-like baseline, and a proposed one employing a novel coupling layer. Synthetic mask perturbations are used for evaluating and fine-tuning the learnt probability densities. Expert annotations on a separate test-set are employed to measure detection performance relative to inter-user variability. Regular coupling-layers tend to focus more on local pixel correlations and to disregard semantic content. Experiments and analyses show that, in contrast, the proposed architecture is capable of capturing semantic content and is therefore better suited for OoD detection of faulty lumen segmentations. When compared against expert consensus, the proposed model achieves an accuracy of 78.6% and a sensitivity of 76%, close to the inter-user mean of 80.9% and 79%, respectively, while the baseline model achieves an accuracy of 64.3% and a sensitivity of 48%.

Keywords: out-of-distribution; normalizing flows; coronary computed tomography angiography; lumen segmentation

1. Introduction

Coronary computed tomography angiography (CCTA) is an effective imaging modality, increasingly accepted as a first-line test to diagnose coronary artery disease (CAD). Advancements in CCTA have allowed for minimal radiation exposure, effective coronary characterization, and detailed imaging of atherosclerosis over time. Due to the increasing body of evidence showing the effectiveness of CCTA [1,2], recent ACC/AHA chest pain guidelines recommend CCTA as a first line test for patients with stable and acute chest pain.

The rapid progress in Artificial Intelligence (AI) approaches for pattern recognition over the last decade has led to several concepts, applications, and products built around the primary goal of augmenting and/or assisting the radiologist in their reading and reporting workflow, mainly focusing on automatic detection and characterization of features and on automatic measurements in the images. The majority of state-of-the-art CCTA image analysis algorithms are powered by artificial intelligence (AI) [3]. The accurate segmentation of the coronary artery lumen on CCTA is a crucial step for the automated detection and assessment of CAD for numerous use cases:

- Anatomical quantification of stenosis—for instance, minimum lumen area, minimum lumen diameter or percentage diameter stenosis [4].
- Morphological quantification: amount and composition of coronary plaques [5].
- Functional quantification of coronary function—for instance, CFD or machine learning based Fractional Flow Reserve (FFR) computation [6,7].

While the performance of AI based methods has improved markedly over the years, given the importance of an accurate lumen detection, semi-automated approaches are currently still being employed. Thus, the lumen is first automatically detected, and then manually inspected and edited by the radiologist if deemed necessary. This process, together with coronary artery centerline editing, required, e.g., between 10 and 60 min in a study assessing the diagnostic performance of ML-based CT-FFR for the detection of functionally obstructive coronary artery disease [6]. One potential approach for significantly reducing the time required for a semi-automated CCTA lumen analysis is to pre-select locations which are likely to require inspection and editing, and to present only those for review to the radiologist. Considering that a Deep Neural Network (DNN) is responsible for generating the lumen segmentation masks, this pre-selection step can be linked to the topic of confidence and out-of-distribution detection in Deep Learning. It is known that the output of classic DNNs may be unreliable when applied on out-of-domain, noisy or uncertain input data. Many methods have been proposed for assessing model output confidence. van Amersfoort et al. [8] uses a bi-Lipschitz deep feature extractor which feeds a sparse Gaussian Process (GP): segmenting an image involves at the lowest level many classification sub-problems, where each pixel is labelled according to the object/class it pertains to. Therefore, a GP can be adapted to model a segmentation task as a classification task, and, hence, the associated output uncertainty value can be extracted for each pixel. Image segmentations which display large mask uncertainties can be flagged as unreliable and proposed for human inspection. Another approach is to employ energy-based models. Liu et al. [9] shows that a model trained with a SoftMax final activation contains implicitly a density estimator. An energy-score can be computed for each pixel and aggregated mask-scores can be compared to predefined thresholds to determine which samples require manual inspection. Within these two methods, the model confidence is shaped during learning the target task. Therefore, in classification problems, the output confidence is low whenever the input sample is either far away from the training distribution or it is placed close to the nonlinear class-separation manifold in the input space.

Regular confidence methods do not provide a reason why the output confidence is low, and the class separation learnt by the model is highly dependent on the target task and on the model architecture. Normalizing Flow (NF) models on the other hand can be trained explicitly to model input data probability densities. Given a downstream target task T, if only its input data is employed for building the NF model, then estimating the likelihood of input samples for the target task can be obtained through the NF model. Input samples with low probabilities can be flagged as out-of-distribution and the target model's output should be considered unreliable, as it would operate outside its training distribution. An NF model can also be built by stacking the input samples with their expected GT output. This way, the NF model can be placed downstream of the target task and act as an Audit model, detecting cases where the previous model provided faulty predictions. In either scenario, the NF is a separate model and therefore imposes no constraints on the model responsible for the target task. NFs are a class of generative models which can perform exact log-likelihood computation. They have been employed in various setups, for instance:

- Image generation[10,11]: Being reversible models, random samples from the prior distribution can be transformed into the data domain, therefore obtaining new synthetic datapoints.
- Prior for variational inference: Instead of employing a fixed distribution (usually the normal distribution) in the KL term for ELBO maximization in variational inference, NF can be employed to model a much more expressive prior distribution. In a variational auto-encoder (VAE), this allows the encoder to better capture input patterns by not placing a fixed constraint on its computed embeddings. Ziegler and Rush [12] employed such a method for character-level language modeling and polyphonic music generation.
- Out-of-Distribution (OoD) detection: As log-likelihood values can be exactly and efficiently computed, NF may be good candidates in outlier detection [13].

NF can usually operate efficiently in both directions: forward (or inference) direction, where input samples x from the input domain X are transformed into embeddings z which are likely under a chosen distribution Z. At each layer, the input is modified towards Z and the *logDet* value (i.e., $\ln\left(\left|\det\left(\frac{\partial f}{\partial x}\right)\right|\right)$, where f is the NF) is summed with the current layer contribution. The backward (or generative) direction employs the bijection property of the NF to transform an embedding z into a synthetic sample x_{new}. Refs. [14,15] offer an introduction and review into the current approaches used in the NF framework.

In this paper, we present an approach based on NF for the OoD detection of coronary lumen segmentations. NF models which are built from coupling layers as proposed in [10,11] tend to focus on local pixel correlations instead of the global semantic meaning [13,16] and, as a result, OoD samples may in fact produce larger log-probability values than in-distribution data. We investigate the usage of a new type of coupling layer, which employs reversible 1×1 convolutions in which the filter parameters are computed based on the passed-through components. We compare the proposed architecture against a Glow-like architecture on the task of detecting mismatched pairs of CCTA lumen images and their corresponding lumen segmentations. The coronary lumen images and masks are 3D volumes stacked along the channel axis. We also employ synthetic perturbations on the binary masks and use the perturbed samples as explicit outliers to further shape the learnt probability density of "correct" image-mask pairs. The end goal is to flag those samples for which the given segmentation does not properly match with the lumen image. Overall, we assess the performance of the NF models as follows: (i) against the synthetic mask perturbations, and (ii) using expert annotations.

2. Methods
2.1. Patients and Imaging Protocol

Two datasets were used for the purpose of this study: a primary dataset as basis for the conventional train\validation\test split and an additional secondary separate test set.

The primary dataset included 560 patients who underwent contrast enhanced CCTA for clinical indications at Das Radiologische Zentrum (Heidelberg, Germany). CCTA was performed on a third generation dual-source CT scanner (SOMATOM Force, Siemens Healthcare GmbH, Erlangen, Germany). Beta-blockers or sublingual nitroglycerin were administered prior to the scan if clinically indicated. Prospective and retrospective gating protocols were utilized with a tube voltage varying between 70–150 kV. Reconstructed matrix size was 512×512 with a pixel size between 0.289–0.496 mm. Slice thickness and increment were 0.6 mm and 0.4 mm, respectively.

The secondary dataset included 53 patients. It was retrospectively collected from patients who underwent contrast enhanced CCTA from an independent test center. CCTA was acquired on a dual-source CT scanner (SOMATOM Force, Siemens Healthcare GmbH, Erlangen, Germany). Beta-blockers were not used since physiological cardiac function was also assessed. Tube voltages for the scans varied between 80–120 kV. Reconstructed matrix size was 512×512 with a pixel size of 0.391 mm. Slice thickness and increment were 0.75 mm and 0.4 mm, respectively.

2.2. CCTA Annotations

Three expert readers performed lumen annotations. These experts received 25 h of theoretical and practical training from a level 3 SCCT certified cardiothoracic imaging radiologist. Lumen annotations were performed on standard Windows workstations with a dedicated in-house annotation tool. The tool has two orthogonal curved multiplanar reconstruction (cMPR) views and one cross-sectional view where the experts can perform drawing and editing. First, coronary centerlines and corresponding lumen boundaries were generated using previously developed methods [17,18]. These automatically generated centerlines and lumen boundaries were then manually edited to obtain the final lumen annotations. Lumen annotations were performed only between the proximal and distal lesion markers defined by the expert readers before starting the lumen annotation process. To this end, annotators placed markers at the start and end of the diseased coronary artery regions to define lesions for every branch. These lesions were then extended proximally and distally by 5 mm to include healthy coronary artery regions. If the lesions overlapped after the extension, the lesions were merged. For each extended lesion, the experts first checked and edited the lumen boundaries in 4 cMPR views. They also reviewed their results in the cross-sectional view and edited the contours if required. The window and level were automatically set using values extracted from the DICOM tags; however, the experts were encouraged to modify these values according to existing guidelines [19] to achieve the best visualization of the coronary lumen.

2.3. Data Preparation for Convolutional Neural Networks

For each extended lesion, the centerline is sampled equidistantly at 0.25 mm intervals. Unit vectors that are tangential to the centerline are computed. To define a 2D local coordinate system along the centerline, two other unit vectors are determined at every centerline point using a rotation minimizing frame technique [20]. Cross-sectional images are then generated by sampling the CCTA volume at regular grid positions around the centerline along the local 2D coordinate system. Distances from the corresponding branch mesh to regular grid positions are also computed to generate corresponding cross-sectional distance maps, which then can be binarized to obtain lumen masks. The resulting cross sectional images and masks have a size of 64 × 64 pixels with 0.125 mm isotropic pixel spacing.

2.4. NF Architectures

We investigated the use of a Glow-style NF architecture, combining layers previously introduced in [10,11], such as checkerboard and channel masking coupling layers, invertible 1 × 1 Convolutions, Split and Squeeze layers. Our baseline network is depicted in Figure 1 and described in Table 1. We employed affine coupling layers as in Equation (1), where x and y are the input and output tensors, respectively. Subscripts a and b typically denote the two halves of the tensors: one which is passed-through unchanged and the other one which is updated in a linear fashion with respect to itself, but in a highly non-linear fashion with respect to the former half, through functions s and t (which are Deep Neural Networks).

$$y_a = x_a \\ y_b = (x_b - t_{DNN}(x_a)) \, s_{DNN}(x_a) \quad (1)$$

Networks s and t are in our case a two-head 3D CNN, with its architecture described in Table 2. The final activation function of head s was chosen as $\exp(\tanh(x))$ in order to easily compute the contribution to logDet (as $\sum \tanh(x)$ across all spatial dimensions and channels) and provide a bound of $[e^{-1}, e^1]$ to the scaling done at each coupling layer, ensuring numerical stability and a bounded global maximal value of logDet.

The input samples consist of chunks of 8 adjacent cross sections (down-sampled to 32 × 32 resolution) and 2 channels (the concatenation of the angiography and the binary mask volumes). There are 3 squeezing operations which contract the input resolution

2^3 times down to $1 \times 4 \times 4$, with increasing number of channels, being the same setup as used in Glow [11] (however we employed fewer layers on each scale due to runtime considerations). The effective receptive field of a coupling layer is given by the receptive field of the s and t network, in this case, $5 \times 5 \times 5$. Stacking coupling layers and using multiple scales (i.e., squeeze layers) increase the final NF receptive field, similar to the operation of classical CNNs.

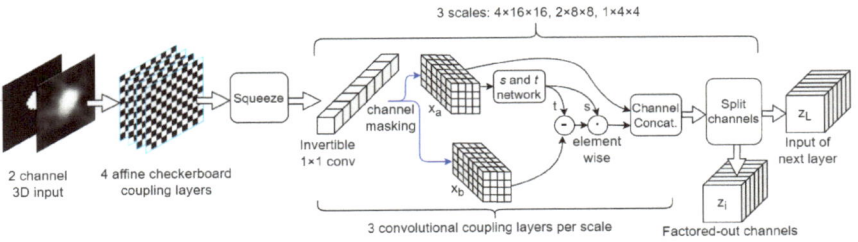

Figure 1. Baseline model architecture. Activation norm not depicted.

Table 1. Glow-style baseline architecture.

Stage	No. Blocks	Block Description	Resolution	No. Channels	Total Number of Parameters
1	4	Affine coupling layer using checkerboard mask; Activation Norm (if not last block)	$8 \times 32 \times 32$	2	~2 millions
2 3 4	1	3D Squeeze operation	Stage 2: $4 \times 16 \times 16$ Stage 3: $2 \times 8 \times 8$ Stage 4: $1 \times 4 \times 4$	Stage 2: 16 Stage 3: 64 Stage 4: 256	
	3	Activation Norm; Invertible 1×1 Convolution; Affine coupling layer using channel-wise masking			
	1	Split channels		After stage 2: 8 After stage 3: 32	

Table 2. s and t network architecture.

Stage	Block	No. Filters	Cumulative Receptive Field
1	Conv3D with $3 \times 3 \times 3$ kernel, stride 1, padding 1; BatchNorm; LeakyReLU	64	$3 \times 3 \times 3$
2	Conv3D with $1 \times 1 \times 1$ kernel, stride 1, padding 0; BatchNorm; LeakyReLU; Dropout	64	$3 \times 3 \times 3$
3	Conv3D with $3 \times 3 \times 3$ kernel, stride 1, padding 1; BatchNorm; LeakyReLU; Dropout	64	$5 \times 5 \times 5$
$4 - s$	Conv3D with $1 \times 1 \times 1$ kernel, stride 1, padding 0	As many as x's channels for checkerboard masking or half for channel masking	$5 \times 5 \times 5$
$4 - t$	Conv3D with $1 \times 1 \times 1$ kernel, stride 1, padding 0		

In [13], it has been shown that NFs which employ affine coupling layers are prone to focus more on local pixel correlations instead of semantic content and exploit coupling layer co-adaptation in order to maximize the final log-probability, i.e., the inductive bias of a stack of affine coupling layers encourages them to encode information about masked pixels in subsequent layers so that t is a good approximation to x_b and thus s can be increased, leading to larger final log-probabilities. Therefore, stacks of affine coupling layers are incentivized to guess local pixel values by exploiting texture correlations and information feed-forward by bypassing the masks, instead of building features increasing in complexity and expressivity as it is happening for classical stacks of convolutional layers. Semantic features can describe higher level characteristics of the modeled objects, such as global shapes (e.g., the relatively circular shape of the lumen with its spatial continuity between slices), global appearance (e.g., how distinguishable is the lumen from the background) and object correlations (e.g., the mask's spatial alignment to the lumen; calcifications, if present, should be around the lumen, etc.).

Behrmann et al. [21] shows that by constraining a general-purpose residual network [22] to be bi-Lipschitz, it can be used as a NF architecture. The expressive power of ResNets is shown to be preserved even with this constraint. On a classification task the invertible ResNet performed better while on a density modeling task it performed similar to affine-coupling networks. However, sampling from a NF consisting of a ResNet is an iterative process at each residual layer and training/inference involves approximating the logDet at each layer through a power series truncation. Therefore, such ResNet-based models are not that straight-forward in their operation as, e.g., Glow-based models. Ref. [23,24] tackle the problem of conditional probability modelling through the use of NF. In coupling layers, instead of using only the passed-through portion of the layer input to compute s and t, an additional embedding (dependent on the conditioning variable) is also employed. In addition, several layers such as Activation Norm (ActNorm) and invertible 1×1 Convolutions no longer have constant (but trainable) parameters for all inputs/samples, but instead their weights are computed based on the conditional embedding, therefore tailoring their effect for each particular pair of (condition, sample).

Inspired by the above research, we propose the use of a novel type of coupling layer, one which can operate efficiently for both NF directions, does not focus on local pixel correlations and has an inductive bias similar to conventional CNNs. The layer resembles a standard Glow-like sequence of 1×1 Invertible Convolution, channel masking, affine coupling layer. However, the last step is replaced with a 1×1 convolution (with applied bias) whose parameters are computed based on the passed-through channels, as in [25]. The applied bias is broadcasted to all spatial positions, therefore is it the same across the width, height and depth of the resulting tensor, meaning that the layer is no longer capable to reproduce masked pixel values as revealed in [13]. The same (sample specific) convolution kernel is applied at all spatial positions, in contrast to the element-wise computation done in (1). This behavior is similar to classical CNNs, with the exception that now the filter weights are not the same for all samples. Equation (2) describes the layer's operation, with simplified notation: $*$ means 1×1 Convolution with kernel k and $+$ is a broadcasting sum. k is computed by a CNN and has shape c_{modif}-by-c_{modif}, where c_{modif} is the number of channels which are updated. b is a vector of c_{modif} elements. The CNN responsible for computing k and b is described in Table 3.

$$y_a = x_a \\ y_b = x_a * k(x_a) + b(x_a) \qquad (2)$$

It is observed that the layer is self-conditioned, i.e., it does not employ an external conditioning network or another parallel flow as in [23,24], since the lumen binary mask and the angiographic image were not treated separately, but were concatenated on the channel axis. This is possible because the mask and the image should be highly correlated spatially in order to achieve high log-probability.

A new NF architecture was designed employing the above coupling layer. The first stage is a sequence of Additive Coupling Layers with checkerboard masking. According to [13], these layers will focus mainly on local pixel correlations, but this is equivalent to the functioning of the first layers in classical CNNs, where the receptive field-of-view is small and the filters tend to search for simple patterns such as corners, edges, textures, etc. As opposed to affine couplings, additive couplings are volume preserving, i.e., they do not contribute directly to logDet and final $\log(p(x))$, but indirectly through the upstream layers.

Table 3. CNN architecture for computation of k and b employed inside the coupling layer.

Stage	Block	No. Filters	Cumulative Receptive Field
1	Conv3D with $3 \times 3 \times 3$ kernel, stride 1, padding 1; BatchNorm; LeakyReLU	64	$3 \times 3 \times 3$
2	MaxPool3D $2 \times 2 \times 2$, stride 2		$4 \times 4 \times 4$
3	Conv3D with $1 \times 1 \times 1$ kernel, stride 1, padding 0; BatchNorm; LeakyReLU; Dropout	64	$4 \times 4 \times 4$
4	Conv3D with $3 \times 3 \times 3$ kernel, stride 1, padding 1; BatchNorm; LeakyReLU; Dropout	64	$8 \times 8 \times 8$
$5-k$	Conv3D with $1 \times 1 \times 1$ kernel, stride 1, padding 0; Average pooling	c_{modif}^2	full
$5-b$	Conv3D with $1 \times 1 \times 1$ kernel, stride 1, padding 0; Average pooling	c_{modif}	full

The next stages consist of cascades of coupling layers, as described in Figure 2 and Table 4. In contrast to a classical CNN, where filters of shape 3×3 (or larger) and strides larger than 1 are used (either in convolutional or max pool layers) to increase the effective field-of-view (FoV), in our architecture the FoV in these stages is increased solely by the squeeze operations. After squeezing, a $1 \times 1 \times 1$ patch of pixels is formed from a patch of $2 \times 2 \times 2$ pixels which were flattened spatially into the channel dimension. Therefore, the FoV doubles on each spatial axis for each squeeze step. This allows 1×1 Convolutions to operate on increasingly larger FoV, similar to the functioning of a classical CNN, while still retaining the capability of efficient forward/backward NF computation. There are enough squeeze operations so that the resolution on the last stage decays to $1 \times 1 \times 1$. Naturally, we restrict the input spatial dimensions to be powers of 2.

One possible disadvantage is that after each squeeze operation, the number of channels c_i (at stage i) increases exponentially with the number of squeezed dimensions (see Table 4). This directly impacts the proposed coupling layer's runtime and complexity, since it must produce matrix k whose size scales with the square of c_i. In addition, inference and sampling involve computing the determinant and inverse of k, respectively. One workaround to alleviate this issue is to modify the splitting layers so that the tensors are not split in half along the channel axis anymore, but instead only a quarter is retained for the rest of the computation graph while the other 75% of channels are factored out. This can be applied especially in the first stages, where the embeddings mostly describe texture. After such a split, the input to the next squeeze has only $c_i/4$ channels, half that of a regular split. Cascading such splits throughout the network can alleviate the effect of the exponentially-increasing c_i, especially for larger resolution inputs. In our experiments, the first split layer only retains $c_i/4$ channels. The net effect is that there are only 512 channels in the last stage, as opposed to the original 1024 (as described in Table 4), resulting in faster runtimes and fewer model parameters.

In this new architecture, BatchNorm was employed instead of ActNorm. In classical CNNs, batch norm acts by computing the batch statistics and then using them to normalize the output. In our approach, two running averages of the batch mean and standard-deviation are employed for normalization and they are updated with current batch statistics after their use, so that the normalization procedure is dependent only on past batches and any cross-talk between samples in the current batch is eliminated. In either CNN and NF cases, batch norm's main purpose is to provide "checkpoints" for activations inside the network, i.e., after each BatchNorm layer the activations have preset statistics (i.e., are centered around 0 with a std. dev. of 1). This has been shown to improve the training process [11,26].

In all our experiments, the network weights are initialized such that the layers are an identity mapping in the beginning of training, as suggested in [11]. We employed the PyTorch DL framework with the Adam optimizer with a learning rate of 1×10^{-4} and trained until the validation loss plateaued.

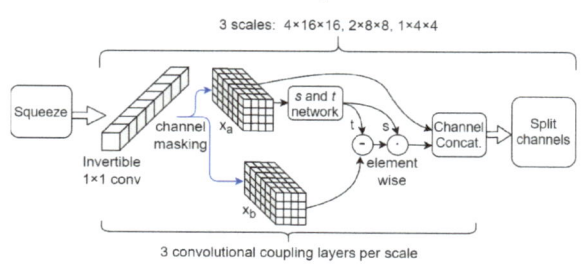

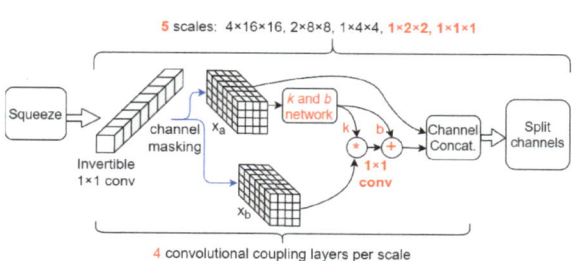

Figure 2. Comparison between baseline inner architecture (**top**) and proposed inner architecture (**bottom**) employing the novel coupling layer. Updated parts are highlighted in red. Normalization layers not depicted.

Table 4. Improved NF architecture employing the novel coupling layers.

Stage	No. Blocks	Block Description	Resolution	No. Channels	Total Number of Parameters
1	4	Additive coupling layer using checkerboard mask; BatchNorm (if not last block)	$8 \times 32 \times 32$	2	
2 3 4 5 6	1	3D Squeeze operation	Stage 2: $4 \times 16 \times 16$ Stage 3: $2 \times 8 \times 8$ Stage 4: $1 \times 4 \times 4$ Stage 5: $1 \times 2 \times 2$ Stage 6: $1 \times 1 \times 1$	Stage 2: 16 Stage 3: 64 Stage 4: 256 Stage 5: 512 Stage 6: 1024	~8.7 millions
	4	BatchNorm; Invertible 1×1 Convolution; convolutional coupling layer using channel-wise masking			
	1	Split channels		After stage 2: 8 After stage 3: 32 After stage 4: 128 After stage 5: 256	

2.5. Synthetic Mask Perturbations

Our application's goal is to detect incorrect pairs of (angiography image, lumen mask), i.e., samples where the segmentation is not in full agreement with the image. To test our models, we devised a method to obtain "wrong" datapoints (or samples which are not in the distribution of "correct" image-mask pairs) starting from our initial data (considered to be "correct").

We augmented the datasets by applying preset perturbations on the lumen segmentation binary mask, while keeping the angiographic image untouched. Three types of mask perturbations were employed:

- zooming: we applied zoom in/out operations on the mask image with respect to the mask center, so that the resulting mask is still aligned with the angiography, but larger/smaller than before. Figure 3 displays an example for various levels of zoom.
- morphing: we applied dilations or erosions along 4 directions on the height * width plane: left-right, top-bottom, topLeft-bottomRight and topRight-bottomLeft. This perturbation only affects one part of the mask (the eroded or dilated part), while the other part is left untouched. Figure 4 displays an example for various levels of morphing. By convention, negative and positive levels refer to the two ways in the selected direction, with zero meaning original mask position (levels are expressed as ratios of the original mask size along the chosen direction). At every level, either dilation (resulting in prolonged masks) or erosion (resulting in shortened masks) can be applied.
- translations: in the same 4 directions on the height * width plane, we translated whole mask images. Each level increment signifies a pixel shift. Figure 5 shows an example for various levels of translation.

For each network architecture, we performed two training procedures: one employing only original (unperturbed) data and one employing a dataset consisting of the original data and its perturbed version. The perturbations are applied during train time, similar to data augmentation techniques, such that each original data sample gets perturbed on all perturbation types, levels and directions over the training epochs. At each epoch, the ratio between untouched and perturbed data is 1-to-1.

When only original data is used, the training loss function consists of maximizing the log-probabilities across the train set. When perturbed data is also employed, we used a train loss function (Equation (3)) similar to the hinge loss introduced in [13,27], where the model tries to maximize predicted log-probs for original (untouched, *in-distrib*) samples and tries to decrease predicted log-probs under a certain threshold T for perturbed samples (OoD). The hinge loss allows us to shape the learnt probability density modeled by the NF, by directly offering supervision in regions around the original samples in the input domain. Therefore, the NF can be trained to be sensitive to the used synthetic perturbations and to mark perturbed samples as OoD

$$L(\theta, x) = \mathbb{E}_{x \in inDistrib}(\ln(p_\theta(x))) - \mathbb{E}_{x \in OoD}(\max(0, \ln(p_\theta(x)) - T))$$
$$\theta_{optimal} = \operatorname{argmax}_\theta L(\theta, x) \quad (3)$$

where p_θ is the probability density modeled by the NF.

Training only on original data and then testing on synthetic perturbations gives insight into the OoD detection capability which stems purely from the inductive bias of the NF architecture. In addition, we argue that NF models, being a class of generative models, provide a form of explainability by being able to produce samples from their learnt probability density. By sampling repeatedly from the model and computing the associated log-probs, one can observe the kind of samples which the model considers to be in-distribution. We believe that this gives insight into the semantic content which is interpreted by the model and into the functioning of the computational chain of layers. Section 3.3 will discuss in further detail.

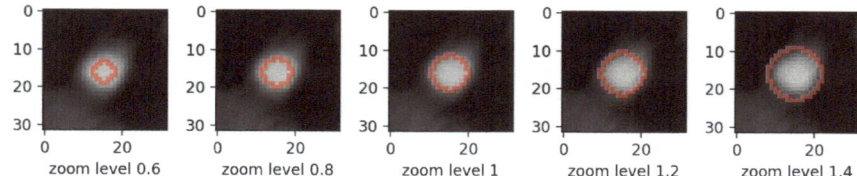

Figure 3. Example of zoom perturbation on lumen mask. Only the mask contour is shown.

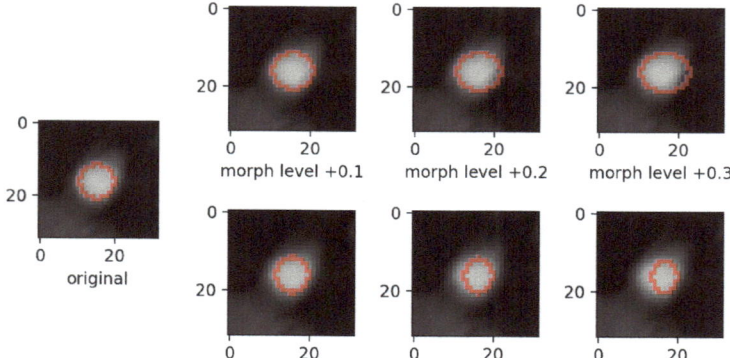

Figure 4. Example of mask morphing perturbation in the left-to-right direction. Top row shows dilations (the mask is prolonged), bottom row shows erosions (the mask is shortened). Only the mask contour is shown.

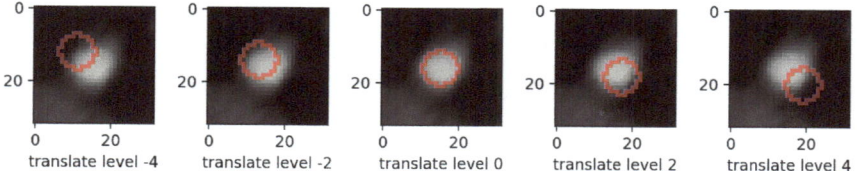

Figure 5. Example of mask translation perturbation in the topLeft-bottomRight direction. Only the mask contour is shown.

3. Results and Discussion

3.1. Evaluation on Synthetic Mask Perturbations

First, we evaluated the baseline and the proposed networks trained on original (unperturbed) data. We applied the synthetic perturbations on the testset in increasing levels of severity and measured how well the models can distinguish between log-probs of original and log-probs of perturbed samples. At each perturbation level, we computed the area under the RoC curve. Figures 6–8 display the AUROC values for translation, zooming and morphing perturbations, respectively. We use AuRoC as a metric for assessing how well two individual data distributions can be separated by using a probability threshold. A value close to 1 indicates that there are probability thresholds which yield near 100% accuracy in detecting perturbations, while values close to 0.5 indicate that the probabilities of the two data distributions have high overlap and are therefore indistinguishable by simple thresholding. Hence, the closer the AuRoC values are to 1, the better is the performance of the method. Zoom level $1.0\times$ and translate level 0 do not have any effect on the test data, so naturally the AuRoC is 0.5 since it is comparing the same distribution against itself.

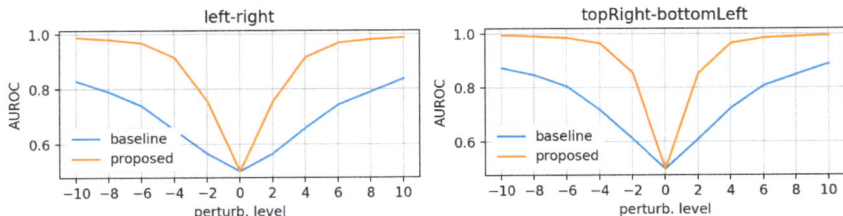

Figure 6. AuRoC performance for the baseline and proposed architecture when tested against mask translation in various directions. Training done only on original data.

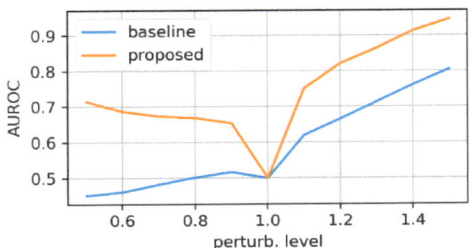

Figure 7. AuRoC performance for the baseline and proposed architecture when tested against mask zooming. Training done only on original data.

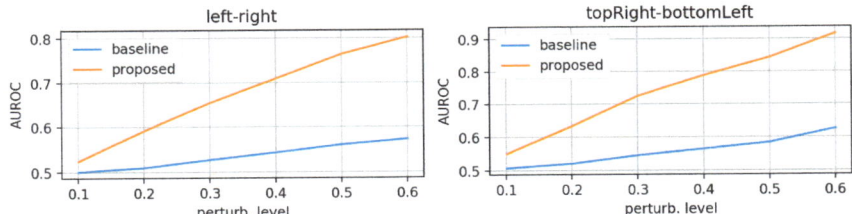

Figure 8. AuRoC performance for the baseline and proposed architecture when tested against mask morphing in various directions. Training done only on original data.

It is observed that the proposed model has superior performance across all perturbation types and levels. Zooming under $1.0\times$ actually yields higher log-probs for the baseline model, resulting in AuRoC values under 0.5. Even at small mask perturbation levels (e.g., $0.9\times/1.1\times$ zooming, ± 2 pixels translation), the proposed model has much larger sensitivity in detecting the mask alterations (even though it was not trained explicitly to do so) in contrast to the baseline model, where the log-probs start to decrease more significantly only at larger perturbation levels. The mask morphing is the hardest to detect since part of the mask remains the same. Thus, the baseline model is largely insensitive to this type of perturbation as the maximum AuRoC at a high perturbation level of 60% is under 0.65. In comparison, the AuRoC for the proposed network has a much faster variation for increasing perturbation severity, achieving values over 0.9 for some directions at 60% morphing.

Next, we evaluated the test-time sensitivity against synthetic perturbations after training using the augmented trainset and loss from Equation (3). We employed the following perturbation levels to generate OoD samples for training:

- translation (in all 4 directions) of ± 3 or ± 4 pixels;
- zooming levels of $0.65\times$, $0.8\times$, $1.2\times$, and $1.35\times$;
- morphing (in all 4 directions, erosions/dilations) levels of 0.2 and 0.35 (ratio of initial mask size).

Any OoD sample had only one type of perturbation applied to it. In addition, given the fact that each sample is a 3D volume consisting of 8 2D slices, the perturbation could be applied on each slice following some preset variation for its severity/level along the slice axis (see Figure 9 for the employed severity variation types). A severity variation type is sampled randomly for each OoD sample and each component 2D slice is perturbed according to its corresponding severity.

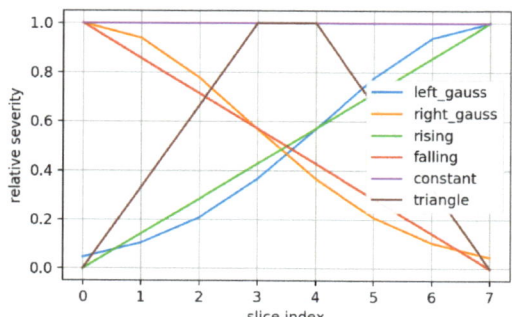

Figure 9. There were 7 perturbation severity variations employed for OoD samples generation. Each slice index has its corresponding relative severity level, according to the chosen severity variation.

Figures 10–12 display the test-set AUROC values for translation, zooming and morphing perturbations, respectively. The two models perform similarly well, except for some morphing directions, where the proposed model has slightly lower AuRoC values for small perturbation levels.

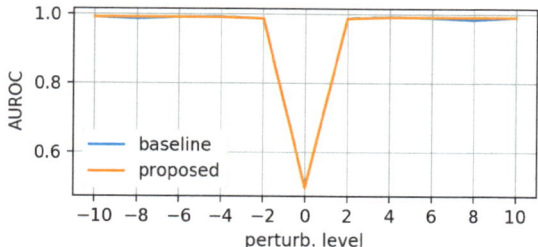

Figure 10. AuRoC performance for the baseline and proposed architecture when tested against mask translation in various directions. Training done on augmented dataset using the hinge loss in Equation (3).

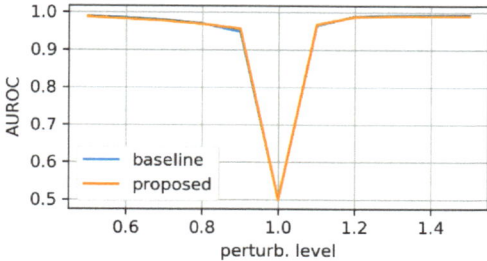

Figure 11. AuRoC performance for the baseline and proposed architecture when tested against mask zooming. Training done on augmented dataset using the hinge loss in Equation (3).

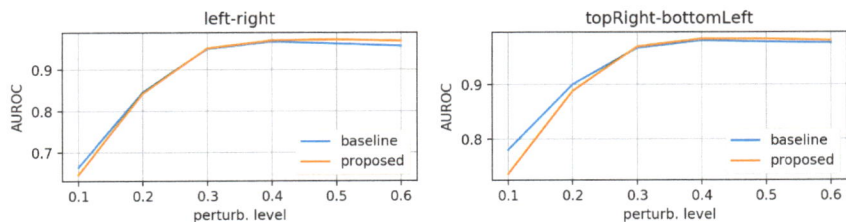

Figure 12. AuRoC performance for the baseline and proposed architecture when tested against mask morphing in various directions. Training done on augmented dataset using the hinge loss in Equation (3).

This training procedure yields models which show high sensitivity towards samples from the outlier distribution used explicitly during training, and which can separate the log-prob distributions even for small levels of perturbations. However, in [13] it is reported that even though NF models can achieve good separation between in-distrib and OoD sets used explicitly during training, there may also be other OoD sets (unseen during training) which still achieve log-probs as high as in-distribution data. Therefore, the inductive bias of the model still plays a major role in the generalization and usability of a NF model in the face of new data, even if it was explicitly tuned to decrease log-probabilities for *some* forms of outliers. The next sections will further inspect the two models trained on augmented data and will provide evidence that the baseline model, even though it can now detect some outliers, does not model a useful probability density which is descriptive of the training data.

To obtain a log-probability signal which describes the likelihood of an entire vessel segment, a sliding window approach was employed in which overlapping chunks of 8 adjacent cross-sections are fed through the NF model to obtain the log-prob values for each chunk. Using this procedure, middle cross-sections can participate in at most 8 chunks, therefore there may be up to eight predicted log-probability values linked to each middle cross-Section. A voting scheme based on averaging is employed, where the final log-prob value for each cross-section is computed by averaging the linked predicted log-probs. Figure 13 depicts such an example, where a synthetic perturbation is applied with a known severity variation. The proposed NF model detects when the perturbation severity is high enough, while outputting high log-prob values when the perturbation is negligible.

3.2. Evaluation on Expert Annotations

The results in the previous section indicate that the herein proposed model is superior to the Glow-like baseline. Hence, we first ran the proposed model (trained on the augmented trainset) on the secondary dataset described in Section 2.1. We employed two relative thresholds of 60% and 90% of the mean log-probability value observed on the primary dataset's test split, when no perturbation was applied. All lesions which had *at least one* cross-Section log-probability value under the 60% threshold were selected as candidates with possibly wrong mask annotations, yielding a total of 31 of these lesions. To construct the bin of candidates with possibly correct mask annotations, we randomly sampled 31 lesions from those for which all cross-Section log-probability values were above the 90% threshold. Thus, we did not consider lesions which had any intermediate values of log-probability (i.e., between the 60% and 90% thresholds) without also having at least one low log-probability cross-section, to avoid the effect of model uncertainty for data which it considers to be near the separation manifold between correctly annotated lesions and faulty ones. We then further excluded lesions which had a reference diameter (computed as average of healthy proximal and distal diameters) lower than 1.5 mm (the typical threshold employed in CCTA based studies assessing CAD anatomically and functionally). As a result, a test set containing 56 lesions from 35 patients was employed for the evaluation.

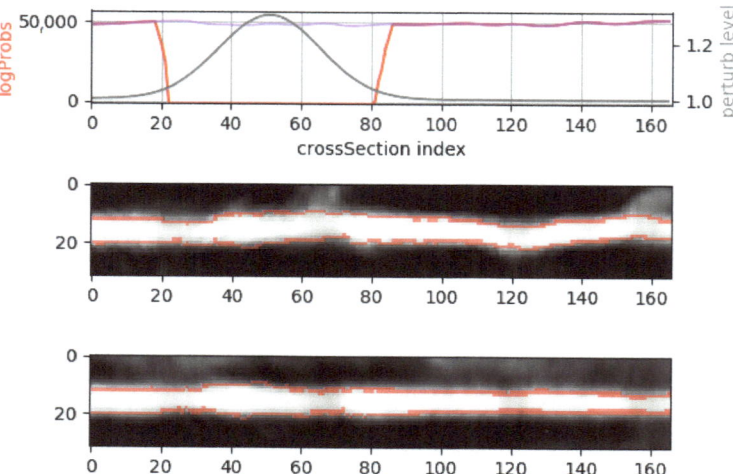

Figure 13. Whole vessel-segment prediction using a sliding window approach and the proposed architecture. A zooming perturbation with a known severity variation (top plot, gray signal) is applied (note that zoom level 1.0× is an identity transform). The resulting log-prob signal (top plot, red signal) dips whenever the perturbation is severe enough, compared to the original log-prob signal in the absence of any perturbation (top plot, purple signal). The bottom 2 plots display two lateral views of the vessel segment (2 projections on different axes), with the perturbed mask contour overlaid.

The selected test set was manually and independently annotated by three expert readers at lesion-level: each lesion was marked as being either "correctly" or "incorrectly annotated", based on the following instructions: a lesion should be marked as "incorrectly annotated" if *at least one* cross-sectional contour would require editing; otherwise, if *no* cross-sectional contour requires editing, then the lesion should be marked as "correctly annotated". The annotator instructions were devised so as to match the procedure employed to construct this separate test set, with the goal of being able to directly compare the labels from the NF model to the ones provided by the annotators.

The Glow-like baseline NF model was also applied on the 56 lesions and the same relative thresholds and criteria were employed to classify each lesion. Evaluating the two NF models against the human annotations was framed as a binary classification problem. Table 5 summarizes relevant metrics (accuracy, sensitivity, specificity, PPV and NPV) for the proposed and baseline models. Annotation consensus was obtained through a majority vote between the three annotators. The mean inter-user metric values were obtained by averaging all 6 possible metric values pertaining to pairs of annotators, e.g., Annotator_1 (as GT) versus Annotator_3 (as Prediction), Annotator_3 (as GT) versus Annotator_1 (as Prediction), etc. When compared against annotation consensus, the proposed model has higher performance than the baseline on all considered metrics.

Of special interest is the sensitivity metric, which measures the percentage of NF-flagged lesions as being incorrect from the set of lesions considered incorrect by the human annotators' consensus. The higher this metric value, the more capable is a NF model in detecting faulty segmentation masks. We observe that the proposed model has sensitivity of 76.0%, close to the inter-user value of 79.0%, while the baseline model only achieves 48%. Overall, according to the majority vote of the expert readers, 25 lesions were annotated as requiring editing, out of which 17 had unanimous annotations and 8 had non-unanimous annotations. The NF model correctly classified 16 out of the 17 unanimously annotated lesions and three of the eight non-unanimously annotated lesions. This indicates a 94.1% sensitivity on the unanimously annotated lesions.

The overall accuracy score also increases to 78.6% (close to the inter-user value of 80.9%) for the proposed model, as compared to an accuracy value of 64.3% for the baseline Glow-like model. These results reinforce the observation that the baseline model is unable to fully capture semantic content while the proposed model does. Similar behavior was observed in the previous section, where the proposed model had better AuRoC values in detecting synthetic perturbations when trained only on original data.

Table 5. Metrics on the secondary dataset for the baseline and the proposed model. The proposed model consistently outperforms the baseline and has metric values close to inter-expert agreement.

Metric	Inter-Expert Agreement Average [Min, Max]	Baseline Model	Proposed Model
Accuracy	0.81 [0.79, 0.86]	0.64	0.79
Sensitivity	0.79 [0.70, 0.87]	0.48	0.76
Specificity	0.83 [0.76, 0.90]	0.77	0.81
PPV	0.79 [0.70, 0.87]	0.63	0.76
NPV	0.83 [0.76, 0.90]	0.65	0.81

3.3. Sampling from the Models

We employed the models trained on the augmented trainset to generate novel samples. Similar to sampling procedures in [11], we employed $\mathcal{N}(0, 0.6 \cdot I)$ instead of the actual prior distribution (i.e., standard normal multivariate distribution) in order to produce samples with larger log-probs and which look more realistic. Each new sample was run back through the model in the forward direction to compute the log-probs, confirming that the sample is in fact in-distribution (the sampling procedure may seldomly generate samples of lower log-probability). Figure 14 shows samples from the two models.

As already revealed in [13], the baseline model tends to focus more on textures and is unable to capture the semantics of the training data. We observed that in most of the generated samples, the segmentation mask is lacking (i.e., only zeros are generated on the mask channel). In addition, the usual round shape of the lumen is not distinguishable in the image channel. In [11], the proposed Glow model can indeed generate realistic samples. That model operates on the same spatial resolution as ours (32×32) and uses the same number of 3 spatial scales (i.e., squeeze operations); however, it employs up to 48 coupling layers per scale (as opposed to ours, which only uses 3 coupling layers per scale due to runtime considerations). We hypothesize that many glow-like layers are required at each scale because of their tendency to disregard semantic content and a deep stack of such layers can approximate some semantic content as very complex textures.

In contrast, the proposed architecture uses a small number of 4 (novel) coupling layers per scale and manages to capture the semantic content of a usual data point: the lumen has the typical shape in the image channel, the segmentation mask is present (with plausible pixel-values, e.g., close to either 0 or 1) and respects the shape and position of the lumen in the image channel.

We argue that inspecting the generated samples is an explainability mechanism which offers insight into the learnt probability distribution, i.e., the model can provide example inputs which are very likely under the learnt density and by repeating the sampling procedure enough times, an approximation of the typical set of the learnt distribution may be constructed. If a generative model consistently produces samples with high associated log-probabilities but which have low quality under manual inspection and are implausible considering the specific topic/domain, then this is proof that the learnt probability density is not a good approximation of the true probability density and, therefore, the model cannot be reliably used for OoD detection (even if its train-time loss function encouraged the separation of *some* OoD data sets). In practice, the amount of available OoD data is usually limited and the inductive bias of the model still holds a huge importance in the quality of the distribution fitting. Zhang et al. [27] shows that "even good generators can still exhibit OoD detection failures", therefore the above condition of good/plausible sample generation is necessary, but not sufficient. It is necessary because the learnt probability density should match the underlying data probability density as closely as possible. However, even for a good generator (with high validation-set likelihoods) there may be small-volume regions of the sample space where the model assigns high-density but low overall probability mass. This faulty assignment of high-density might be caused by model estimation error [27], but because the overall probability mass may be negligible (due to the small volumes of poorly modeled regions), it does not affect the generation of synthetic samples. Still, training a generative model for OoD detection can require accurate estimation in regions which are unimportant for good generation [27]. Using an ensemble of generative models may alleviate the effect of model estimation errors, as each model instance may mis-predict on different regions of the sample space and thus errors could be averaged out.

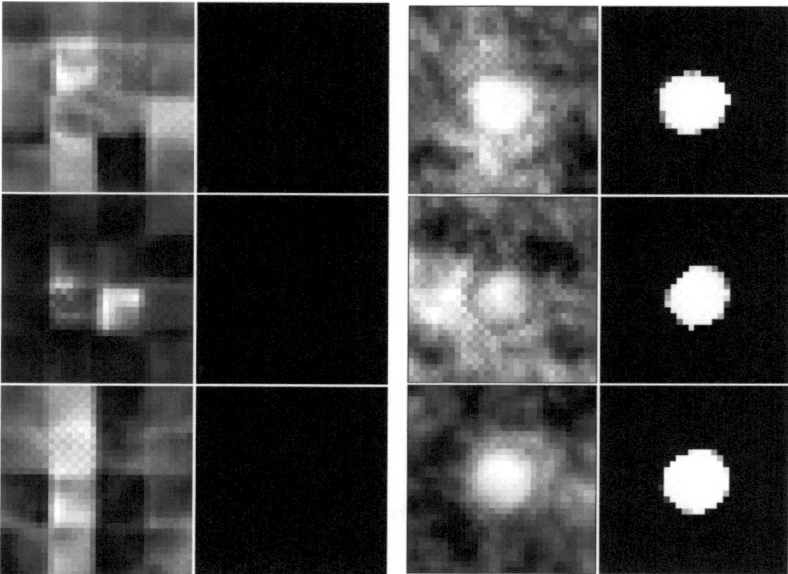

Figure 14. Three samples (pairs of lumen image and segmentation mask) generated by the proposed network (**right**) and by the baseline model (**left**).

3.4. Inspecting the Flows

The main hypothesis in [28] is that, in hierarchical VAEs, the lowest latent variables "learn generic features that can be used to describe a wide range of data" and thus OoD data can achieve high likelihoods "as long as the learned low-level features are appropriate". It is further suggested that "OOD data are in-distribution with respect to these low-level features, but not with respect to semantic ones".

Inspired by the hierarchical likelihood bounds approach in [28], we inspected the progressive transformation of mask-perturbed samples as they are passed through the sequence of coupling layers inside the NF models (i.e., going in the forward direction from $x \in X$ to $z \in Z$). We recorded the likelihood of the first factored out embeddings (termed z_{bottom}) after the first splitting operation and their associated logDet values at that stage. We computed pseudo-likelihood values, associated to these "bottom features", by ignoring the rest of the computational chain and the part of the embeddings which were not factored out (termed z_{top}). We also computed pseudo-likelihood values associated with the "top features", i.e., by summing the likelihood (under the prior) of embeddings z_{top} and the updates to logDet done at stages after the first splitting operation.

Intuitively, the "bottom features" operate at a smaller field-of-view and tend to capture more local patterns, while the "top features" are built based on the "bottom features" and at larger fields-of-views, therefore being able to access the global semantic content of a sample. Formally, the two pseudo-likelihoods are computed as in Equation (4):

$$\begin{aligned} \mathcal{L}_{bottom} &= \ln(p(z_{bottom})) + \ln(|\det(\nabla(f_1 \circ f_2 \circ \ldots \circ f_k))|) \\ \mathcal{L}_{top} &= \ln(p(z_{top})) + \ln(|\det(\nabla(f_{k+1} \circ f_{k+2} \circ \ldots \circ f_M))|) \end{aligned} \quad (4)$$

where M is the number of layers in the NF model and k is the index of the splitting layer which factored out z_{bottom} and retained z_{top} for the upstream layers. Because the chosen prior distribution is a diagonal multivariate Gaussian, the sum of $\mathcal{L}_{bottom}$ and $\mathcal{L}_{top}$ yields exactly $\ln(p(x))$.

Employing models trained only on original (unperturbed) data, we computed AuRoC values for the two pseudo-likelihoods when applying perturbations of increasing severity and compared performances to the standard case where the regular log-probabilities are used to discriminate OoD samples. Figure 15 shows the AuRoC values for the baseline and proposed models when detecting mask translations in various directions. For the baseline model, the bottom features have worse performance in detecting outliers, while the top features perform better than regular log-probs. This observation is in line with the hypothesis in [28], that higher level latent variables can better discriminate through semantic content, while lower level latents would yield similar likelihoods for OoD data if textures appear to be in-distribution.

However, the proposed model shows consistent performance across the top and bottom level features and regular log-probabilities. This suggests that even at the first spatial scale, the novel coupling layers try to capture semantic features instead of local spatial correlations and that the inductive bias of this coupling layer is better suited for OoD detection than regular affine coupling layers (as used in [10] or [11]).

Inspecting the proposed model also on mask-morphing or mask-zooming perturbations reveals the same behavior. However, despite the fact that pseudo-likelihoods of bottom-features exhibit similar OoD detection performance as regular log-probs, the network architecture cannot be truncated to use a smaller number of spatial scales. An experiment where only four spatial scales were employed (instead of the original 5) was conducted. The resolution on the final spatial scale was $1 \times 2 \times 2$ instead of $1 \times 1 \times 1$. The experiment revealed that the network in this configuration is unable to ensure spatial coherency across the entirety of the input image (of resolution $8 \times 32 \times 32$) but only on patches of $8 \times 16 \times 16$, since each pixel position in the $1 \times 2 \times 2$ map has a field-of-view of 16 pixels. Even though the k-and-b network has a full receptive view, the k kernel is applied on a 1×1 basis and cannot semantically link adjacent spatial sections in the x_b tensor. Therefore, the proposed network decays the spatial resolution down to $1 \times 1 \times 1$, where the computed k kernel can operate on the entire receptive field of the input. Figure 16 shows samples generated from a model with only four spatial scales instead of five. The samples reveal that the 4 quadrants are not semantically connected.

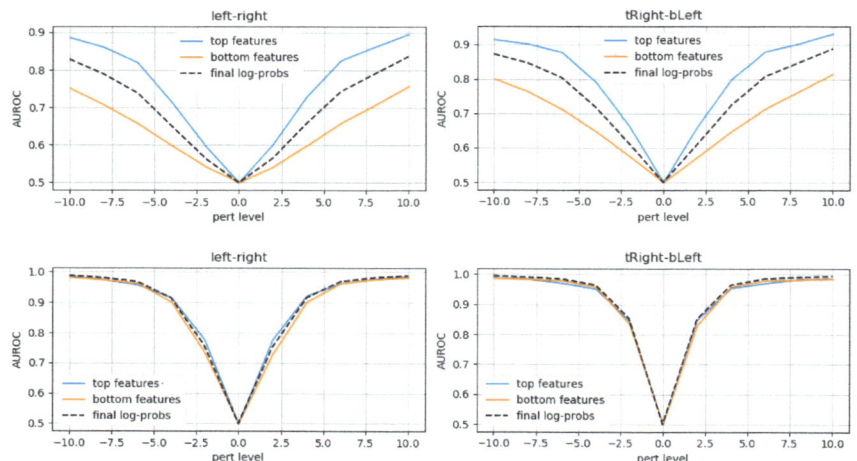

Figure 15. Baseline model (**first row**) compared to proposed network (**last row**): AuRoC values for separating OoD mask-translated samples, using either top-features or bottom-features. Dashed black lines represent default model performance using regular log-probs.

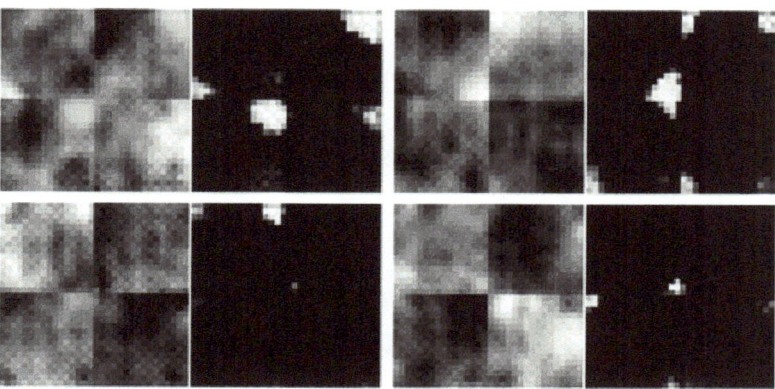

Figure 16. Generated samples are not coherent across the entire 32 × 32 spatial resolution, but only on the 16 × 16 quadrants.

4. Conclusions

While the performance of AI based methods has improved markedly over the past few years, semi-automated approaches are currently still being employed. One potential approach for significantly reducing the processing time is to pre-select regions of interest which are likely to require manual inspection and editing. Herein, we linked this pre-selection step to the topic of confidence and out-of-distribution detection, based on NF. The usage of a novel coupling layer which exhibits an inductive bias favoring the exploitation of semantical features instead of local pixel correlations was investigated on the task of detecting mismatched pairs of CCTA lumen images and their corresponding lumen segmentations. A network architecture employing such layers was tested against a Glow-like baseline. The proposed network showed better performance in OoD detection when tested against synthetic perturbations, while the sensitivity of detecting faulty annotations was close to inter-expert agreement. Samples from the model confirm that the learnt probability density managed to capture the relevant informational content from the training samples, instead of just modelling plain textures.

During model development and testing stages, only manual annotations were employed in the input mask-channel; therefore, the use of the investigated models is not tied to any specific segmentation model. Thus, any model-specific segmentation artefacts, which could possibly alter the observed probability density of correct image-mask pairs, are avoided by learning only from manual annotations (e.g., possible segmentation failure modes are not included in the learnt probability density). In deployment scenarios, the mask channel would be fed by a separate segmentation model and the proposed NF can therefore act as an independent audit model, detecting cases where the mask is not in full agreement with the underlying lumen images. As the failure modes of the NF model would be uncorrelated with the failure modes of the segmentation model, the proposed setup is better suited for robust pre-selection of vessel locations which are likely to require inspection and editing, leading to time savings when performing semi-automated CCTA lumen analysis.

CCTA is a powerful non-invasive test for ruling out CAD, i.e., avoiding unnecessary invasive coronary angiography (ICA). A recent review has summarized the latest aspects addressing the CCTA suitability for selecting patients for invasive coronary angiography (ICA) and subsequent revascularization [29]. Clinical trials have shown that performing CCTA in patients receiving a clinical indication for ICA results in lower costs and more effective patient care [30].

However, small errors in the CCTA interpretation (e.g., minimal lumen area or diameter) can have a significant influence on the interpretation of the anatomical and/or functional significance of a stenosis. A large grey zone of uncertainty in the clinical interpretation may be the consequence.

The method proposed herein allows for more confident decision making using CCTA imaging alone. Using the proposed out-of-domain detection method, the gray zone in the clinical interpretation can potentially be narrowed down.

Author Contributions: C.F.C.: Methodology, Software, Validation, Formal analysis, Writing—Original Draft, Visualization. L.M.I.: Conceptualization, Methodology, Writing—Original Draft, Writing—Review & Editing, Visualization, Project administration, Funding acquisition. S.C.: Conceptualization, Investigation, Resources, Data Curation. C.S. and M.W.: Conceptualization, Writing—Review and Editing, Supervision. P.F., S.S., F.A. and S.J.B.: Writing—Review and Editing. P.S.: Conceptualization, Supervision, Project administration, Funding acquisition. S.R.: Conceptualization, Methodology, Supervision, Project administration. All authors have read and agreed to the published version of the manuscript.

Funding: The research leading to these results has received funding from the EEA Grants 2014–2021, under Project contract No. 33/2021. This work was partially supported by a grant of the Romanian National Authority for Scientific Research and Innovation, CCCDI–UEFISCDI, project number ERANETPERMED-HeartMed, within PNCDI III.

Institutional Review Board Statement: Not applicable.

Informed Consent Statement: Not applicable.

Data Availability Statement: The data has been acquired as part of the project acknowledged in the manuscript, and cannot be made public, considering GDPR regulations and the content of the informed consent signed by the patients.

Acknowledgments: The concepts and information presented in this paper are based on research results that are not commercially available. Future commercial availability cannot be guaranteed.

Conflicts of Interest: Costin Florian Ciușdel and Lucian Mihai Itu are employees of Siemens SRL, Advanta, Brasov, Romania. Chris Schwemmer and Michael Wels are employees of Siemens Healthcare GmbH, Computed Tomography-Research & Development, Forchheim, Germany. Serkan Cimen, Puneet Sharma and Saikiran Rapaka are employees of Siemens Healthineers, Digital Technology & Innovation, Princeton NJ, USA. Philipp Fortner, Sebastian Seitz, Florian André and Sebastian Johannes Buß are employees of Das Radiologische Zentrum—Radiology Center, Sinsheim-Eberbach-Walldorf-Heidelberg, Germany.

Abbreviations

The following abbreviations are used in this manuscript:

CCTA	Coronary computed tomography angiography
OoD	Out-of-Distribution
NF	Normalizing Flows
AI	Artificial Intelligence
CNN	Convolutional Neural Network
GT	Ground truth
CAD	Coronary artery disease
AuRoC	Area under the Receiver operating Characteristics
FFR	Fractional Flow Reserve
AHA	American Heart Association
CFD	Computational fluid dynamics
KL	Kullback–Leibler divergence
logDet	Logarithm of determinant
cMPR	curved Multiplanar Reconstruction
VAE	Variational Auto-Encoder

References

1. Mark, D.B.; Federspiel, J.J.; Cowper, P.A.; Anstrom, K.J.; Hoffmann, U.; Patel, M.R.; Davidson-Ray, L.; Daniels, M.R.; Cooper, L.S.; Knight, J.D.; et al. Economic Outcomes With Anatomical Versus Functional Diagnostic Testing for Coronary Artery Disease. *Ann. Intern. Med.* **2016**, *165*, 94–102. [CrossRef] [PubMed]
2. Levin, D.C.; Parker, L.; Halpern, E.J.; Rao, V.M. Coronary CT Angiography: Reversal of Earlier Utilization Trends. *J. Am. Coll. Radiol.* **2019**, *16*, 147–155. [CrossRef]
3. Han, D.; Liu, J.; Sun, Z.; Cui, Y.; He, Y.; Yang, Z. Deep learning analysis in coronary computed tomographic angiography imaging for the assessment of patients with coronary artery stenosis. *Comput. Methods Programs Biomed.* **2020**, *196*, 105651. [CrossRef] [PubMed]
4. Muscogiuri, G.; Van Assen, M.; Tesche, C.; De Cecco, C.N.; Chiesa, M.; Scafuri, S.; Guglielmo, M.; Baggiano, A.; Fusini, L.; Guaricci, A.I.; et al. Artificial Intelligence in Coronary Computed Tomography Angiography: From Anatomy to Prognosis. *BioMed Res. Int.* **2020**, *2020*, 6649410. [CrossRef] [PubMed]
5. Williams, M.C.; Earls, J.P.; Hecht, H. Quantitative assessment of atherosclerotic plaque, recent progress and current limitations. *J. Cardiovasc. Comput. Tomogr.* **2022**, *16*, 124–137. [CrossRef] [PubMed]
6. Coenen, A.; Kim, Y.H.; Kruk, M.; Tesche, C.; Geer, J.D.; Kurata, A.; Lubbers, M.L.; Daemen, J.; Itu, L.; Rapaka, S.; et al. Diagnostic Accuracy of a Machine-Learning Approach to Coronary Computed Tomographic Angiography-Based Fractional Flow Reserve. *Circ. Cardiovasc. Imaging* **2018**, *11*, e007217. [CrossRef] [PubMed]
7. Sankaran, S.; Grady, L.; Taylor, C.A. Fast Computation of Hemodynamic Sensitivity to Lumen Segmentation Uncertainty. *IEEE Trans. Med Imaging* **2015**, *34*, 2562–2571. [CrossRef] [PubMed]
8. van Amersfoort, J.; Smith, L.; Jesson, A.; Key, O.; Gal, Y. Improving Deterministic Uncertainty Estimation in Deep Learning for Classification and Regression. *arXiv* **2021**, arXiv:2102.11409.
9. Liu, W.; Wang, X.; Owens, J.; Li, Y. Energy-based Out-of-distribution Detection. In *Advances in Neural Information Processing Systems*; Larochelle, H., Ranzato, M., Hadsell, R., Balcan, M.F., Lin, H., Eds.; Curran Associates, Inc.: Red Hook, NY, USA, 2020; Volume 33, pp. 21464–21475.
10. Dinh, L.; Sohl-Dickstein, J.; Bengio, S. Density estimation using Real NVP. *arXiv* **2017**, arXiv:1605.08803.
11. Kingma, D.P.; Dhariwal, P. Glow: Generative Flow with Invertible 1x1 Convolutions. In *Advances in Neural Information Processing Systems*; Bengio, S., Wallach, H., Larochelle, H., Grauman, K., Cesa-Bianchi, N., Garnett, R., Eds.; Curran Associates, Inc.: Red Hook, NY, USA, 2018; Volume 31.
12. Ziegler, Z.; Rush, A. Latent Normalizing Flows for Discrete Sequences. In Proceedings of the 36th International Conference on Machine Learning, Long Beach, CA, USA, 9–15 June 2019; Volume 97, pp. 7673–7682.
13. Kirichenko, P.; Izmailov, P.; Wilson, A.G. Why Normalizing Flows Fail to Detect Out-of-Distribution Data. In *Advances in Neural Information Processing Systems*; Larochelle, H., Ranzato, M., Hadsell, R., Balcan, M.F., Lin, H., Eds.; Curran Associates, Inc.: Red Hook, NY, USA, 2020; Volume 33, pp. 20578–20589.
14. Papamakarios, G.; Nalisnick, E.; Rezende, D.J.; Mohamed, S.; Lakshminarayanan, B. Normalizing Flows for Probabilistic Modeling and Inference. *J. Mach. Learn. Res.* **2021**, *22*, 1–64.
15. Kobyzev, I.; Prince, S.J.; Brubaker, M.A. Normalizing Flows: An Introduction and Review of Current Methods. *IEEE Trans. Pattern Anal. Mach. Intell.* **2021**, *43*, 3964–3979. [CrossRef] [PubMed]
16. Nalisnick, E.; Matsukawa, A.; Teh, Y.W.; Gorur, D.; Lakshminarayanan, B. Do Deep Generative Models Know What They Don't Know? In Proceedings of the International Conference on Learning Representations, New Orleans, LA, USA, 6–9 May 2019.

17. Zheng, Y.; Tek, H.; Funka-Lea, G. Robust and Accurate Coronary Artery Centerline Extraction in CTA by Combining Model-Driven and Data-Driven Approaches. In Proceedings of the Medical Image Computing and Computer-Assisted Intervention—MICCAI 2013, Nagoya, Japan, 22–26 September 2013; Mori, K., Sakuma, I., Sato, Y., Barillot, C., Navab, N., Eds.; Springer: Berlin/Heidelberg, Germany, 2013; pp. 74–81.
18. Lugauer, F.; Zheng, Y.; Hornegger, J.; Kelm, B.M. Precise Lumen Segmentation in Coronary Computed Tomography Angiography. In Proceedings of the Medical Computer Vision: Algorithms for Big Data, Cambridge, MA, USA, 18 September 2014; Menze, B., Langs, G., Montillo, A., Kelm, M., Müller, H., Zhang, S., Cai, W.T., Metaxas, D., Eds.; Springer International Publishing: Cham, Switzerland, 2014; pp. 137–147.
19. Leipsic, J.; Abbara, S.; Achenbach, S.; Cury, R.; Earls, J.P.; Mancini, G.J.; Nieman, K.; Pontone, G.; Raff, G.L. SCCT guidelines for the interpretation and reporting of coronary CT angiography: A report of the Society of Cardiovascular Computed Tomography Guidelines Committee. *J. Cardiovasc. Comput. Tomogr.* **2014**, *8*, 342–358. [CrossRef] [PubMed]
20. Poston, T.; Fang, S.; Lawton, W. Computing and Approximating Sweeping Surfaces Based on Rotation Minimizing Frames. In Proceedings of the 4th International Conference on CAD/CG, Wuhan, China, 23–25 October 1995.
21. Behrmann, J.; Grathwohl, W.; Chen, R.T.Q.; Duvenaud, D.; Jacobsen, J.H. Invertible Residual Networks. In Proceedings of the 36th International Conference on Machine Learning, Long Beach, CA, USA, 9–15 June 2019; Volume 97, pp. 573–582.
22. He, K.; Zhang, X.; Ren, S.; Sun, J. Deep Residual Learning for Image Recognition. In Proceedings of the 2016 IEEE Conference on Computer Vision and Pattern Recognition (CVPR), Las Vegas, NV, USA, 27–30 June 2016; pp. 770–778. [CrossRef]
23. Sorkhei, M.; Henter, G.E.; Kjellström, H. Full-Glow: Fully Conditional Glow for More Realistic Image Generation. In Proceedings of the Pattern Recognition, Bonn, Germany, 28 September–1 October 2021; Bauckhage, C., Gall, J., Schwing, A., Eds.; Springer International Publishing: Cham, Switzerland, 2021; pp. 697–711.
24. Lu, Y.; Huang, B. Structured Output Learning with Conditional Generative Flows. *Proc. AAAI Conf. Artif. Intell.* **2020**, *34*, 5005–5012. [CrossRef]
25. Karami, M.; Schuurmans, D.; Sohl-Dickstein, J.; Dinh, L.; Duckworth, D. Invertible Convolutional Flow. In *Advances in Neural Information Processing Systems*; Wallach, H., Larochelle, H., Beygelzimer, A., d' Alché-Buc, F., Fox, E., Garnett, R., Eds.; Curran Associates, Inc.: Red Hook, NY, USA, 2019; Volume 32.
26. Ioffe, S.; Szegedy, C. Batch Normalization: Accelerating Deep Network Training by Reducing Internal Covariate Shift. In Proceedings of the 32nd International Conference on International Conference on Machine Learning—ICML'15, Lille, France, 7–9 July 2015; Volume 37, pp. 448–456.
27. Zhang, L.; Goldstein, M.; Ranganath, R. Understanding Failures in Out-of-Distribution Detection with Deep Generative Models. In Proceedings of the 38th International Conference on Machine Learning, Virtual Event, 18–24 July 2021; Volume 139, pp. 12427–12436.
28. Havtorn, J.D.; Frellsen, J.; Hauberg, S.; Maaløe, L. Hierarchical VAEs Know What They Don't Know. In Proceedings of the 38th International Conference on Machine Learning, Virtual Event, 18–24 July 2021; Volume 139, pp. 4117–4128.
29. van den Hoogen, I.J.; van Rosendael, A.R.; Lin, F.Y.; Bax, J.J.; Shaw, L.J.; Min, J.K. Coronary Computed Tomography Angiography as a Gatekeeper to Coronary Revascularization: Emphasizing Atherosclerosis Findings Beyond Stenosis. *Curr. Cardiovasc. Imaging Rep.* **2019**, *12*, 24. [CrossRef] [PubMed]
30. Chang, H.J.; Lin, F.Y.; Gebow, D.; An, H.Y.; Andreini, D.; Bathina, R.; Baggiano, A.; Beltrama, V.; Cerci, R.; Choi, E.Y.; et al. Selective Referral Using CCTA Versus Direct Referral for Individuals Referred to Invasive Coronary Angiography for Suspected CAD: A Randomized, Controlled, Open-Label Trial. *JACC Cardiovasc. Imaging* **2019**, *12*, 1303–1312. [CrossRef] [PubMed]

Article

Towards a Deep-Learning Approach for Prediction of Fractional Flow Reserve from Optical Coherence Tomography

Cosmin-Andrei Hatfaludi [1,2,*], Irina-Andra Tache [1,3], Costin Florian Ciușdel [1], Andrei Puiu [1,2], Diana Stoian [1,2], Lucian Mihai Itu [1,2], Lucian Calmac [4,5], Nicoleta-Monica Popa-Fotea [4,5], Vlad Bataila [4] and Alexandru Scafa-Udriste [4,5]

1. Advanta, Siemens SRL, 15 Noiembrie Bvd, 500097 Brasov, Romania; irina.tache@upb.ro (I.-A.T.); costin.ciusdel@siemens.com (C.F.C.); andrei.puiu@siemens.com (A.P.); diana.stoian@siemens.com (D.S.); lucian.itu@siemens.com (L.M.I.)
2. Automation and Information Technology, Transilvania University of Brasov, Mihai Viteazu nr. 5, 5000174 Brasov, Romania
3. Department of Automatic Control and Systems Engineering, University Politehnica of Bucharest, 014461 Bucharest, Romania
4. Department of Cardiology, Emergency Clinical Hospital, 8 Calea Floreasca, 014461 Bucharest, Romania; lcalmac@gmail.com (L.C.); fotea.nicoleta@yahoo.com (N.-M.P.-F.); vladbataila@yahoo.co.uk (V.B.); alexscafa@yahoo.com (A.S.-U.)
5. Department Cardio-Thoracic, University of Medicine and Pharmacy "Carol Davila", 8 Eroii Sanitari, 050474 Bucharest, Romania
* Correspondence: cosmin.hatfaludi@unitbv.ro

Abstract: Cardiovascular disease (CVD) is the number one cause of death worldwide, and coronary artery disease (CAD) is the most prevalent CVD, accounting for 42% of these deaths. In view of the limitations of the anatomical evaluation of CAD, Fractional Flow Reserve (FFR) has been introduced as a functional diagnostic index. Herein, we evaluate the feasibility of using deep neural networks (DNN) in an ensemble approach to predict the invasively measured FFR from raw anatomical information that is extracted from optical coherence tomography (OCT). We evaluate the performance of various DNN architectures under different formulations: regression, classification—standard, and few-shot learning (FSL) on a dataset containing 102 intermediate lesions from 80 patients. The FSL approach that is based on a convolutional neural network leads to slightly better results compared to the standard classification: the per-lesion accuracy, sensitivity, and specificity were 77.5%, 72.9%, and 81.5%, respectively. However, since the 95% confidence intervals overlap, the differences are statistically not significant. The main findings of this study can be summarized as follows: (1) Deep-learning (DL)-based FFR prediction from reduced-order raw anatomical data is feasible in intermediate coronary artery lesions; (2) DL-based FFR prediction provides superior diagnostic performance compared to baseline approaches that are based on minimal lumen diameter and percentage diameter stenosis; and (3) the FFR prediction performance increases quasi-linearly with the dataset size, indicating that a larger train dataset will likely lead to superior diagnostic performance.

Keywords: deep-learning; few-shot learning; ensemble models; coronary artery disease; optical coherence tomography; fractional flow reserve

1. Introduction

Cardiovascular disease (CVD) is the number one cause of death worldwide, and coronary artery disease (CAD) is the most prevalent CVD, accounting for 42% of these deaths. In CAD patients, plaque builds up in the coronary arteries and limits the blood flow to the myocardium, especially when the demand is increased (exercise, stress). In severe cases, this can lead to myocardial infarction, or even death.

X-ray coronary angiography (XA) represents the gold standard in CAD imaging [1]. Optical coherence tomography (OCT) is used in certain scenarios in conjunction with XA.

OCT has the highest resolution among all invasive imaging modalities, allowing for a precise intra-vascular evaluation of stent apposition and expansion [2–4], thus, representing a paramount tool for PCI (percutaneous coronary intervention) optimization [5]. Nonetheless, its ability to assess the functional significance of a stenosis is not negligible [6].

The purely anatomical assessment of CAD, independent from the medical imaging modality, does not fully capture the functional significance of coronary stenoses. In view of the limitations of the anatomical evaluation of CAD, Fractional Flow Reserve (FFR) has been introduced as a functional index. FFR is defined as the ratio of flow in the stenosed branch at hyperemia—a condition of stress, with maximum coronary blood flow—to the hypothetical hyperemic flow in the same branch under healthy conditions. This can be shown to be closely approximated by the ratio of hyperemic cycle-averaged pressure distal to the stenosis to the cycle-averaged aortic pressure [7]. An FFR value ≤ 0.8 is considered to be positive, i.e., the patient requires invasive treatment, such as percutaneous coronary intervention (PCI-stenting) or coronary artery bypass graft (CABG). An FFR value > 0.8 is considered to be negative, i.e., typically only optimal medical therapy is prescribed. Several clinical trials have demonstrated the superiority of FFR-guided decision-making [8], which represents the current gold standard. However, although providing obvious advantages, studies indicate that the use of FFR is still relatively low due to the need to administer hyperemia-inducing drugs, additional costs, and the extended duration and invasive nature of the procedure [9]. Hence, computational approaches for FFR prediction have been introduced, relying either on computational fluid dynamics (CFD) or on artificial intelligence (AI).

Blood-flow computations, performed using CFD, when used in conjunction with patient-specific anatomical models that are extracted from medical images, have been proposed for diagnosis, risk stratification, and surgical planning [4]. Model-based assessment of coronary stenoses has been previously performed using such techniques in several clinical studies, based on anatomical models that are reconstructed from coronary computed tomography angiography (CCTA) [10–13], XA [14–18], or OCT [19–22]. Computed FFR has been the main quantity of interest in these studies, all of which showed that computed FFR has good diagnostic accuracy compared to invasively measured FFR. The CFD models consist of partial differential equations, which can be only numerically solved, leading to a large number of algebraic equations. Due to the time-consuming process that is employed for reconstructing the anatomical model, and the computationally intensive aspect of the CFD models [23,24], they are not used for intra-operative assessment and planning, where near real-time performance is required.

Alternatively, artificial intelligence-based solutions may be employed that are capable of providing results in real-time. To develop such solutions, a large database is required for the training phase, containing pairs of input-output data. The input data are represented by the anatomical information, while the output are invasive FFR [25]. Once the training phase has been finalized, the online usage provides results instantaneously. Such supervised machine learning (ML) algorithms are routinely employed in medical imaging applications, e.g., organ segmentation [26]. Moreover, machine learning models can also be employed to reproduce the behavior of non-linear computational models [27,28].

Recently, machine learning models for the prediction of FFR based on CCTA [29], XA [30], OCT [31], and intravascular ultrasound (IVUS) [32] have been introduced. All these approaches rely on the extraction of features describing the vascular geometry, specifically the arterial lumen, and, in some studies, also on patient features.

The goal of the present study is to evaluate the feasibility of using deep neural networks (DNN) to predict the invasively measured FFR from the radius of the coronary lumen that is extracted along the centerline of the coronary artery of interest. The starting point is represented by OCT images, the coronary lumen is then automatically extracted for each cross-section and subsequently processed to determine an equivalent radius value. The radius values are then arranged in a one-dimensional (1D) sequence, to be fed as input to the DNN. Our approach is in contrast to previous ML-based approaches for FFR prediction, since we use as input raw, reduced-order anatomical data instead of hand-crafted features. The second important aspect of the study is that we focus on intermediate lesions, for which the visual anatomical assessment of CAD based on XA does not allow for a clear clinical decision. As a result, the dataset contains a large number of lesions having an FFR value that is close to the cut-off of 0.8, making the prediction task more challenging.

Deep-learning (DL) is a class of machine learning algorithms that uses multiple layers to extract higher level features from the raw input [33]. The FFR prediction task can be formulated either as a regression problem (predict the exact value of FFR) or as a classification problem (predict the FFR class, e.g., binary classification: ≤0.8 or >0.8). There are several types of neural networks that are suitable for the FFR prediction, amongst others:

- fully connected neural network, commonly referred to as artificial neural networks (ANNs). Potential disadvantages of ANNs are the large number of trainable parameters, which leads to the requirement of large training datasets, and the difficulty in capturing the inherent properties in 1D/2D/3D data structures
- convolutional neural networks (CNNs). Compared to ANNs, CNNs can capture the inherent properties in 1D/2D/3D data structures, but still require relatively large training sets. Also, fixed size input data are required if the network is not fully convolutional.
- recurrent neural networks (RNNs) [34]. RNNs have the advantage that a variable length input sequence can be processed, but they may be affected by vanishing and exploding gradient issues.

Few-shot learning (FSL) is a type of learning where the prediction is performed based on a limited number of samples [35]. In a study that was published by Yang et al., the models that were used for FSL were classified into four categories: multitask learning, embedding learning, learning with external memory, and generative modeling.

OCT images were previously used in a variety of DL-based applications: stent strut detection [36,37], stent strut segmentation [38–40], coronary calcification segmentation [41,42], atherosclerotic plaque characterization [43], and lumen segmentation [44]. Furthermore, DL-based approaches were employed also in studies addressing other types of optical signals [45,46].

Herein, we evaluate the performance of ANNs, CNNs, and RNNs in both regression and classification formulations. Additionally, we also consider the use of FSL, focusing specifically on prototypical networks [47], a subcategory of the embedding learning models, considered the state of the art for classification tasks. More details that are related to prototypical networks are included in Appendix A.1.

2. Materials and Methods

2.1. Data Set

2.1.1. Study Design

This was a single-center, retrospective study that was carried out at the Clinical Emergency Hospital, Bucharest, Romania. The study complied with the Declaration of Helsinki for investigation in human beings. The study protocol was approved by the local ethics committee and each patient signed an informed consent form before the enrolment in the study.

2.1.2. Study Population

Patients at least 18 years old, with stable angina, and an indication for diagnostic XA due to intermediate or high likelihood of obstructive coronary artery disease, were considered. Further inclusion criteria were: at least one lesion with 40% to 80% diameter stenosis by visual assessment, and invasive FFR measurement considered required by the operator for clinical decision-making. Patients were excluded if they were unable to provide informed consent, had significant arrhythmia (heart rate over 120 bpm), suspected acute coronary syndrome, atrial fibrillation, low systolic pressure (below 90 mmHg), contraindication to beta blockers, nitroglycerin or adenosine, a non-cardiac illness with a life expectancy of less than 2 years, pathological aortic valve, rest state angina, or myocardial infarct during the last 6 months. Additionally, aorto-ostial lesions were excluded from the study. A total of 80 patients were included in the study.

2.1.3. Procedure Protocol

Coronary angiography (Siemens Artis Zee, Forchheim, Germany) was performed after iso-centering in posterior-anterior and lateral planes, via a transradial (preferred) or transfemoral approach. In all cases, a 6 French diagnostic catheter was used after intracoronary injection of glyceryl trinitrate according to routine practice in the hospital, with manual contrast injection and cine acquisition at a frame rate of 15 frames/second. OCT imaging was performed using a frequency-domain OCT systems (St. Jude Medical/Abbott, St. Paul, MN, USA). The fiber probe was pulled back at a constant speed and cross-sectional images were generated with a spacing of 0.2 mm.

The acquisition of physiological data for FFR calculation was performed according to conventional practice [48] with a commercially available FFR measurement system (PressureWire Aeris; St. Jude Medical, Minneapolis, MN, USA). The 0.014 coronary wire with a pressure tip was advanced until the pressure sensor passed the orifice of the guiding catheter. Transcatheter aortic and intracoronary pressure tracings were equalized. Subsequently, the guidewire was advanced into the respective coronary artery until the pressure sensor passed the index lesion. Hyperemia was induced by the administration of adenosine either intravenously at a constant rate of 140 µg/kg/min, or as an intracoronary bolus (100 µg for the right and 200 µg for the left coronary artery); the pressure recording was started, and the FFR was determined. A total of 102 coronary lesions in 80 patients underwent FFR analysis. This invasively measured FFR represents the ground truth that is used during the training of the deep neural networks, as described in the following.

2.2. Data Pre-Processing

The OCT data were exported from the OCT workstation available onsite. All OCT slices are RGB images, and the exported data contains the automatically detected coronary lumen, which is overlaid on the image and depicted in green. The spacing between the slices is 0.2 mm, and the number of slices per acquisition is constant at 376. Figure 1 displays the data pre-processing workflow starting from the exported OCT images with automatically detected lumen contour. First, the contours are automatically extracted by processing the green channel as follows: a threshold representing 90% of the maximum intensity value is used to create a binary image, and all the contours are extracted [49]. We then retain the contour which surrounds the center of the image: if there are multiple such contours, we pick the one with the largest area. Next, we use an in-house developed application to collect manual input that is provided by the clinical expert:

- selection of the proximal start and distal end slice, which define the coronary artery region of interest. Slices representing the catheter are excluded, alongside other slices with sub-optimal image quality (e.g., blood artifacts);
- rejecting/correcting erroneous contours within the selected slice-range: the automatically detected contours may be incorrect on certain slices, typically in bifurcation regions and/or if the lumen has a profoundly non-circular shape (e.g., concave shape). Erroneous bifurcation contours are rejected, while erroneous contours in the stenosis region are corrected (required in less than 10% of the OCT acquisitions).

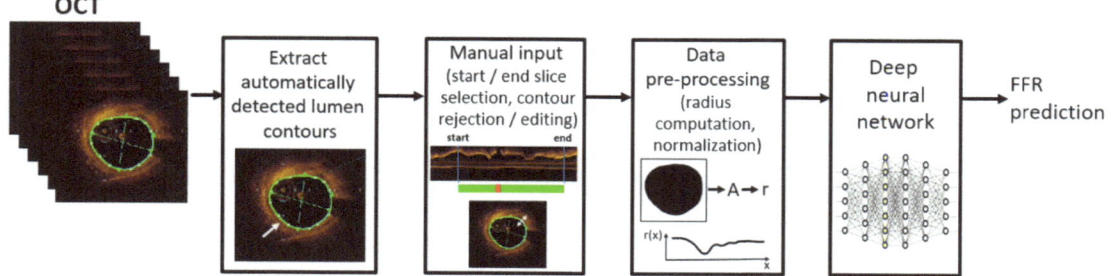

Figure 1. OCT data processing workflow, including FFR prediction using a deep neural network.

Next, the data are pre-processed: the inside area of each non-rejected lumen contour in the selected slice-range is computed and the effective radius is determined (considering an equivalent circular contour with identical area). The radius of rejected contours is set using linear interpolation that is applied on the radiuses of the closest neighboring contours that have not been rejected. The radiuses are then arranged in a 1D sequence, starting with the proximal slice of the selected slice-range. Since the OCT slices are equidistant, only the radius values are used as input. For the further processing using deep neural networks, the 1D radius sequence is padded to a size of 376 (maximum length of an OCT sequence), and z-score normalization is performed [50]. The mean and standard deviation of each acquisition are computed, and then a global mean and global standard deviation are computed for the training set by averaging the mean and standard deviation values of the acquisitions that are included in the training set. The acquisitions in the validation/test split are normalized using the values that are employed for the training set. The 1D sequence of normalized radius values is used as input for the deep neural network predicting FFR.

2.3. Deep Neural Network Based FFR Prediction

Different types of neural network models are considered for the prediction of the invasively measured FFR, ANNs, CNNs, and RNNs, applied with different approaches:

- a regression approach: models predict a rational number representing invasive FFR
- a classification approach: models predict the class of the FFR value (positive, i.e., FFR ≤ 0.8, or negative, i.e., FFR > 0.8)
- a FSL approach: similar to the classification approach.

As ANN, we used a fully connected neural network with 4 hidden layers, and the rectified linear unit (ReLU) [51] as the activation function for the hidden layers. The details of the ANN architecture are included in Appendix A (Table A1).

As CNN, we used a fully convolutional neural network (1D convolutions) with eight layers. For the hidden layers we used ReLU as activation function, and batch normalization was employed [52]. For the regression and the classification approach we added a final fully connected layer to perform the prediction. For the FSL approach, this layer is not required. The details of the CNN architectures are included in Appendix A (Tables A2 and A3).

As RNN, we included a bidirectional gated recurrent unit (GRU) [53] layer on top of the previously described fully convolutional neural network (referred to as CNN + RNN in the Appendix A). This avoids the padding requirement. The CNN layers learn the relevant features from the input, and then the RNN performs the final prediction based on those features. Training a fully RNN network was not possible considering the small size of the available dataset. For the regression and the classification approach we added a fully connected layer after the bidirectional GRU to perform the prediction. For the bidirectional GRU, we used ReLU as the activation function. The details of the RNN architecture are included in Appendix A (Table A4).

No activation function was used on the last layer for the regression approach, and the sigmoid function [54] was chosen for the classification approach. For the FSL approach, the output of the network is represented by the features from the last hidden layer. The class is then determined by the smallest Euclidean distance between the output of the network and the two class clusters. These are defined by the mean features of the training set samples of each class.

For the classification and FSL approaches, all the samples with invasive FFR ≤ 0.8 represent the positive class and all the samples with invasive FFR > 0.8 represent the negative class. Since the dataset consists of only 102 invasive values, the models are evaluated using the leave-one-out cross validation strategy that is applied at the patient level [55]. For each fold, the samples of one patient are moved to a validation set, while the model is trained for a fixed number of epochs (300) on the samples of the remaining patients. The classification accuracy is computed for each epoch, and the epoch leading to the highest accuracy on the entire dataset, i.e., all folds, is chosen for reporting the statistics. Additionally, only during training of the classification-based approaches, we also ignored the samples with invasive FFR values in the range 0.79–0.81 (six samples). By removing these samples that are close to the cut-off point, the model is able to learn to better discriminate between the classes. For all the models we used the Adam optimizer [56], mean squared error as a loss function for the regression approach, and cross entropy [57] for the classification and the FSL approach (more details are included in Appendix A.2). All the architectures were optimized using grid search [58], applied for: number of layers, number of neurons per layer, dropout percentage, and the learning rate. The implementation is based on Python, and the PyTorch [59] library for DL model training and inference.

To allow for a fair assessment of the performance, an ensemble approach is considered for each configuration: each of the proposed models is trained 20 times using different random seeds. For each configuration, the 20 models are then combined into one ensemble model. For regression approaches, the ensemble prediction for one sample is the mean value of the predictions of all 20 models. For classification and FSL approaches, the ensemble prediction for one sample is the mean value of the probabilities of all 20 models. This allows for a more robust assessment of the model performance, which is independent from the random seed that is used during training. The value 20 was chosen following experiments which indicated that the ensemble model performance did not change when using larger values.

For all the ensemble models, we performed the receiver operating characteristic (ROC) analysis [60] and we computed the area under the curve (AUC) score [61]. Based on the ROC curves, we selected for each ensemble model the optimal cut-off point as being the point closest to the point (0, 1) [62]. The reported model performance metrics are based on the optimal cut-off point. The formula that is used to determine the point closest to (0, 1) is [63]:

$$ER(c) = \sqrt{(1 - Se(c))^2 + (1 - Sp(c))^2} \qquad (1)$$

where ER is the closest point to (0, 1), c is a cut-point, Se is sensitivity, and Sp is specificity.

Similar to other studies, we further consider the minimum lumen diameter (MLD) and percentage diameter stenosis (%DS) as simple baseline references to assess the performance of the DL models. The %DS is computed as follow:

$$DS = (1 - r_{min}/r_{avg}) \times 100 \qquad (2)$$

where r_{min} is the minimum radius of the sequence, r_{avg} is the average of the proximal and distal reference radius values of the lesion, as extracted from the OCT data.

For both MLD and %DS, we also apply the leave-one-out cross validation strategy at the patient level, as follows: for each fold, a threshold value is chosen which balances sensitivity and specificity on the respective training set, and then this threshold is applied to classify the test sample(s).

To evaluate the results, we computed the diagnostic statistics (accuracy, sensitivity, specificity, negative predictive value (NPV), and positive predictive value (PPV) [64]) for all approaches, and additionally the mean absolute error (MAE), mean error (ME), and the mean squared error (MSE) for the regression approach. For the diagnostic statistics we additionally computed the 95% confidence intervals.

3. Results

3.1. Population Characteristics

Baseline patient and lesion characteristics are summarized in Tables 1 and 2: 80 patients (66 male, 14 female) with 102 lesions were included in this study. The mean patient age was 60.5 ± 11.2 years. The mean FFR was 0.80 ± 0.08, and 48 of the lesions were hemodynamically significant according to the criterion FFR ≤ 0.80.

Table 1. Baseline patient characteristics and risk factors ($n = 80$).

Male	66 (82%)
Female	14 (18%)
Age (years)	60.5 ± 11.2 years
Race	All Caucasian
Weight	81.93 ± 16.15 kg
Height	172.13 ± 8.05 cm
Diabetes	27 (33.75%)
Hypertension	60 (75%)
Hypercholesterolemia	62 (77.5%)
Smoking history	42 (52.5%)
Family history of CAD	3 (2.9%)
Previous myocardial infarction	46 (45%)
Previous Angina	64 (80%)
Ejection fraction	$48.28 \pm 6.31\%$

Table 2. Baseline lesion characteristics ($n = 102$).

Index Artery	
Left Anterior Descending artery (LAD)	57
Left Circumflex artery (LCx)	20
Right Coronary Artery (RCA)	25
Fractional Flow Reserve	
Mean $\pm$ SD	0.80 ± 0.08
Median (IQR)	0.83 (0.75–0.86)
FFR ≤ 0.80	48
FFR < 0.75	25
$0.75 \leq$ FFR ≤ 0.85	47
FFR > 0.85	30

3.2. Invasive FFR Prediction Performance

Figure 2 displays the ROC curve, the AUC scores including their 95% confidence intervals (CI), and the closest point to (0, 1) for all the approaches. The best three approaches based on AUC score are regression CNN, FSL RNN, and FSL CNN. Interestingly, the AUC score is superior for the regression CNN approach, but the FSL CNN approach has the closest point to (0, 1), i.e., the best diagnostic performance statistics, as shown below.

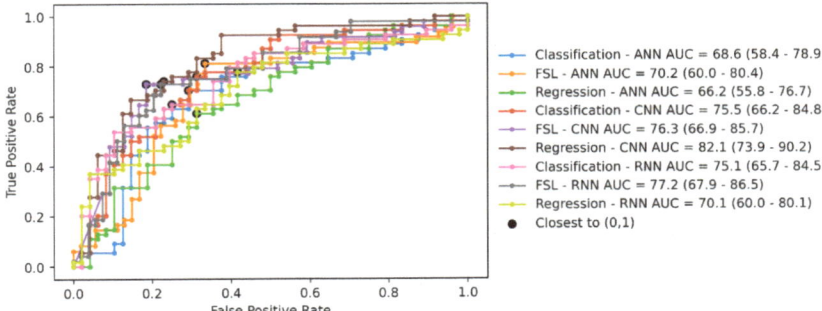

Figure 2. The ROC curve, AUC score, and the closest point to (0, 1) for all approaches. Values in the parentheses represent the 95% confidence intervals computed as in [65].

The performance and statistics of the various ensemble DL models and approaches considered herein are displayed in Table 3.

Table 3. Diagnostics and performance statistics of the considered ensemble DL models and approaches. Values in the parentheses represent the 95% confidence intervals.

Approach	Ensemble Arch.	Train_Accuracy [%]	Validation									
			Accuracy [%]	Sensitivity [%]	Specificity [%]	NPV [%]	PFV [%]	AUC [%]	MAE	ME	MSE	Corr.
Regression	ANN	73.7	64.7 (55.1–73.3)	61.1 (47.8–80.1)	68.8 (54.7–80.1)	61.1 (47.8–73.0)	68.8 (54.7–80.1)	66.2 (55.8–76.7)	0.062	0.007	0.105	0.273
	CNN	85.9	75.5 (66.3–82.8)	74.1 (61.1–86.7)	77.1 (63.5–86.7)	72.5 (59.1–82.9)	78.4 (65.4–87.5)	82.1 (73.9–90.2)	0.082	−0.008	0.015	0.342
	RNN	69.7	68.6 (59.1–76.8)	77.8 (65.1–71.2)	58.3 (44.3–71.2)	70.0 (54.6–81.9)	67.7 (55.4–78.0)	70.1 (60–80.1)	0.072	0.022	0.011	0.261
Classification	ANN	78.4	70.6 (61.1–78.6)	70.4 (57.2–81.8)	70.8 (56.8–81.8)	68.0 (54.2–79.2)	73.1 (59.7–83.2)	68.6 (58.4–78.9)	-	-	-	-
	CNN	98.7	72.5 (63.2–80.3)	75.9 (63.1–80.1)	68.8 (54.7–80.1)	71.7 (57.5–82.7)	73.2 (60.4–83.0)	75.5 (66.2–84.8)	-	-	-	-
	RNN	73.8	69.6 (60.1–77.7)	64.8 (51.5–85.1)	75.0 (61.2–85.1)	65.5 (52.3–76.6)	74.5 (60.5–84.7)	75.1 (65.7–74.5)	-	-	-	-
FSL	ANN	78.9	72.5 (63.2–80.3)	79.2 (65.7–77.8)	66.7 (53.4–77.8)	78.3 (64.4–87.7)	67.9 (54.8–78.6)	70.2 (60–80.4)	-	-	-	-
	CNN	78.6	**77.5 (68.4–84.5)**	72.9 (59.0–89.6)	81.5 (69.2–89.6)	77.2 (64.8–86.2)	77.8 (63.7–87.5)	76.3 (66.9–85.7)	-	-	-	-
	RNN	75.6	75.5 (66.3–82.8)	72.9 (59.0–86.8)	77.8 (65.1–86.8)	76.4 (63.7–85.6)	74.5 (60.5–84.7)	77.2 (60–80.1)	-	-	-	-

In terms of diagnostic performance, the FSL approach is performing better than classical regression and classification, while in terms of AUC, the CNN regression is superior to other methods. Since the 95% confidence intervals overlap, the differences are statistically not significant. FSL algorithms have been designed for optimal performance on small datasets where they tend to perform better than classic models. The best performing architecture is the one that is based on CNN. Furthermore, the training accuracy suggests that overfitting is not present for eight of the nine approaches. For the classic CNN-based classification, the model seems to overfit, even though different attempts were made to address this: L2 regularization and dropout. The confusion matrix for the best approach is depicted in Table 4.

Table 4. The confusion matrix for the FSL-CNN approach.

Predicted Values	Actual Values	
	Positive (1)	Negative (0)
Positive (1)	35	13
Negative (0)	11	44

For comparison, MLD has an accuracy of 67.64%, a sensitivity of 64.81%, and a specificity of 70.83%. The %DS has an accuracy of 63.72%, a sensitivity of 62.96%, and a specificity of 64.58%.

Each ensemble model consists of 20 models that were trained with different seed values. Table 5 displays the mean accuracy, the standard deviation (std) of the accuracy, the minimum accuracy (min), and the maximum accuracy (max) for the validation dataset when employing the default operating points/thresholds of 0.8 for regression and 0.5 for classification. While all variations are quite small, the smallest std is obtained for the models that are based on FSL, which further underlines the robustness of this approach. Additionally, we computed the ensemble model mean uncertainty by averaging the uncertainty of the ensemble model for each examination [66]. The ensemble model uncertainty for regression approaches is the standard deviation of the predictions of all models for one sample. An intuitive approximation for the ensemble model's uncertainty for classification and FSL approaches was chosen as:

$$Mean\ ensamble\ uncertainty = \sum_i^N \frac{abs(round(y(i)) - y(i))}{N}, \quad (3)$$

where $y(i)$ is the ensemble model prediction for each sample and N is the number of samples; this uncertainty measure is the distance between the output probability and the predicted class label (0 or 1), therefore, predictions such as 0.1 or 0.9 are considered "confident" while others such as 0.4 or 0.6 are considered more "uncertain". This approximation is feasible since ensemble models usually have well-calibrated outputs [66]. The ensemble uncertainty results of the regression approaches are not directly comparable to the ensemble uncertainty results for the classification and FSL approaches, and it has been also shown [66] that regression-based uncertainty that is computed as the ensemble predictions' standard deviation is not well-calibrated as the MSE training loss "is not a scoring rule that captures predictive uncertainty" [66]. For the regression approaches, RNNs tend to have the smallest uncertainty. For classification and FSL approaches the uncertainty is similar for five of the approaches, while FSL CNN has a much smaller uncertainty.

Table 5. Diagnostic performance statistics of the considered ensemble DL models and approaches.

		Accuracy				
Approach	Ensemble Arch.	Mean [%]	Std [%]	Min [%]	Max [%]	Uncertainty [%]
Regression	ANN	61.57	4.55	53.92	70.59	4.48
	CNN	61.76	2.65	55.88	65.69	12.91
	RNN	63.19	3.82	54.9	71.57	2.25
Classification	ANN	68.43	1.69	65.69	72.55	32.55
	CNN	67.75	3.1	63.73	73.53	32.9
	RNN	68.04	1.71	64.71	71.57	31.69
FSL	ANN	66.67	3.34	59.8	72.55	30.9
	CNN	75.59	1.2	72.55	76.47	2.77
	RNN	74.46	1.37	71.57	76.47	34.71

The reason the default thresholds were employed in Table 5 is that selecting a best-operating-point with respect to some metrics and some held-out test-set is part of a post-processing stage; uncertainty estimates, however, depend solely on two factors: the input samples (i.e., input noise, out-of-distribution, etc.) and the learned model (here, the training procedure, the network architecture, and especially the training set have a large influence); the ground-truth label of a test input sample has no influence on the prediction uncertainty. Therefore, for an unbiased assessment, uncertainty measures of all the approaches were computed from the raw ensemble predictions and compared with the mean accuracy that was obtained from using the default thresholds.

Figure 3 displays four sample cases: one for each of the categories true positive (TP), true negative (TN), false positive (FP), and false negative (FN). A representative angiographic frame is displayed, indicating the invasive FFR value and the coronary vessel and region of interest that is visualized using OCT. Further, the longitudinal OCT view and the radius profile that were used as input to the DNNs are displayed.

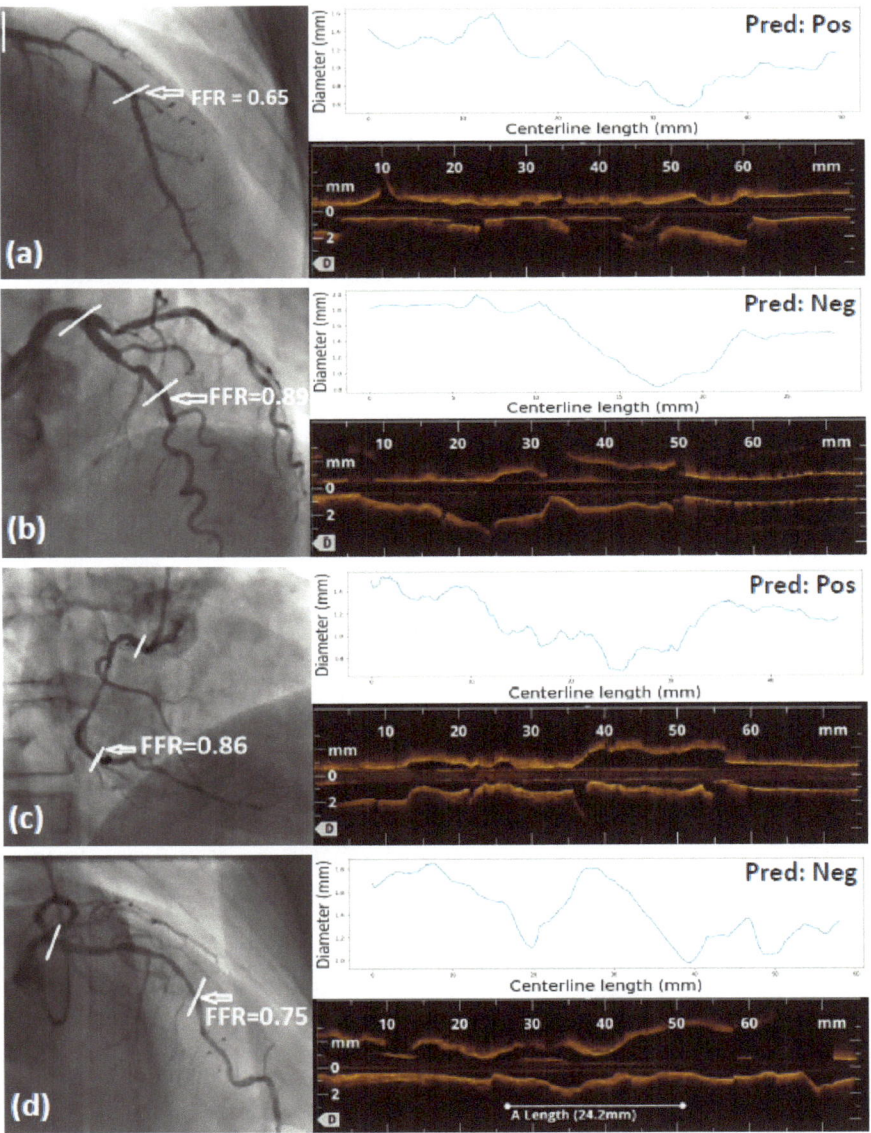

Figure 3. Four sample cases: one for each of the categories: (**a**) TP, (**b**) TN, (**c**) FP, and (**d**) FN. A representative angiographic frame is displayed indicating the invasive FFR value and the coronary vessel and region of interest visualized using OCT. The longitudinal OCT view and the radius profile used as input to the DNNs are also displayed.

3.3. Subgroup Analyses

In the following, we use the best performing model according to the results in Table 3 (FSL-CNN) to perform a series of subgroup analyses.

As detailed in Section 2.1, the dataset contains a large number of samples in the interval 0.75–0.85 (46%). Hence, we have computed the statistics separately for lesions with FFR < 0.75, lesions with FFR > 0.85, and for the lesions with intermediate values. The results are displayed in Table 6. As expected, the accuracy of the model increases in the two bins at the extremes.

Table 6. Diagnostic performance and 95% CI of the model for lesions with FFR < 0.75, lesions with FFR > 0.85, and for the lesions with intermediate values.

FFR Interval	Accuracy [%]	Sensitivity [%]	Specificity [%]
FFR > 0.85	86.6 (70.3–94.6)	N/A	86.6 (70.3–94.6)
0.75–0.85	68.0 (53.8–79.6)	60.8 (40.7–77.8)	75.0 (55.1–88.0)
FFR < 0.75	84.0 (65.3–93.6)	84.0 (65.3–93.6)	N/A

In another analysis, we assessed the performance as a function of the vessel on which the measurement was performed. The results are displayed in Table 7 and indicate a higher accuracy on the LCx, compared to the other two main coronary arteries. The literature suggests that the LCx has typically a smaller baseline and hyperemic flow velocity compared to the LAD and RCA, which impacts the FFR measurements [67]. In other words, the same radius profile will lead to different invasive FFR values on different arteries. Since the type of artery is not used as an input to the DNN, a performance difference is expected.

Table 7. Diagnostic performance and 95% CI of the model for the three main coronary arteries.

Coronary Artery	Accuracy [%]	Sensitivity [%]	Specificity [%]
LAD	75.4 (62.8–84.7)	76.4 (60.0–87.5)	73.9 (53.5–87.4)
LCX	85.0 (58.3–91.9)	80.0 (37.5–96.3)	86.6 (54.8–92.9)
RCA	76.0 (56.5–88.5)	55.5 (26.6–81.1)	87.5 (63.9–96.5)

Most of the measurements in the study were performed in the LAD. The clinical literature suggests that proximal LAD lesions are of particular interest for long-term patient outcome [68]. Hence, we have divided LAD lesions into proximal lesions and others (mid or distal lesion). The results are displayed in Table 8 and indicate a similar performance in terms of accuracy, but the sensitivity is slightly lower for proximal lesions. This is an expected outcome since literature indicates that a lesion with a certain severity will lead to smaller FFR values when it is located in the proximal LAD, compared to the mid and distal LAD. Hence, the model slightly underestimates the severity of proximal LAD lesions.

Table 8. Diagnostic performance and 95% CI of the model for different lesion locations on the LAD.

LAD Lesions Location	Accuracy [%]	Sensitivity [%]	Specificity [%]
proximal LAD	74.1 (56.7–86.2)	70.5 (46.8–86.7)	78.5 (52.4–92.4)
mid/distal LAD	76.9 (57.9–88.9)	82.3 (58.9–93.8)	66.6 (35.4–87.9)

In another analysis, we assessed the prediction performance for male and female patients. The results in Table 9 indicate that the model performs slightly better for male patients. This is an expected outcome since the vast majority of lesions are from male patients (82%).

Table 9. Diagnostic performance and 95% CI of the model as a function of patient sex.

Gender	Accuracy [%]	Sensitivity [%]	Specificity [%]
Male	78.8 (67.7–85.1)	73.8 (58.9–84.6)	83.7 (67.3–90.2)
Female	70.5 (46.8–86.7)	66.6 (29.9–90.3)	72.7 (43.4–90.2)

The age of the patient can be another important factor in the clinical decision-making. We have divided the data at the patient level into three equally large bins. The results in Table 10 indicate a marked difference between the three subgroups. The intermediate bin has a slightly larger number of intermediate lesions (18 vs. 15/14), partially explaining the difference in diagnostic performance.

Table 10. Diagnostic performance and 95% CI of the model as a function of age.

Age Interval	Accuracy [%]	Sensitivity [%]	Specificity [%]
<58	81.2 (64.6–91.1)	75.0 (53.2–88.8)	91.6 (64.6–98.5)
58–66	69.2 (50.9–79.3)	60.0 (35.7–80.1)	75.0 (50.8–85.0)
>66	83.8 (67.3–92.9)	84.6 (57.7–95.6)	83.3 (60.7–94.1)

Finally, in another subgroup analysis we have considered the centerline length of the input data and have divided the samples into three equally sized bins. The results in Table 11 display a balanced performance, i.e., the considered length has no major influence on the model performance.

Table 11. Diagnostic performance and 95% CI of the model as a function of the OCT sequence length.

Vessel Length [cm]	Accuracy [%]	Sensitivity [%]	Specificity [%]
<4.74	77.1 (57.9–85.8)	53.8 (29.1–76.7)	90.9 (66.6–92.5)
4.74–5.74	75.0 (57.8–86.7)	78.5 (52.4–92.4)	72.2 (49.1–87.5)
>5.74	79.4 (63.2–89.6)	80.9 (59.9–92.3)	76.9 (49.7–91.8)

3.4. Effect of Dataset Size

To assess the impact of the number of samples on the performance, we trained the best performing approach (CNN architecture with FSL) on datasets containing only a part of the original dataset. We started with 30% of the original dataset, and then increased the size in increments of 10%, until reaching 100%, i.e., the original dataset. The smaller datasets were set up by random sampling from the original dataset. To limit the selection bias, for each percentage we ran 20 experiments, where for each experiment a new random sampling was performed, and the CNN was initialized with a new random seed. The accuracy and the standard deviation for all the considered experiments is displayed in Figure 4.

As expected, the dataset size has an important impact on the accuracy. Encouragingly, a relatively linear increase in performance can be observed, indicating that with larger datasets, the performance should further increase. Moreover, the variation, i.e., standard deviation, decreases as the dataset size increases. This is motivated by two aspects. First, the smaller the percentage of data are, the larger is the variability of the actual dataset that is employed for the leave-one-out cross-validation. When 100% of the data are employed, the variability stems only from the random seed that is used for the initialization. Secondly, the larger the dataset, the more robust the prediction will be, i.e., with a smaller variability.

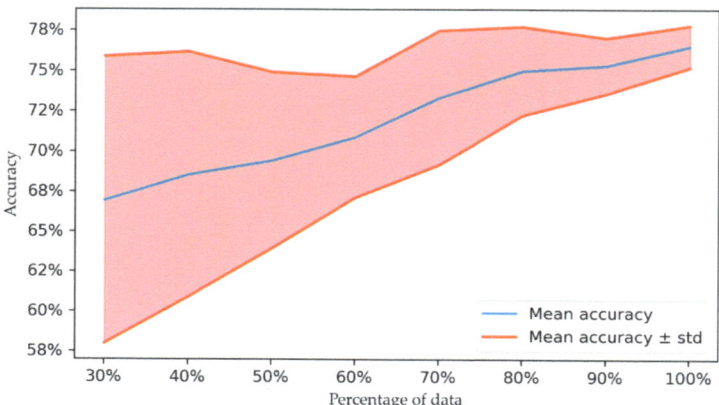

Figure 4. Model accuracy as a function of the dataset size.

3.5. Saliency Maps and Runtime

To analyze the features that the model is focusing on, we computed the saliency maps [69] for the best ensemble model (CNN-FSL). To obtain the saliency map for the ensemble model, we computed the derivative of the output with respect to the input for each individual model and then we averaged all saliency maps (see Figure 5). As expected, the output of the ensemble CNN-FSL model is influenced by all coronary diameters, but the gradient is larger in the stenosis area, which is known as the main determinant for the measured FFR values.

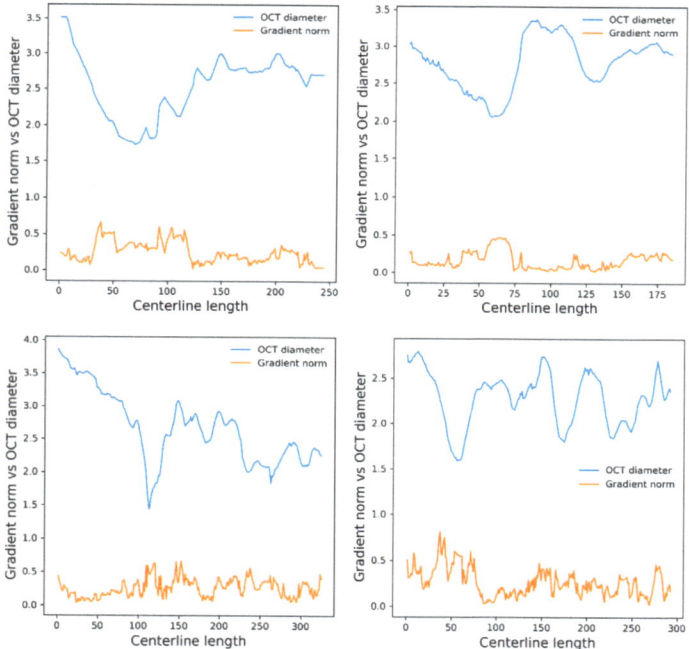

Figure 5. The saliency map that was computed for the ensemble CNN-FSL model. The saliency maps on the top correspond to samples with an invasive FFR > 0.8, and the saliency maps on the bottom correspond to samples with an invasive FFR < 0.8.

The training time for one-fold and one epoch is approximately 1050 ms for all the described approaches, the inference time for regression and classification approaches for one sample is approximately 2 ms, and the inference time for FSL approaches for one sample is approximately 25 ms. This difference of one order of magnitude is determined by the necessity of determining the classification clusters. All experiments were run on a desktop computer with AMD Ryzen 9 5900X CPU, 128 GB of RAM, and an NVIDIA RTX 3060 graphics card.

4. Discussion and Conclusions

4.1. Deep Learning-Based Prediction of FFR

As more data are emerging from studies that are based on artificial intelligence and computational modelling, the incremental diagnostic value of predicted coronary functional diagnostic indices over the traditional XA-based visual or quantitative lesion grading is becoming more evident.

We have introduced a method for the deep-learning-based prediction of FFR from routine optical coherence tomography. No specific requirements were formulated for the OCT acquisition. We demonstrated that this approach has a high potential in assessing functionally significant stenoses. Different models and approaches were proposed and evaluated. The experiments indicated the superiority of the FSL-based approach, a type of DL formulation that is specialized for small datasets. However, given the large overlap in the 95% confidence intervals, the differences between the methods are statistically not significant.

Thus, the main findings of this study can be summarized as follows: (1) DL-based FFR prediction from reduced-order raw anatomical data is feasible in a dataset that is focused on intermediate lesions for which the visual anatomical assessment of CAD based on XA does not allow for a clear clinical decision, and with no restriction on the type of lesions that were included in the study, and on the OCT acquisition; (2) DL-based FFR prediction provides superior diagnostic performance compared to baseline approaches based on MLD or %DS; (3) the FFR prediction performance increases quasi-linearly with the dataset size, indicating that a larger training dataset will likely lead to superior diagnostic performance.

The diagnostic accuracy of 77.5% achieved herein is lower compared to that of other studies focusing on FFR prediction from OCT, which reported an accuracy ranging between 88% and 95% [21,22,31,70,71]. There are two main aspects that are responsible for this difference. First, the complexity of the dataset that is processed herein is higher than that of other studies: 46% of the samples have an invasive FFR value ranging between 0.75 and 0.85, while in other studies these grey zone lesions represented between 20% and 44% of the entire dataset [21,22,31,70,71].

Secondly, past studies focusing on FFR prediction from OCT either rely on computational fluid dynamics (CFD) [21,22,70,71], or on ML-based approaches including hand-crafted features [31]. By applying a deep neural network directly on the raw data that are represented by the effective radius along the centerline of the vessel of interest, we allow the model to automatically learn powerful features for FFR prediction. The results that were obtained in other application areas (healthcare or others) demonstrate that classic machine learning (ML) techniques and hand-crafted features typically outperform DL-based approaches when the training set is small, but, conversely, the DL-based approaches outperform classic ML-based approaches when the size of the trainset increases significantly [70]. The results in Figure 4, depicting the accuracy as a function of the dataset size, confirm that a larger dataset will enable a better performance: the performance of the DL model increases quasi-linearly with the dataset size. As shown in Table 3, the diagnostic performance of the proposed model is already considerably higher outside of the 0.75–0.85 FFR value interval.

To increase the prediction performance of DL models, different types of regularization are employed in the literature: mathematical expressions added to the loss function (L1, L2 regularization) [71], dropout (used to randomly drop out neurons during training) [72], and data augmentation [73]. Herein, we have used L2 regularizations and dropout. Data augmentation, i.e., generating new samples by perturbing the input data, is difficult to perform when training against invasively measured FFR, since the approximation of the ground truth values is not straightforward. We have considered data augmentation by adding a small amount of noise to the 1D radius sequence used as input, but the results have not improved.

A DL- or ML-based prediction of FFR was considered also in studies relying on other types of medical images (CCTA, XA). Kumamaru et al. [74] proposed a DL model to estimate invasive FFR from CCTA. They had a dataset containing 207 measurements from 131 patients and have obtained an accuracy of 75.9% in predicting an abnormal invasive FFR (≤ 0.8). Another interesting approach was proposed by Zreik et al. [75], they used DL in an unsupervised manner and obtained an overall accuracy of 78% on CCTA data. They obtained an accuracy of 66% for FFR < 0.7, 75% for an FFR between 0.7 and 0.8, 79% for an FFR between 0.8 and 0.9, and 73% for an FFR > 0.9. Itu et al. [29] proposed a DL model that was trained on ground truth values computed with a CFD-based approach on a database of synthetically-generated coronary anatomies. They achieved an accuracy of 83.2% on CCTA data.

4.2. Clinical Impact

Despite the overwhelming clinical evidence that an FFR-guided revascularization strategy improves patient outcome, still the number of coronary interventions preceded by FFR measurements is relatively low due to the limitations of invasive pressure measurements [76]. Hence, a virtual functional index would increase the adoption of physiology-guided coronary interventions, while drastically reducing the requirement for invasive pressure measurements.

The proposed method is potentially well suited for a clinical setting, given the real-time prediction performance of the DL model. Certain manual steps are required in the current pipeline, but these can be automated using algorithms for image quality assessment, e.g., to exclude slices with blood artifacts, and more accurate lumen contour detection [77]. The approach only requires knowledge of the coronary luminal geometry, which can be extracted directly from OCT.

4.3. Limitations

The motivation to perform invasive FFR was clinical, which resulted in a large proportion of anatomically borderline lesions in a population with extensive atherosclerotic disease. No cases were excluded, and the results should be interpreted with the consideration that this was a retrospective single-center study.

The anatomical data that was used as input to the DL model may not always accurately reflect the true luminal geometry due to limitations of the OCT acquisition itself (heart motion during automatic pullback, sub-optimal calibration), and small errors that are introduced by the linear interpolation of radius values for the rejected contours. Furthermore, by using the effective radius information as input, we neglect the actual three-dimensional shape of the coronary lumen. The literature suggests that this has a small impact [78], but in certain samples, with non-circular lumen geometry, e.g., concave shape, the impact may not be negligible.

Moreover, the manual editing steps limit the real-time capabilities of the algorithm and introduce intra- and inter-observer variability.

While the subgroup analyses indicate that the length of the considered segment does not influence the results, the maximal length of 7.5 cm may represent a limitation in the case of serial stenoses. For example, if lesions are present in the proximal and distal segment of a vessel, a processed vessel length that is larger than the limit of 7.5 cm would be required to accurately predict FFR.

Finally, to validate our findings and to provide more representative results, the proposed method requires further validation in larger, prospective studies, that are conducted at multiple clinical sites.

4.4. Future Work

Multiple future directions can be defined, given also the current limitations that are listed above. First, the size of the training set should be increased to exploit the capabilities of a deep neural network-based approach. To limit the complexity of the input data, we currently use the effective radius, however, we envision the use of the coronary lumen mask as input, which may then allow the model to consider lumen non-circularities for the prediction. The dimensionality of the input data would increase from 1D to 3D, which would require a larger training set for enabling an accurate prediction. Furthermore, with the increase in the dataset size, other deep-learning approaches (evaluated herein or others) might lead to the best FFR prediction performance.

When employing a classification-based approach, another possible future direction is to increase the number of output classes. For example, a three-class approach would predict lesions as being functionally significant, functionally non-significant, or intermediate/uncertain. This would allow for the definition of hybrid decision-making strategy, where lesions which are not in the intermediate, i.e., uncertain class, can be confidently diagnosed, while for the ones in the intermediate class further aspects may be considered for the final decision, potentially even performing the invasive FFR measurement. The invasive FFR cut-off values for distinguishing the three classes may be chosen based on the performance of the model, e.g., to ensure a sensitivity/specificity of at least 95% for the lesions which are not in the intermediate class. The better the performance of the model, the closer the cut-off values may be to 0.8, i.e., the fewer lesions would be predicted as being uncertain.

Herein, we have considered only the coronary lumen information as input. Previous studies have demonstrated that FFR is influenced also by other patient characteristics (demographics, other pathologies, etc.) [31]. The results of the sub-group analyses have shown the patient sex and age and the vessel of interest may influence the prediction. Additional features may be considered directly as input into the deep neural network, or a cascaded modeling approach may be designed: the first model processes only the coronary lumen information, while the second model, which takes as input the output of the first model, processes all additional features to perform a final and more accurate prediction.

Standard OCT acquisitions have been used for obtaining the input data for the FFR prediction. OCT acquisition guidelines containing specific requirements (e.g., include the entire stenosis in the OCT sequence) may likely improve the prediction accuracy. Such an approach was successfully applied in a previous study [79].

The method that is described herein may be applied similarly on coronary lumen information that is extracted from other imaging modalities (XA, CCTA, IVUS). Since the image resolution, especially on XA and CCTA, is lower than on intra-vascular images, the coronary lumen information may be less accurate. However, XA and CCTA allow for a more complete evaluation of the coronary tree since the vessel of interest can be assessed in all its segments, alongside large side branches. A different methodology might lead to the optimal performance in that case, e.g., based on graph neural networks [80].

Finally, the approach can also be extended to predict other hemodynamic quantities, such as coronary flow reserve (CFR), rest Pd/Pa [81], the instantaneous wave-free ratio (iFR) [82], or hyperemic/basal stenosis resistance (HSR/BSR) [83,84], each of which can be used as a ground-truth during training.

Author Contributions: Conceptualization, C.-A.H., I.-A.T. and L.M.I.; methodology, C.-A.H., C.F.C., I.-A.T. and A.P.; software, C.-A.H., A.P. and D.S.; validation, L.C., N.-M.P.-F., V.B. and A.S.-U.; formal analysis, C.F.C.; investigation, L.C., N.-M.P.-F. and V.B.; resources, I.-A.T. and L.M.I.; data curation, I.-A.T. and L.C.; writing—original draft preparation, C.-A.H., I.-A.T. and L.M.I.; writing—review and editing, C.F.C.; visualization, C.-A.H. and L.M.I.; supervision, L.M.I. and A.S.-U.; project administration, L.M.I. and A.S.-U.; funding acquisition, L.M.I. and A.S.-U. All authors have read and agreed to the published version of the manuscript.

Funding: The research leading to these results has received funding from the EEA Grants 592 2014-2021, under Project contract no. 33/2021. This work was supported by a grant of the Romanian Ministry of Education and Research, CCCDI-UEFISCDI, project number PN-III-P2-2.1-PED-2019-2434, within PNCDI III.

Institutional Review Board Statement: Not applicable.

Informed Consent Statement: Not applicable.

Data Availability Statement: The data has been acquired as part of the projects acknowledged in the manuscript, and cannot be made public, considering GDPR regulations and the content of the informed consent signed by the patients.

Acknowledgments: The concepts and information presented in this paper are based on research results that are not commercially available. Future commercial availability cannot be guaranteed.

Conflicts of Interest: Irina-Andra Tache, Costin Florian Ciusdel, Andrei Puiu, Diana Stoian, and Lucian Mihai Itu are employees of Siemens SRL, Advanta, Brasov, Romania. Cosmin-Andrei Hatfaludi receives a scholarship from Siemens SRL, Advanta, Brasov, Romania. The other authors declare no conflict of interest.

Abbreviations

The following abbreviations are used in the manuscript:

CVD	Cardiovascular disease
CAD	Coronary artery disease
XA	X-ray coronary Angiography
OCT	Optical coherence tomography
PCI	Percutaneous coronary intervention
FFR	Fractional flow reserve
CABG	Coronary artery bypass graft
CFG	Computational fluid dynamics
CCTA	Coronary computed tomography angiography
ML	Machine Learning
IVUS	Intravascular ultrasound
DNN	Deep neural network
DL	Deep learning
ANN	Artificial neural network
CNN	Convolutional neural network
RNN	Recurrent neural network
FSL	Few-shot learning
ReLU	Rectified linear unit
GRU	Gated recurrent unit
MLD	Minimum lumen diameter
%DS	Percentage diameter stenosis
NPV	Negative predictive value
PPV	Positive predictive value
MAE	Mean absolute error
ME	Mean error
MSE	Mean squared error
LAD	Left Anterior Descending artery
LCx	Left Circumflex artery
RCA	Right Coronary Artery
Arch.	Architecture
Corr.	Correlation
TP	True positive

TN	True negative
FP	False positive
FN	False negative
CFR	Coronary flow reserve
iFR	Instantaneous wave-free ratio
HSR	Hyperemic stenosis resistance
BSR	Basal stenosis resistance
FC	Fully connected
BCE	Binary cross entropy
FoV	Field of view

Appendix A.

Table A1. ANN architecture. The layer that is highlighted in purple is used only for the regression and the classification approaches (not for the FSL approach). For the regression approach, we used no activation function and the activation function that is highlighted in green is used for the classification approach (not for the FSL approach).

Layer Index	Layer	Input Features	Output Features	Activation Function	Regularization
1	FC	376	32	ReLU	-
2	FC	32	64	ReLU	-
3	FC	64	128	ReLU	-
4	FC	128	256	ReLU	Dropout
5	FC	256	1	Sigmoid	-

Table A2. CNN architecture that is used for the FSL approach.

Layer Index	Layer	Kernel Size	Input Channels	Output Channels	Stride	Activation Function	Regularization	Normalization	Receptive FoV
1	Conv1D	3	1	64	2	ReLU	-	Batch norm	3
2	Conv1D	3	64	128	2	ReLU	-	Batch norm	7
3	Conv1D	3	128	256	2	ReLU	-	Batch norm	15
4	Conv1D	3	256	512	2	ReLU	-	Batch norm	31
5	Conv1D	3	512	512	2	ReLU	-	Batch norm	63
6	Conv1D	3	512	512	1	ReLU	-	Batch norm	127
7	Conv1D	3	512	512	1	ReLU	-	Batch norm	191
8	Conv1D	3	512	512	1	ReLU	-	Batch norm	255

Table A3. The fully connected layers that were added on top of the architecture that is presented in Table A2, for the CNN-based regression and classification. For the regression approach, we used no activation function and the activation function that is highlighted in green is used for the classification approach (not for the FSL approach).

Layer	Input Features	Output Features	Activation	Regularization
FC	2048	1024	ReLU	Dropout
FC	1024	1	Sigmoid	-

Table A4. The bidirectional GRU that was added on top of the architecture that is presented in Table A2, used for CNN + RNN approach. The layer that is highlighted in purple is only used for the regression and the classification approach (not for the FSL approach). For the regression approach, we used no activation function and the activation function that is highlighted in green is used for the classification approach (not for the FSL approach).

Layer	Input Features	Hidden Size	Output Features	Activation	Regularization
Bidirectional GRU	512	512	1024	-	Dropout
FC	1024	-	1	Sigmoid	-

Appendix A.1. Prototypical Networks

Prototypical networks [37] are a subcategory of the embedding learning models. Prototypical networks are used mainly for classification tasks, in both few-shot learning and zero-shot learning scenarios. These neural networks learn a metric space from the data, and then the classification is performed by computing distances to the prototype representations of each class, which are M-dimensional representations of each class cluster, based on an embedding function. They are computed by averaging the embedding vectors of all the training samples of a class (i.e., the neural network features predicted for the input data):

$$v_k = \frac{1}{N_s}\sum_{i=1}^{N_s} f_\phi(x_i), \tag{A1}$$

where v_k is the prototype of each class, f_ϕ is the embedding function, and x_i are the support images. The next step consists of classifying the query images. This is performed by computing the distance between each image and the prototypes:

$$p_\phi(y = k|x) = \frac{exp(-d(f_\phi(x), v_k))}{\sum_{k'} exp(-d(f_\phi(x), v_{k'}))} \tag{A2}$$

During training, the loss is computed using:

$$J(\phi) = -log\left(p_\phi(y = k|x)\right) \tag{A3}$$

where k is the true class.

Appendix A.2. Loss Functions

The loss function used for the regression approaches [46] is:

$$MSE = \frac{1}{n}\sum_{i=1}^{n}(Y_i - Y'_i)^2, \tag{A4}$$

where Y_i is the ground truth value and Y'_i is the predicted value.

The loss function used for the classification approach [46] (not for FSL the approach):

$$BCE = -\frac{1}{n}\sum_{i=1}^{n}(Y_i \cdot \ln Y'_i + (1 - Y_i) \cdot \ln(1 - Y'_i)), \tag{A5}$$

where Y_i is the ground truth value and Y'_i is the predicted value.

The loss function used for the FSL approaches is described in Equation (A1).

References

1. Ryan, T.J. The coronary angiogram and its seminal contributions to cardiovascular medicine over five decades. *Circulation* **2002**, *106*, 752–756. [CrossRef] [PubMed]
2. Gutierrez-Chico, J.L.; Alegría-Barrero, E.; Teijeiro-Mestre, R.; Chan, P.H.; Tsujioka, H.; de Silva, R.; Viceconte, N.; Lindsay, A.; Patterson, T.; Foin, N. Optical coherence tomography: From research to practice. *Eur. Heart J.-Cardiovasc. Imaging* **2012**, *13*, 370–384. [CrossRef]
3. Gutiérrez-Chico, J.L.; Regar, E.; Nüesch, E.; Okamura, T.; Wykrzykowska, J.; di Mario, C.; Windecker, S.; van Es, G.-A.; Gobbens, P.; Jüni, P. Delayed coverage in malapposed and side-branch struts with respect to well-apposed struts in drug-eluting stents: In vivo assessment with optical coherence tomography. *Circulation* **2011**, *124*, 612–623. [CrossRef] [PubMed]
4. Gutiérrez-Chico, J.L.; Wykrzykowska, J.; Nüesch, E.; van Geuns, R.J.; Koch, K.T.; Koolen, J.J.; di Mario, C.; Windecker, S.; van Es, G.-A.; Gobbens, P. Vascular tissue reaction to acute malapposition in human coronary arteries: Sequential assessment with optical coherence tomography. *Circ. Cardiovasc. Interv.* **2012**, *5*, 20–29. [CrossRef] [PubMed]
5. Ali, Z.A.; Maehara, A.; Généreux, P.; Shlofmitz, R.A.; Fabbiocchi, F.; Nazif, T.M.; Guagliumi, G.; Meraj, P.M.; Alfonso, F.; Samady, H. Optical coherence tomography compared with intravascular ultrasound and with angiography to guide coronary stent implantation (ILUMIEN III: OPTIMIZE PCI): A randomised controlled trial. *Lancet* **2016**, *388*, 2618–2628. [CrossRef]

6. Gonzalo, N.; Escaned, J.; Alfonso, F.; Nolte, C.; Rodriguez, V.; Jimenez-Quevedo, P.; Bañuelos, C.; Fernández-Ortiz, A.; Garcia, E.; Hernandez-Antolin, R. Morphometric assessment of coronary stenosis relevance with optical coherence tomography: A comparison with fractional flow reserve and intravascular ultrasound. *J. Am. Coll. Cardiol.* **2012**, *59*, 1080–1089. [CrossRef] [PubMed]
7. Pijls, N.H.; de Bruyne, B.; Peels, K.; van der Voort, P.H.; Bonnier, H.J.; Bartunek, J.; Koolen, J.J. Measurement of fractional flow reserve to assess the functional severity of coronary-artery stenoses. *N. Engl. J. Med.* **1996**, *334*, 1703–1708. [CrossRef] [PubMed]
8. Tonino, P.A.; De Bruyne, B.; Pijls, N.H.; Siebert, U.; Ikeno, F.; vant Veer, M.; Klauss, V.; Manoharan, G.; Engstrøm, T.; Oldroyd, K.G. Fractional flow reserve versus angiography for guiding percutaneous coronary intervention. *N. Engl. J. Med.* **2009**, *360*, 213–224. [CrossRef]
9. Tu, S.; Bourantas, C.V.; Nørgaard, B.L.; Kassab, G.S.; Koo, B.K.; Reiber, J. Image-based assessment of fractional flow reserve. *EuroIntervention J. EuroPCR Collab. Work. Group Interv. Cardiol. Eur. Soc. Cardiol.* **2015**, *11*, V50–V54. [CrossRef]
10. Yang, D.H.; Kim, Y.-H.; Roh, J.H.; Kang, J.-W.; Ahn, J.-M.; Kweon, J.; Lee, J.B.; Choi, S.H.; Shin, E.-S.; Park, D.-W. Diagnostic performance of on-site CT-derived fractional flow reserve versus CT perfusion. *Eur. Heart J.-Cardiovasc. Imaging* **2017**, *18*, 432–440. [CrossRef]
11. Coenen, A.; Lubbers, M.M.; Kurata, A.; Kono, A.; Dedic, A.; Chelu, R.G.; Dijkshoorn, M.L.; Gijsen, F.J.; Ouhlous, M.; van Geuns, R.-J.M. Fractional flow reserve computed from noninvasive CT angiography data: Diagnostic performance of an on-site clinician-operated computational fluid dynamics algorithm. *Radiology* **2015**, *274*, 674–683. [CrossRef] [PubMed]
12. Renker, M.; Schoepf, U.J.; Wang, R.; Meinel, F.G.; Rier, J.D.; Bayer II, R.R.; Möllmann, H.; Hamm, C.W.; Steinberg, D.H.; Baumann, S. Comparison of diagnostic value of a novel noninvasive coronary computed tomography angiography method versus standard coronary angiography for assessing fractional flow reserve. *Am. J. Cardiol.* **2014**, *114*, 1303–1308. [CrossRef] [PubMed]
13. Koo, B.-K.; Erglis, A.; Doh, J.-H.; Daniels, D.V.; Jegere, S.; Kim, H.-S.; Dunning, A.; DeFrance, T.; Lansky, A.; Leipsic, J. Diagnosis of ischemia-causing coronary stenoses by noninvasive fractional flow reserve computed from coronary computed tomographic angiograms: Results from the prospective multicenter DISCOVER-FLOW (Diagnosis of Ischemia-Causing Stenoses Obtained Via Noninvasive Fractional Flow Reserve) study. *J. Am. Coll. Cardiol.* **2011**, *58*, 1989–1997.
14. Tu, S.; Westra, J.; Yang, J.; von Birgelen, C.; Ferrara, A.; Pellicano, M.; Nef, H.; Tebaldi, M.; Murasato, Y.; Lansky, A. Diagnostic accuracy of fast computational approaches to derive fractional flow reserve from diagnostic coronary angiography: The international multicenter FAVOR pilot study. *Cardiovasc. Interv.* **2016**, *9*, 2024–2035.
15. Tröbs, M.; Achenbach, S.; Röther, J.; Redel, T.; Scheuering, M.; Winneberger, D.; Klingenbeck, K.; Itu, L.; Passerini, T.; Kamen, A. Comparison of fractional flow reserve based on computational fluid dynamics modeling using coronary angiographic vessel morphology versus invasively measured fractional flow reserve. *Am. J. Cardiol.* **2016**, *117*, 29–35. [CrossRef]
16. Papafaklis, M.I.; Muramatsu, T.; Ishibashi, Y.; Lakkas, L.S.; Nakatani, S.; Bourantas, C.V.; Ligthart, J.; Onuma, Y.; Echavarria-Pinto, M.; Tsirka, G. Fast virtual functional assessment of intermediate coronary lesions using routine angiographic data and blood flow simulation in humans: Comparison with pressure wire-fractional flow reserve. *EuroIntervention* **2014**, *10*, 574–583. [CrossRef]
17. Tu, S.; Barbato, E.; Köszegi, Z.; Yang, J.; Sun, Z.; Holm, N.R.; Tar, B.; Li, Y.; Rusinaru, D.; Wijns, W. Fractional flow reserve calculation from 3-dimensional quantitative coronary angiography and TIMI frame count: A fast computer model to quantify the functional significance of moderately obstructed coronary arteries. *JACC Cardiovasc. Interv.* **2014**, *7*, 768–777. [CrossRef]
18. Morris, P.D.; Ryan, D.; Morton, A.C.; Lycett, R.; Lawford, P.V.; Hose, D.R.; Gunn, J.P. Virtual fractional flow reserve from coronary angiography: Modeling the significance of coronary lesions: Results from the VIRTU-1 (VIRTUal Fractional Flow Reserve From Coronary Angiography) study. *JACC Cardiovasc. Interv.* **2013**, *6*, 149–157. [CrossRef] [PubMed]
19. Seike, F.; Uetani, T.; Nishimura, K.; Kawakami, H.; Higashi, H.; Aono, J.; Nagai, T.; Inoue, K.; Suzuki, J.; Kawakami, H. Intracoronary optical coherence tomography-derived virtual fractional flow reserve for the assessment of coronary artery disease. *Am. J. Cardiol.* **2017**, *120*, 1772–1779. [CrossRef]
20. Jang, S.-J.; Ahn, J.-M.; Kim, B.; Gu, J.-M.; Sung, H.J.; Park, S.-J.; Oh, W.-Y. Comparison of accuracy of one-use methods for calculating fractional flow reserve by intravascular optical coherence tomography to that determined by the pressure-wire method. *Am. J. Cardiol.* **2017**, *120*, 1920–1925. [CrossRef]
21. Yu, W.; Huang, J.; Jia, D.; Chen, S.; Raffel, O.C.; Ding, D.; Tian, F.; Kan, J.; Zhang, S.; Yan, F. Diagnostic accuracy of intracoronary optical coherence tomography-derived fractional flow reserve for assessment of coronary stenosis severity. *EuroIntervention J. EuroPCR Collab. Work. Group Interv. Cardiol. Eur. Soc. Cardiol.* **2019**, *15*, 189. [CrossRef] [PubMed]
22. Ha, J.; Kim, J.-S.; Lim, J.; Kim, G.; Lee, S.; Lee, J.S.; Shin, D.-H.; Kim, B.-K.; Ko, Y.-G.; Choi, D. Assessing computational fractional flow reserve from optical coherence tomography in patients with intermediate coronary stenosis in the left anterior descending artery. *Circ. Cardiovasc. Interv.* **2016**, *9*, e003613. [CrossRef] [PubMed]
23. Itu, L.; Sharma, P.; Mihalef, V.; Kamen, A.; Suciu, C.; Lomaniciu, D. A patient-specific reduced-order model for coronary circulation. In Proceedings of the 2012 9th IEEE International Symposium on Biomedical Imaging (ISBI), Barcelona, Spain, 2–5 May 2012; pp. 832–835.
24. Deng, S.-B.; Jing, X.-D.; Wang, J.; Huang, C.; Xia, S.; Du, J.-L.; Liu, Y.-J.; She, Q. Diagnostic performance of noninvasive fractional flow reserve derived from coronary computed tomography angiography in coronary artery disease: A systematic review and meta-analysis. *Int. J. Cardiol.* **2015**, *184*, 703–709. [CrossRef]
25. Bishop, C.M.; Nasrabadi, N.M. *Pattern Recognition and Machine Learning*; Springer: Berlin/Heidelberg, Germany, 2006; Volume 4.

26. Zheng, Y.; Comaniciu, D. *Marginal Space Learning for Medical Image Analysis*; Springer: Berlin/Heidelberg, Germany, 2014; Volume 2, p. 6.
27. Mansi, T.; Georgescu, B.; Hussan, J.; Hunter, P.J.; Kamen, A.; Comaniciu, D. Data-driven reduction of a cardiac myofilament model. In Proceedings of the International Conference on Functional Imaging and Modeling of the Heart, London, UK, 20–22 June 2013; pp. 232–240.
28. Tøndel, K.; Indahl, U.G.; Gjuvsland, A.B.; Vik, J.O.; Hunter, P.; Omholt, S.W.; Martens, H. Hierarchical Cluster-based Partial Least Squares Regression (HC-PLSR) is an efficient tool for metamodelling of nonlinear dynamic models. *BMC Syst. Biol.* **2011**, *5*, 90. [CrossRef] [PubMed]
29. Itu, L.; Rapaka, S.; Passerini, T.; Georgescu, B.; Schwemmer, C.; Schoebinger, M.; Flohr, T.; Sharma, P.; Comaniciu, D. A machine-learning approach for computation of fractional flow reserve from coronary computed tomography. *J. Appl. Physiol.* **2016**, *121*, 42–52. [CrossRef] [PubMed]
30. Cho, H.; Lee, J.G.; Kang, S.J.; Kim, W.J.; Choi, S.Y.; Ko, J.; Min, H.S.; Choi, G.H.; Kang, D.Y.; Lee, P.H. Angiography-based machine learning for predicting fractional flow reserve in intermediate coronary artery lesions. *J. Am. Heart Assoc.* **2019**, *8*, e011685. [CrossRef] [PubMed]
31. Cha, J.-J.; Son, T.D.; Ha, J.; Kim, J.-S.; Hong, S.-J.; Ahn, C.-M.; Kim, B.-K.; Ko, Y.-G.; Choi, D.; Hong, M.-K. Optical coherence tomography-based machine learning for predicting fractional flow reserve in intermediate coronary stenosis: A feasibility study. *Sci. Rep.* **2020**, *10*, 20421. [CrossRef] [PubMed]
32. Lee, J.-G.; Ko, J.; Hae, H.; Kang, S.-J.; Kang, D.-Y.; Lee, P.H.; Ahn, J.-M.; Park, D.-W.; Lee, S.-W.; Kim, Y.-H. Intravascular ultrasound-based machine learning for predicting fractional flow reserve in intermediate coronary artery lesions. *Atherosclerosis* **2020**, *292*, 171–177. [CrossRef]
33. Deng, L.; Yu, D. Deep learning: Methods and applications. *Found. Trends Signal Processing* **2014**, *7*, 197–387. [CrossRef]
34. LeCun, Y.; Bengio, Y.; Hinton, G. Deep learning. *Nature* **2015**, *521*, 436–444. [CrossRef]
35. Wang, Y.; Yao, Q.; Kwok, J.T.; Ni, L.M. Generalizing from a few examples: A survey on few-shot learning. *ACM Comput. Surv.* **2020**, *53*, 1–34. [CrossRef]
36. Jiang, X.; Zeng, Y.; Xiao, S.; He, S.; Ye, C.; Qi, Y.; Zhao, J.; Wei, D.; Hu, M.; Chen, F. Automatic detection of coronary metallic stent struts based on YOLOv3 and R-FCN. *Comput. Math. Methods Med.* **2020**, *2020*, 1793517. [CrossRef]
37. Wang, Z.; Jenkins, M.W.; Linderman, G.C.; Bezerra, H.G.; Fujino, Y.; Costa, M.A.; Wilson, D.L.; Rollins, A.M. 3-D stent detection in intravascular OCT using a Bayesian network and graph search. *IEEE Trans. Med. Imaging* **2015**, *34*, 1549–1561. [CrossRef] [PubMed]
38. Wu, P.; Gutiérrez-Chico, J.L.; Tauzin, H.; Yang, W.; Li, Y.; Yu, W.; Chu, M.; Guillon, B.; Bai, J.; Meneveau, N. Automatic stent reconstruction in optical coherence tomography based on a deep convolutional model. *Biomed. Opt. Express* **2020**, *11*, 3374–3394. [CrossRef] [PubMed]
39. Yang, G.; Mehanna, E.; Li, C.; Zhu, H.; He, C.; Lu, F.; Zhao, K.; Gong, Y.; Wang, Z. Stent detection with very thick tissue coverage in intravascular OCT. *Biomed. Opt. Express* **2021**, *12*, 7500–7516. [CrossRef] [PubMed]
40. Lau, Y.S.; Tan, L.K.; Chan, C.K.; Chee, K.H.; Liew, Y.M. Automated segmentation of metal stent and bioresorbable vascular scaffold in intravascular optical coherence tomography images using deep learning architectures. *Phys. Med. Biol.* **2021**, *66*, 245026. [CrossRef]
41. Lee, J.; Gharaibeh, Y.; Kolluru, C.; Zimin, V.N.; Dallan, L.A.P.; Kim, J.N.; Bezerra, H.G.; Wilson, D.L. Segmentation of Coronary Calcified Plaque in Intravascular OCT Images Using a Two-Step Deep Learning Approach. *IEEE Access* **2020**, *8*, 225581–225593. [CrossRef]
42. Gharaibeh, Y.; Prabhu, D.S.; Kolluru, C.; Lee, J.; Zimin, V.; Bezerra, H.G.; Wilson, D.L. Coronary calcification segmentation in intravascular OCT images using deep learning: Application to calcification scoring. *J. Med. Imaging* **2019**, *6*, 045002. [CrossRef]
43. Abdolmanafi, A.; Duong, L.; Ibrahim, R.; Dahdah, N. A deep learning-based model for characterization of atherosclerotic plaque in coronary arteries using optical coherence tomography images. *Med. Phys.* **2021**, *48*, 3511–3524. [CrossRef]
44. Pociask, E.; Malinowski, K.P.; Ślęzak, M.; Jaworek-Korjakowska, J.; Wojakowski, W.; Roleder, T. Fully automated lumen segmentation method for intracoronary optical coherence tomography. *J. Healthc. Eng.* **2018**, *2018*, 1414076. [CrossRef]
45. Jiao, C.; Xu, Z.; Bian, Q.; Forsberg, E.; Tan, Q.; Peng, X.; He, S. Machine learning classification of origins and varieties of Tetrastigma hemsleyanum using a dual-mode microscopic hyperspectral imager. *Spectrochim. Acta Part A Mol. Biomol. Spectrosc.* **2021**, *261*, 120054. [CrossRef] [PubMed]
46. Wang, T.; Shen, F.; Deng, H.; Cai, F.; Chen, S. Smartphone imaging spectrometer for egg/meat freshness monitoring. *Anal. Methods* **2022**, *14*, 508–517. [CrossRef] [PubMed]
47. Snell, J.; Swersky, K.; Zemel, R. Prototypical networks for few-shot learning. *Adv. Neural Inf. Processing Syst.* **2017**, *30*.
48. Kern, M.J.; Lerman, A.; Bech, J.-W.; De Bruyne, B.; Eeckhout, E.; Fearon, W.F.; Higano, S.T.; Lim, M.J.; Meuwissen, M.; Piek, J.J. Physiological assessment of coronary artery disease in the cardiac catheterization laboratory: A scientific statement from the American Heart Association Committee on Diagnostic and Interventional Cardiac Catheterization, Council on Clinical Cardiology. *Circulation* **2006**, *114*, 1321–1341. [CrossRef] [PubMed]
49. Bradski, G. The openCV library. *Dr. Dobb's J. Softw. Tools Prof. Program.* **2000**, *25*, 120–123.
50. Patro, S.; Sahu, K.K. Normalization: A preprocessing stage. *arXiv* **2015**, arXiv:1503.06462. [CrossRef]
51. Agarap, A.F. Deep learning using rectified linear units (relu). *arXiv* **2018**, arXiv:1803.08375.

52. Santurkar, S.; Tsipras, D.; Ilyas, A.; Madry, A. How does batch normalization help optimization? *Adv. Neural Inf. Processing Syst.* **2018**, *31*.
53. Dey, R.; Salem, F.M. Gate-variants of gated recurrent unit (GRU) neural networks. In Proceedings of the 2017 IEEE 60th International Midwest Symposium on Circuits and Systems (MWSCAS), Boston, MA, USA, 6–9 August 2017; pp. 1597–1600.
54. Han, J.; Moraga, C. The influence of the sigmoid function parameters on the speed of backpropagation learning. In Proceedings of the International Workshop on Artificial Neural Networks, Malaga-Torremolinos, Spain, 7–9 June 1995; pp. 195–201.
55. Wong, T.-T. Performance evaluation of classification algorithms by k-fold and leave-one-out cross validation. *Pattern Recognit.* **2015**, *48*, 2839–2846. [CrossRef]
56. Zhang, Z. Improved adam optimizer for deep neural networks. In Proceedings of the 2018 IEEE/ACM 26th International Symposium on Quality of Service (IWQoS), Banff, AB, Canada, 4–6 June 2018; pp. 1–2.
57. Kline, D.M.; Berardi, V.L. Revisiting squared-error and cross-entropy functions for training neural network classifiers. *Neural Comput. Appl.* **2005**, *14*, 310–318. [CrossRef]
58. Liashchynskyi, P.; Liashchynskyi, P. Grid search, random search, genetic algorithm: A big comparison for NAS. *arXiv* **2019**, arXiv:1912.06059.
59. Paszke, A.; Gross, S.; Massa, F.; Lerer, A.; Bradbury, J.; Chanan, G.; Killeen, T.; Lin, Z.; Gimelshein, N.; Antiga, L. Pytorch: An imperative style, high-performance deep learning library. *Adv. Neural Inf. Processing Syst.* **2019**, *32*.
60. Hoo, Z.H.; Candlish, J.; Teare, D. What is an ROC curve? *Emerg. Med. J.* **2017**, *34*, 357–359. [CrossRef] [PubMed]
61. Lobo, J.M.; Jiménez-Valverde, A.; Real, R. AUC: A misleading measure of the performance of predictive distribution models. *Glob. Ecol. Biogeogr.* **2008**, *17*, 145–151. [CrossRef]
62. Unal, I. Defining an optimal cut-point value in ROC analysis: An alternative approach. *Comput. Math. Methods Med.* **2017**, *2017*, 3762651. [CrossRef] [PubMed]
63. Youden, W.J. Index for rating diagnostic tests. *Cancer* **1950**, *3*, 32–35. [CrossRef]
64. Wong, H.B.; Lim, G.H. Measures of diagnostic accuracy: Sensitivity, specificity, PPV and NPV. *Proc. Singap. Healthc.* **2011**, *20*, 316–318. [CrossRef]
65. Genders, T.S.; Spronk, S.; Stijnen, T.; Steyerberg, E.W.; Lesaffre, E.; Hunink, M.M. Methods for calculating sensitivity and specificity of clustered data: A tutorial. *Radiology* **2012**, *265*, 910–916. [CrossRef]
66. Lakshminarayanan, B.; Pritzel, A.; Blundell, C. Simple and scalable predictive uncertainty estimation using deep ensembles. *Adv. Neural Inf. Processing Syst.* **2017**, *30*.
67. Wieneke, H.; Von Birgelen, C.; Haude, M.; Eggebrecht, H.; Mohlenkamp, S.; Schmermund, A.; Bose, D.; Altmann, C.; Bartel, T.; Erbel, R. Determinants of coronary blood flow in humans: Quantification by intracoronary Doppler and ultrasound. *J. Appl. Physiol.* **2005**, *98*, 1076–1082. [CrossRef]
68. Kobayashi, Y.; Johnson, N.P.; Berry, C.; De Bruyne, B.; Gould, K.L.; Jeremias, A.; Oldroyd, K.G.; Pijls, N.H.; Fearon, W.F.; Investigators, C.S. The influence of lesion location on the diagnostic accuracy of adenosine-free coronary pressure wire measurements. *JACC Cardiovasc. Interv.* **2016**, *9*, 2390–2399. [CrossRef] [PubMed]
69. Simonyan, K.; Vedaldi, A.; Zisserman, A. Deep inside convolutional networks: Visualising image classification models and saliency maps. *arXiv* **2013**, arXiv:1312.6034.
70. Bote-Curiel, L.; Munoz-Romero, S.; Gerrero-Curieses, A.; Rojo-Álvarez, J.L. Deep learning and big data in healthcare: A double review for critical beginners. *Appl. Sci.* **2019**, *9*, 2331. [CrossRef]
71. Demir-Kavuk, O.; Kamada, M.; Akutsu, T.; Knapp, E.-W. Prediction using step-wise L1, L2 regularization and feature selection for small data sets with large number of features. *BMC Bioinform.* **2011**, *12*, 412. [CrossRef] [PubMed]
72. Srivastava, N.; Hinton, G.; Krizhevsky, A.; Sutskever, I.; Salakhutdinov, R. Dropout: A simple way to prevent neural networks from overfitting. *J. Mach. Learn. Res.* **2014**, *15*, 1929–1958.
73. Wong, S.C.; Gatt, A.; Stamatescu, V.; McDonnell, M.D. Understanding data augmentation for classification: When to warp? In Proceedings of the 2016 International Conference on Digital Image Computing: Techniques and Applications (DICTA), Gold Coast, Australia, 30 November–2 December 2016; pp. 1–6.
74. Kumamaru, K.K.; Fujimoto, S.; Otsuka, Y.; Kawasaki, T.; Kawaguchi, Y.; Kato, E.; Takamura, K.; Aoshima, C.; Kamo, Y.; Kogure, Y. Diagnostic accuracy of 3D deep-learning-based fully automated estimation of patient-level minimum fractional flow reserve from coronary computed tomography angiography. *Eur. Heart J.-Cardiovasc. Imaging* **2020**, *21*, 437–445. [CrossRef] [PubMed]
75. Zreik, M.; van Hamersvelt, R.W.; Khalili, N.; Wolterink, J.M.; Voskuil, M.; Viergever, M.A.; Leiner, T.; Išgum, I. Deep learning analysis of coronary arteries in cardiac CT angiography for detection of patients requiring invasive coronary angiography. *IEEE Trans. Med. Imaging* **2019**, *39*, 1545–1557. [CrossRef]
76. Petraco, R.; Park, J.J.; Sen, S.; Nijjer, S.; Malik, I.; Pinto, M.E.; Asrress, K.; Nam, C.W.; Foale, R.; Sethi, A. Hybrid iFR-FFR decision-making strategy: Implications for enhancing universal adoption of physiology-guided coronary revascularization. *Am. J. Cardiol.* **2013**, *111*, 54B. [CrossRef]
77. Guo, X.; Tang, D.; Molony, D.; Yang, C.; Samady, H.; Zheng, J.; Mintz, G.S.; Maehara, A.; Wang, L.; Pei, X. A machine learning-based method for intracoronary oct segmentation and vulnerable coronary plaque cap thickness quantification. *Int. J. Comput. Methods* **2019**, *16*, 1842008. [CrossRef]
78. Lyras, K.G.; Lee, J. An improved reduced-order model for pressure drop across arterial stenoses. *PLoS ONE* **2021**, *16*, e0258047. [CrossRef]

79. Gutiérrez-Chico, J.L.; Chen, Y.; Yu, W.; Ding, D.; Huang, J.; Huang, P.; Jing, J.; Chu, M.; Wu, P.; Tian, F. Diagnostic accuracy and reproducibility of optical flow ratio for functional evaluation of coronary stenosis in a prospective series. *Cardiol. J.* **2020**, *27*, 350–361. [CrossRef] [PubMed]
80. Georgousis, S.; Kenning, M.P.; Xie, X. Graph deep learning: State of the art and challenges. *IEEE Access* **2021**, *9*, 22106–22140. [CrossRef]
81. Kern, M.J. Coronary physiology revisited: Practical insights from the cardiac catheterization laboratory. *Circulation* **2000**, *101*, 1344–1351. [CrossRef] [PubMed]
82. Sen, S.; Escaned, J.; Malik, I.S.; Mikhail, G.W.; Foale, R.A.; Mila, R.; Tarkin, J.; Petraco, R.; Broyd, C.; Jabbour, R. Development and validation of a new adenosine-independent index of stenosis severity from coronary wave–intensity analysis: Results of the ADVISE (ADenosine Vasodilator Independent Stenosis Evaluation) study. *J. Am. Coll. Cardiol.* **2012**, *59*, 1392–1402. [CrossRef]
83. Meuwissen, M.; Siebes, M.; Chamuleau, S.A.; van Eck-Smit, B.L.; Koch, K.T.; de Winter, R.J.; Tijssen, J.G.; Spaan, J.A.; Piek, J.J. Hyperemic stenosis resistance index for evaluation of functional coronary lesion severity. *Circulation* **2002**, *106*, 441–446. [CrossRef]
84. van de Hoef, T.P.; Nolte, F.; Damman, P.; Delewi, R.; Bax, M.; Chamuleau, S.A.; Voskuil, M.; Siebes, M.; Tijssen, J.G.; Spaan, J.A. Diagnostic accuracy of combined intracoronary pressure and flow velocity information during baseline conditions: Adenosine-free assessment of functional coronary lesion severity. *Circ. Cardiovasc. Interv.* **2012**, *5*, 508–514. [CrossRef]

Article

Balancing Data through Data Augmentation Improves the Generality of Transfer Learning for Diabetic Retinopathy Classification

Zahra Mungloo-Dilmohamud [1,*], Maleika Heenaye-Mamode Khan [1], Khadiime Jhumka [1], Balkrish N. Beedassy [2], Noorshad Z. Mungloo [2] and Carlos Peña-Reyes [3,4]

[1] Faculty of Information, Communication and Digital Technologies, University of Mauritius, Réduit 80837, Mauritius; m.mamodekhan@uom.ac.mu (M.H.-M.K.); khadhiime@gmail.com (K.J.)
[2] Ministry of Health and Wellness, Quatre Bornes 72259, Mauritius; drkrisb12@gmail.com (B.N.B.); noorshad@hotmail.com (N.Z.M.)
[3] School of Management and Engineering Vaud (HES-SO), University of Applied Sciences and Arts Western Switzerland Vaud, 1400 Yverdon-les-Bains, Switzerland; carlos.pena@heig-vd.ch
[4] CI4CB—Computational Intelligence for Computational Biology, SIB—Swiss Institute of Bioinformatics, 1015 Lausanne, Switzerland
* Correspondence: z.mungloo@uom.ac.mu

Citation: Mungloo-Dilmohamud, Z.; Heenaye-Mamode Khan, M.; Jhumka, K.; Beedassy, B.N.; Mungloo, N.Z.; Peña-Reyes, C. Balancing Data through Data Augmentation Improves the Generality of Transfer Learning for Diabetic Retinopathy Classification. *Appl. Sci.* **2022**, *12*, 5363. https://doi.org/10.3390/app12115363

Academic Editors: Lucian Mihai Itu, Constantin Suciu and Anamaria Vizitiu

Received: 30 March 2022
Accepted: 10 May 2022
Published: 25 May 2022

Publisher's Note: MDPI stays neutral with regard to jurisdictional claims in published maps and institutional affiliations.

Copyright: © 2022 by the authors. Licensee MDPI, Basel, Switzerland. This article is an open access article distributed under the terms and conditions of the Creative Commons Attribution (CC BY) license (https:// creativecommons.org/licenses/by/ 4.0/).

Abstract: The incidence of diabetes in Mauritius is amongst the highest in the world. Diabetic retinopathy (DR), a complication resulting from the disease, can lead to blindness if not detected early. The aim of this work was to investigate the use of transfer learning and data augmentation for the classification of fundus images into five different stages of diabetic retinopathy. The five stages are No DR, Mild nonproliferative DR, Moderate nonproliferative DR, Severe nonproliferative DR and Proliferative. To this end, deep transfer learning and three pre-trained models, VGG16, ResNet50 and DenseNet169, were used to classify the APTOS dataset. The preliminary experiments resulted in low training and validation accuracies, and hence, the APTOS dataset was augmented while ensuring a balance between the five classes. This dataset was then used to train the three models, and the best three models were used to classify a blind Mauritian test datum. We found that the ResNet50 model produced the best results out of the three models and also achieved very good accuracies for the five classes. The classification of class-4 Mauritian fundus images, severe cases, produced some unexpected results, with some images being classified as mild, and therefore needs to be further investigated.

Keywords: deep learning; diabetic retinopathy; retinal fundus images; transfer learning; data augmentation

1. Introduction

Diabetes is one of the most challenging health problems in the world, impacting roughly 537 million individuals according to the IDF Diabetes Atlas Tenth edition 2021 (Diabetes Atlas, 2021). According to the same atlas, countries have spent over USD 966 billion on diabetes patients worldwide, a 316 percent increase over the previous 15 years, and yet diabetes will be responsible for 6.7 million deaths in 2021, or 1 death every 5 s. Diabetes poses a danger to the health-care systems of low- and middle-income nations, which account for 75 percent of the world's diabetic population, resulting in many cases going undetected. The most common complication in advanced or uncontrolled diabetic patients is diabetic retinopathy, one of the leading cause of vision loss worldwide, accounting for 21.8 percent of patients across the globe [1]. With Mauritius currently ranking fifth in the global standardized diabetes prevalence among ages 20–79 in 2019 and predicted to reach the second position in 2030 [2], diabetic retinopathy is a serious threat to Mauritians. This is especially true for people in their working years, since this group is more susceptible as per the article "Global estimates of the prevalence of diabetes for 2010 and 2030 in Diabetes Atlas". Patients who have had vision loss as a result of this condition typically have a late diagnosis of diabetes or are unaware that they have diabetes and eye difficulties. A recent study [3] found that diagnosing retinopathy early can prevent or delay a substantial amount of vision loss. This can also help to speed up the healing process or halt disease development. However, establishing a precise diagnosis and the stage of the disease is difficult. Ophthalmologists conduct screenings by visually inspecting the fundus and evaluating colour images. They rely on detecting the presence of microaneurysms, small saccular outpouching of capillaries, retinal haemorrhages and ruptured blood vessels, among many indicators, in the fundoscopic images. This manual method, however, results in inconsistency among readers [4] and is costly and time-consuming. To address the growing number of undiagnosed retinal patients, early disease identification and treatment are critical.

Advancements in convolutional neural networks (CNNs), a type of deep learning, has motivated researchers to use them in medical image analysis for different tasks, amongst which is image classification of diabetic retinopathy. CNNs exhibit a better performance, but they also need a lot of computing resources and large datasets to train. Transfer learning (TL) strategies have been proposed to solve this problem [5–7]. It involves using a previously learned model, on different images, to train a new model. The traits learned by pre-training on the large dataset can be transferred to the new network, where only the classification component is trained on the new smaller dataset, to fine-tune the new data [7]. TL reduces the amount of time spent constructing and training a deep CNN model as well as the computing resources needed. The visual geometry group (VGG) [8], inception modules (GoogleNet) [9], residual neural network (ResNet) [10] and neural architecture search network (NasNetLarge) [10] are examples of the many high-performing pre-trained models found in the literature. In 2017, Masood et al. [11] applied a pre-trained Inception V3 model on the Eye-PACS fundus dataset and achieved an accuracy of 48.2%. Meanwhile, Li et al. [12] investigated the use of transfer learning for identifying DR by comparing several network topologies, such as AlexNet, VGG-S, VGG16 and VGG19, to two datasets: the Messidor and DR1 datasets. With an area under the curve (AUC) of 98.34%, the VGG-S architecture scored the best AUC for the Messidor dataset while an AUC score of 97.86% was obtained for the DR1 dataset. Similarly, in 2019, using the EYE-PACS dataset, Challa et al. [13] proposed a deep All-CNN architecture for DR classification. The model obtained an accuracy of 86.64%, a loss of 0.46 and an average F1 score of 0.6318. Meanwhile, using the Asia Pacific Tele-Ophthalmology Society 2019 Blindness Detection (APTOS 2019 BD) dataset [14], Kassani et al. [15] described a classification method using a modified Xception architecture model, which is an extension of the Inception architecture, on the dataset and obtained an accuracy of 83.09%, a sensitivity of 88.24% and a specificity of 87.00%. Khalifa et al. [16] implemented transfer learning using four pre-trained models, namely AlexNet, Res-Net18, SqueezeNet and GoogleNet. AlexNet obtained the highest accuracy of 97.9%. In Hagos et al. [17], a pre-trained Inception V3 model was applied to a subset of the APTOS dataset for DR classification, and the accuracy was 90.9% and the loss was 3.94%. Sikder et al. [18] presented a method incorporating the ExtraTree classifier, which is a popular ensemble learning algorithm based on decision trees and bagging learning techniques, and achieved a classification accuracy of 91%. In 2020, Shaban et al. [19] proposed a modified version of the VGG-19 that achieved an accuracy of 88%–89% when both 5-fold, and 10-fold cross validation methods were used, respectively. Using the same APTOS 2019 BD dataset, Mushtaq et al. [20] achieved a classification accuracy of 90% using a pre-trained Dense169 model. Before they trained the images, the latter were pre-processed by removing the black border and applying Gaussian blur filter. Moreover, Thota et al. [21] fine-tuned a pre-trained VGG16 model for classifying the severity of DR. An average class accuracy of 74%, sensitivity of 80%, specificity of 65% and AUC of 0.80 were achieved. Gangwar et al. [22] developed a novel deep learning hybrid model with pre-trained Inception-ResNet-v2 as a base model and it obtained a test accuracy of 72.33%

on Messidor-1 and 82.18% on the APTOS dataset. On the other hand, Dai et al. [23] used a deep learning model based on the ResNet architecture to classify fundus images into five different classes. Images were obtained from the Shanghai Integrated Diabetes Prevention and Care System study. Firstly, the different features, such as microaneurysm, hard exudate and haemorrhage were detected, and then they concatenated the model used and the base model for DR classification. The model achieved AUCs of 0.943, 0.955, 0.960 and 0.972, for mild, moderate, severe and proliferative cases. Benson et al. [24] discussed the usage of transfer learning by using a pre-trained Inception V3 on the DR dataset obtained from the VisionQuest Biomedical database. The model classified fundus images into six classes including identifying scars, and it achieved a sensitivity and specificity of 90%, with an AUC of 95%.

The reviews described above highlight the fact that all work carried out to date was for images from a specific country, and hence they were not targeted at a local multiracial population such as Mauritius [25,26]. Therefore, this research work makes the following contributions:

(1) Application of three pre-trained models, VGG16, DenseNet169 and ResNet50, on a publicly available diabetic retinopathy dataset and the data-augmented version of the dataset to solve the class imbalance problem;
(2) Enhance the pre-trained models to improve the performance obtained in (1);
(3) Apply the enhanced models on a blind Mauritian local cohort to predict the different stages of diabetic retinopathy;
(4) Compare the predicted results obtained for the Mauritian dataset using the enhanced models to an actual ophthalmologist's diagnosis.

The paper is structured as follows. Section 2 presents the proposed solution and describes the different components. Section 3 discusses the experimental results. Finally, Section 4 concludes the paper.

2. Materials and Methods

This section highlights the methodology used in implementing deep transfer learning for classification.

2.1. Proposed Workflow and Components

Figure 1 shows the proposed workflow for the system, which can accept different datasets. For this work, two datasets, the APTOS original dataset and a constructed Mauritian dataset, were used. The data were first pre-processed, and data augmentation was applied to the APTOS dataset only. Next, three pre-trained models were applied to the original and augmented APTOS dataset. The results were analyzed, and the models were tuned to reach their ideal minima. The enhanced models were then applied to the blind testing data from the APTOS dataset and the labelled Mauritian dataset, which was not used for the training phase.

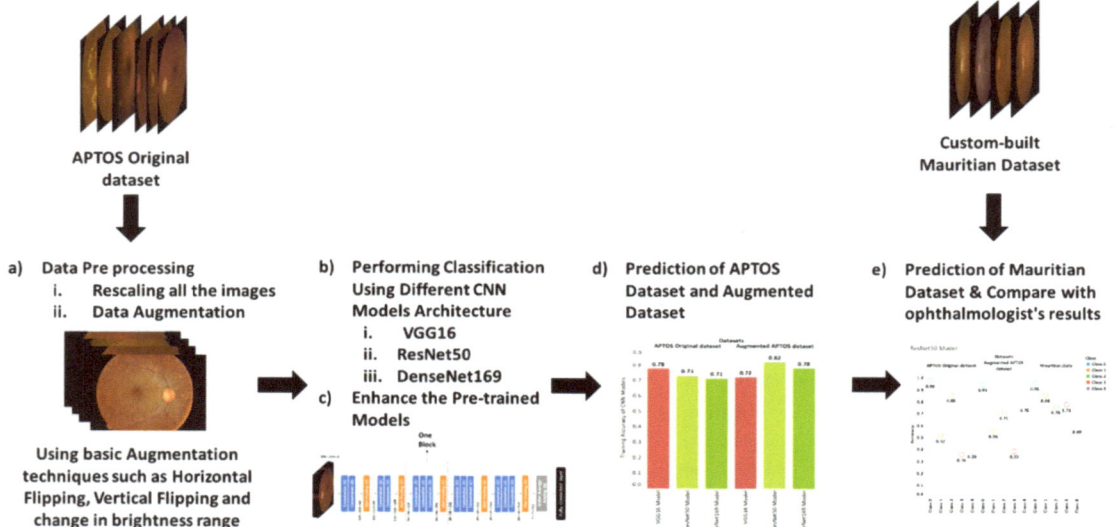

Figure 1. Workflow of our proposed system.

The workflow shown in Figure 1 is as follows: (1) data pre-processing and augmentation (for the APTOS dataset only); (2) training and enhancing the CNN models using the original and augmented APTOS dataset; (3) analyzing results; and (4) classification of the images for the 3 datasets and comparison to actual data.

2.2. Datasets

In this research work, two fundus image datasets were used. The first dataset was the APTOS 2019 diabetic retinopathy dataset, which is publicly available online on Kaggle (https://www.kaggle.com/c/aptos2019-blindness-detection/data, accessed on 17 February 2022). This dataset was selected among the other publicly available datasets since it is from India, which is close to the Mauritian population in terms of ethnicity. The second dataset was created locally from the images obtained from the hospitals in Mauritius. Each image in the APTOS 2019 dataset was assigned a class label of 0–4 according to the severity of the disease, as shown in Figure 2. Each image from the local cohort was also assigned a class label of 0–4 by a local doctor. The original dataset obtained from Kaggle is termed as the original APTOS dataset. The class distribution of the original APTOS dataset is illustrated in Figure 2.

Figure 2 reveals that, despite the data belonging to five different classes, the number of samples in each class varied substantially, resulting in an unbalanced dataset. As discussed in [27–29], an unbalanced dataset leads to a high misclassification rate and sub-optimal performance. To mitigate this challenge, we applied data augmentation, which is one possible solution to this problem. Traditional data augmentation techniques, namely horizontal and vertical flipping and changes in the brightness range [30], were applied to the original APTOS datasets to produce the augmented APTOS dataset.

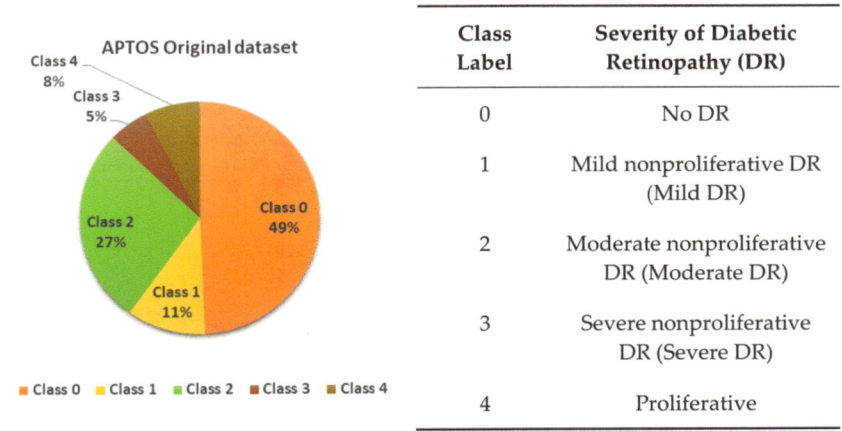

Figure 2. Original APTOS dataset.

Table 1 shows the total number of images for each class in the original APTOS dataset, the augmented APTOS dataset and the local Mauritian dataset. We divided both the APTOS dataset and the augmented APTOS dataset into a training set and testing set. There were 3662 images in the original Aptos dataset, whereby 70% (2563 images) were considered for training and 30% (1099 images) were taken for the testing phase. For the augmented APTOS dataset, data augmentation was performed on the training set only as performed by Gangwar et al. [22]. Only the data from classes 1, 3 and 4 were augmented since the model could not correctly classify these 3 classes in the original APTOS dataset. All the images in these 3 classes were augmented. In this paper, we used two sets for testing data, one which is made up of fundus images from the APTOS 2019 dataset (the remaining 30% of which were not used as training data) and the second being the Mauritian dataset composed of fundus images obtained from a local hospital in Mauritius. Table 1 presents the image count for each class in the training and testing data for the original and augmented APTOS datasets as well as the Mauritian dataset.

Table 1. Number of images class-wise in the 3 datasets.

Training Data	Number of Images in Training/Validation Dataset					Number of Images in Testing Dataset				
	Class 0	Class 1	Class 2	Class 3	Class 4	Class 0	Class 1	Class 2	Class 3	Class 4
Original APTOS dataset	1265	272	697	138	191	540	98	302	55	104
	Total images—2563					Total images—1099				
Augmented APTOS dataset	1265	1306	697	935	1264	540	98	302	55	104
	Total images—5467					Total images—1099				
Mauritian dataset	No training performed using Mauritian data					54	62	45	12	33
						Total images—208				

Figure 3 presents the number of images in each of the 5 classes after the application of data augmentation on the original APTOS dataset. It can be observed that the augmented dataset was more balanced.

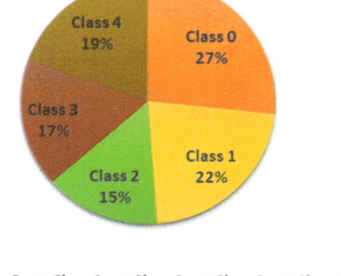

Figure 3. Augmented APTOS dataset.

2.3. Data Pre-Processing

The images were subjected to a pre-processing phase to improve their quality. They were resized as each model accepted images of different resolutions. For the ResNet50 Model, the images were resized to 512 × 512 pixels, whereas they were resized to 224 × 224 pixels for the VGG16 and DenseNet169 models. Another reason for performing pre-processing was the varying size and resolution of photos collected from the Kaggle website. These pictures ranged from 474 × 358 pixels to 3388 × 2588 pixels in width and height. After pre-processing the images, the different CNN models were applied to the training data of the two APTOS datasets to perform classification.

2.4. Transfer Learning Using ResNet50, VGG16 and DenseNet169

In this paper, transfer learning (TL) using the architectures of the three CNNs models, ResNet50, VGG16 and DenseNet169, was applied to the diabetic retinopathy images. In TL, learned features from one task are applied to a different task without having to learn from scratch. This is commonly used when building CNN models since the process of training from scratch requires a lot of computational resources, large datasets and a lot of time [31]. CNN models consist of multiple layers, namely: the convolution layer, pooling layer and fully connected layer. CNN models employ multiple perceptrons to evaluate picture inputs and eventually extract different patterns from the images to output to the fully connected layer. Our CNN models extracted representative patterns to form the feature maps. A 3 × 3 kernel was passed over the input matrix of the diabetic retinopathy image, as illustrated in Figure 4.

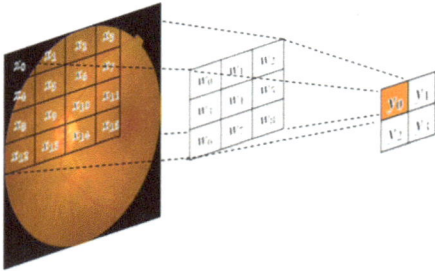

Figure 4. Convolution layer.

The classification function, which is the output of the fully connected layer, plays an important role in the process, whereby the different patterns of the five stages of diabetic retinopathy, learnt by the feature extraction layers, are used to perform the multiclass classification.

The VGG16 model, a CNN architecture pre-trained on the ImageNet dataset, was adopted for the development of our diabetic retinopathy application as it has been fully tested in a similar domain, achieving good performance [32,33]. VGG16 consists of 13 convolutional layers and 3 fully connected layers. There are 5 blocks each containing 2 or 3 convolution layers and ending with a max-pooling layer, as illustrated by Figure 5. A fixed-size image of dimensions (224, 224, 3) is the input to the VGG16 model.

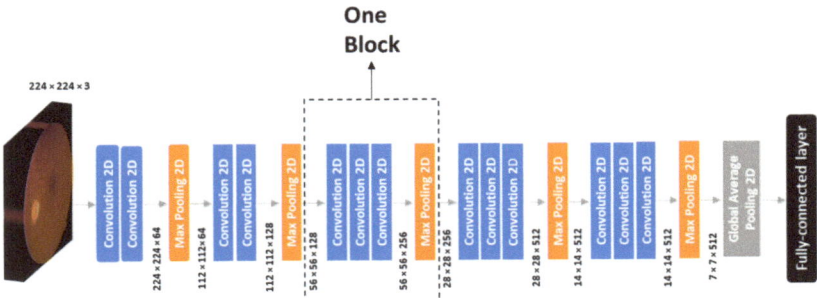

Figure 5. Architecture of the VGG16 model.

ResNet50, another popular CNN architecture, consists of 50 layers organized in so-called residual blocks [9]. It is known for its skip connection approach, which eventually solves the vanishing gradient problem. ResNet50 contains 48 convolution layers along with 1 MaxPool and 1 AveragePool layer. This was desired in our diabetic retinopathy application as it allows the later layers to learn lesser semantic information that was captured in the early layers. A 3×3 filter was used to perform the spatial convolution, which was eventually reduced using the max-pooling method. Figure 6 illustrates the ResNet 50 model with the 48 convolution layers and the 16 skip connections.

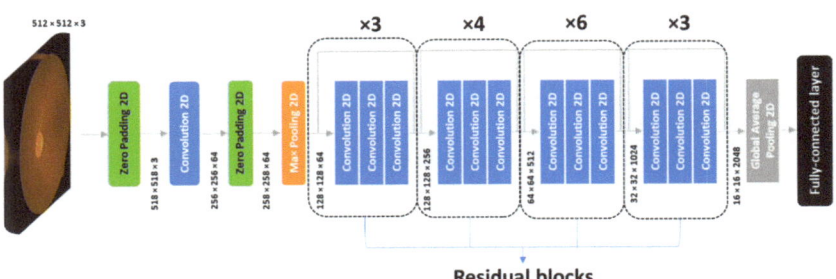

Figure 6. Architecture of ResNet50 model.

The third model that was considered was the DenseNet169 model [34]. Compared to the ResNet50 model, it has more layers. However, it contains a similar block to skip connections called the dense block. With the increase in the number of layers, it gives the model the opportunity to learn more distinctive features. In fact, the architecture consists of four dense blocks with varying numbers of layers as illustrated in Figure 7. Our design for this model consisted of the 2D average pooling, which is in the original architecture, where a dropout layer set to 0.5 was added.

Figure 7. Architecture of DenseNet169 model.

2.5. Enhanced CNN Models

Initially, the architectures of VGG16, ResNet50 and DenseNet169 were applied to the APTOS dataset. To be able to use these architectures in transfer learning and for classifying the diabetic retinopathy images into five classes, fully connected layers were added. The 3-dimensional feature map obtained from the last convolutional layer was converted to one dimension by using global average pooling 2D and passed to a series of a dropout layer, a dense layer and a dropout layer and finally to a dense layer with five nodes, representing the normal and the DR grades. The fully connected layers were selected as in the ResNet model in Taormina et al. [35], and Zhang et al. [36] shows that adding fully connected layers yields better results. The activation function used in the last dense layer was Softmax, as used in ElBedwehy et al. [37] for face detection classification. The Adam optimizer was applied to the 3 models with a learning rate of 1×10^{-3}, and the loss entropy used was the categorical cross entropy. In this work, data balancing was performed using basic image manipulation techniques [38]. In the deep neural network, the Adam optimizer was used instead of the stochastic gradient descent (SGD) since the former is computationally more efficient. The Adam optimizer has been found to be faster in converging the algorithm to the minima, hence reducing the training time [39]. The use of the SGD and other approaches will be explored in future works. Here, only the last 5 layers, namely the global average pooling 2D, dropout, dense, dropout and dense layers, were trained. The other layers were frozen as we were only extracting the features from the base model. These steps resulted in the models producing the relevant learnable parameters during the training process. For example, for the ResNet model, out of the 27,794,309 parameters, 4,206,597 were trainable. In this work, the sequential modelling approach was adopted for adding and customizing the convolution, dropout, dense and optimizer layers. The sequential model is appropriate for a plain stack of layers whereby each layer has exactly one input tensor and one output tensor, which was the case in this application.

To improve the performance of the models and cater for underfitting/overfitting, the 3 models were fine-tuned. The Adam optimizer was again used but this time with a learning rate of 10^{-4}. The learning rate was decremented by 10 as this has been shown to both reduce the risk of overfitting [40] and to improve classification [41]. When the validation loss metric stopped improving, the learning rate was halved as in [42]. Several parameters were changed and added to the models for fine-tuning. Firstly, the loss function was changed to binary cross entropy. Using the latter along with a SoftMax classifier helped the model in reducing the cross entropy loss of each iteration in multiclass classification [43]. Afterwards, an early stopping feature was added to end training when the network began to overfit the data according to the validation loss [44]. Eventually, all the convolutional layers were unfrozen, and the models were set to be trained.

The enhanced transfer learning model that was trained on the augmented APTOS dataset was tested on APTOS test data and on a blind Mauritian test datum annotated by a medical practitioner.

3. Results and Discussions

To evaluate the trained models both before and after fine-tuning, the accuracy regarding training, validation and test sets was calculated. Classification accuracy is the fraction of predictions that a given model predicted correctly. Firstly, a custom-built CNN model similar to that developed by Jayalakshmi et al. [45] was used. The same fully connected layers as in the case of our pre-trained models were joined, and the hyperparameters were tuned to obtain the optimal accuracy. A classification accuracy of 0.73 was obtained here. The model was only able to correctly predict classes 0 and 2. Although the accuracy is quite satisfactory for a binary classification of DR and NoDR, this custom-built model is very limited in the case of a multiclass DR classification. Next, pre-trained networks were implemented. The training and validation accuracy obtained before fine-tuning of the pre-trained networks are illustrated in Figure 8. From the results, it was found that the accuracies were quite low for the models ResNet50 and DenseNet169. Hence, it was deduced that these models were underfitting.

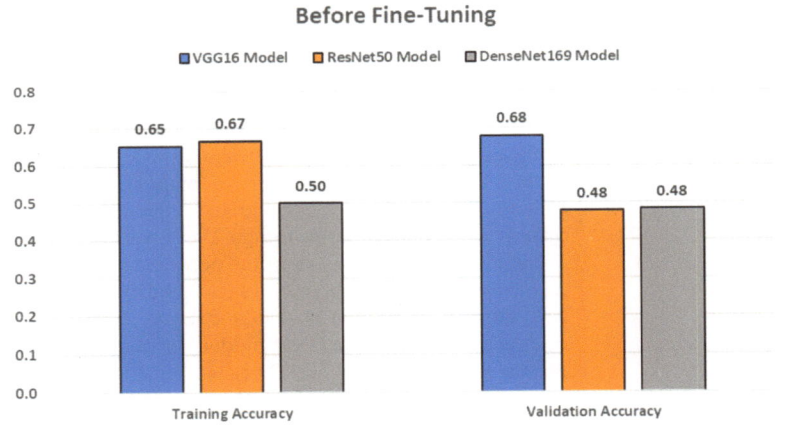

Figure 8. Overall training and validation accuracy before fine-tuning for the original APTOS dataset (after 2 epochs).

Consequently, the models were enhanced, and the weights were adjusted. Different learning rates were applied and evaluated to reach the minima. In addition, the number of epochs were adjusted while analyzing the different accuracies, thus fine-tuning the models. Each model was trained on the same training set used in the previous process. Figure 9 shows the results obtained for training and validation accuracy for each of the three models after fine-tuning.

From Figures 8 and 9, it can be clearly seen that fine-tuning the models improved both the training and the validation classification accuracy of the three models for the original APTOS dataset. We also noticed that using the augmented data improved the generality of transfer learning for the models for both the training and validation data. This can be deduced from the accuracy for the augmented dataset being maintained or increasing across all models compared to the original dataset. Furthermore, ResNet, with the highest accuracy in all cases, showed a better generalization. In parallel, it was also observed that the time taken to train the model decreased considerably (by at least 3 h).

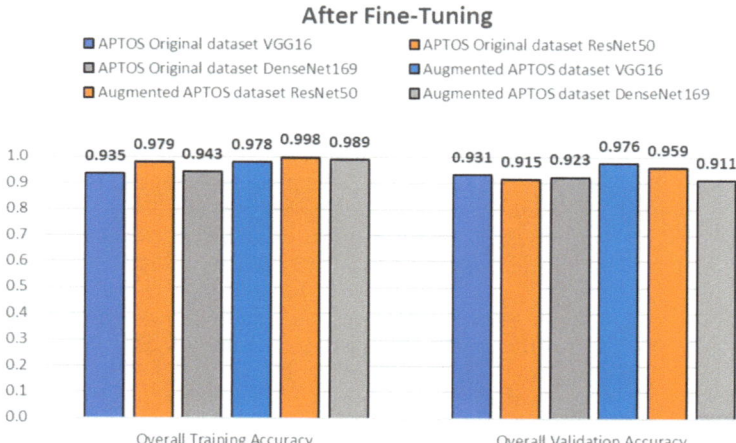

Figure 9. Overall training and validation accuracy of the CNN models after fine-tuning.

Both the overall training accuracy and the validation accuracy were above 90, which is a good indication that the six trained models were able to guess the label for nearly all of the training and validation sets of images. In three out of the six different CNN model training, with ResNet50 using both the original APTOS dataset and the augmented APTOS dataset as the training data, and the DenseNet169 model using the original APTOS dataset as the training data, early stopping occurred to prevent the models from overfitting.

Next, the six models were used to predict the class of the images in the testing data of both the APTOS and the Mauritian datasets. Figure 10 shows the overall testing accuracy obtained with the three CNN models for the original and augmented APTOS datasets. For the ResNet50 model and DenseNet169 model, increases of 9% and 7% were observed, respectively, when dealing with the augmented and balanced dataset. As for the VGG16 model, a decrease of 6.9% was noted for the augmented APTOS dataset.

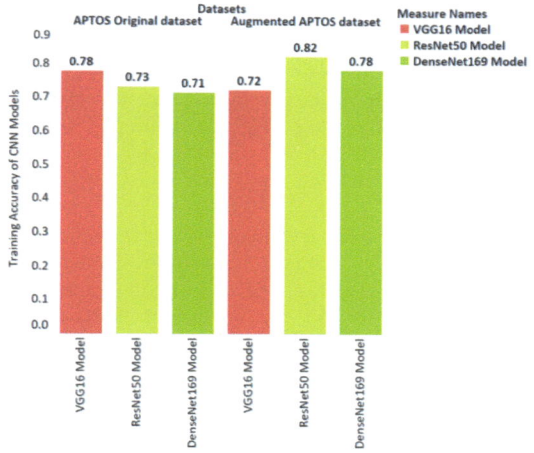

Figure 10. Testing accuracy of the CNN models for the APTOS dataset.

However, this overall testing accuracy for the data is not a good indicator of performance as the proportion of classes in the datasets was different. For example, in the original APTOS dataset, the number of images belonging to class 0 makes up nearly half of the original data, whereas the images in the augmented APTOS dataset are more or less equally distributed among the different classes. Hence, the models will exhibit bias towards class 0 when they are applied to the original APTOS dataset, whereas for the augmented APTOS dataset, the proportion is nearly the same, so comparing the overall testing accuracy between the two datasets is not recommended. To address this issue, the class-wise accuracy was calculated for the three datasets and plotted as shown in Figure 11.

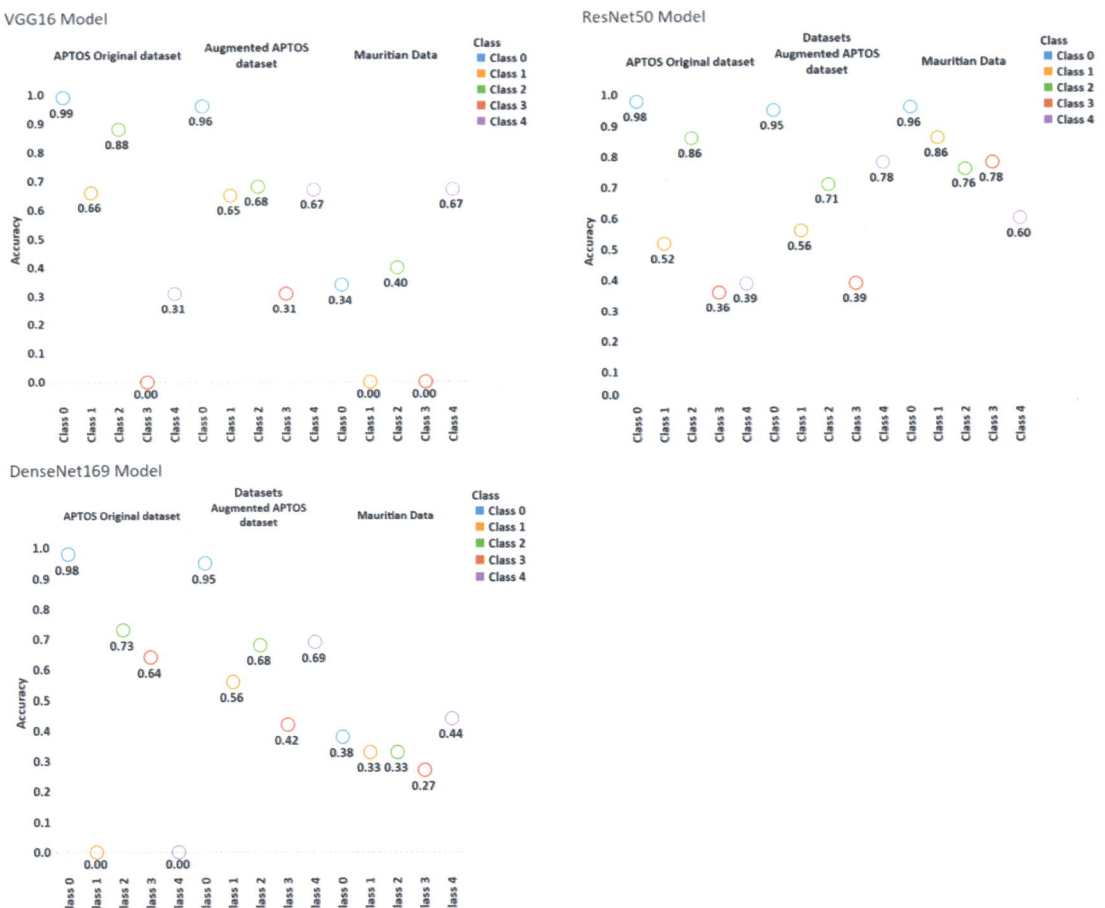

Figure 11. Detailed testing accuracy for each class for the 3 datasets and the 3 models.

A closer study of the plots in Figure 11 shows that the three models were able to predict class 0, "No DR" cases, quite easily for both the original and augmented APTOS datasets; however, only the ResNet50 model was able to classify "No DR" cases for the Mauritius dataset. This is to be expected since class 0 is quite distinct from the other classes given the absence of DR features such as microaneurysms.

For the VGG16 model, class 3 was the one that achieved the lowest accuracy out of all three datasets with none of the 55 cases being correctly classified for the original APTOS dataset. We also noted that none of the cases of class 1 for the Mauritian dataset were correctly identified. This shows that the model was unable to learn to distinguish the features of these two classes. A closer look at the results obtained shows that most of the cases for class 3 were misclassified as class 2 and a few cases as classes 1 and 4. Class 3 represents the moderate cases, which fall between the mild and proliferative cases and therefore may be difficult to identify. There may be intraretinal haemorrhage, which also complicates the task.

For the ResNet50 model, classes 1, 3 and 4 were the most difficult to classify for the original dataset, classes 1 and 3 were the most difficult to classify for the augmented dataset, and class 4 was the most difficult class to classify for the Mauritian dataset. The difficulty in the classification of class 4 for the Mauritian cohort may be due to choroidal fronts and troughs being more pronounced in the local dataset due to presence of pigments. This is due to the local population having different skin colours.

For the DensetNet169 model, the results obtained for the three datasets are variable with classes 1 and 4 being the less distinctive for the original dataset, classes 1 and 3 being less distinctive for the augmented dataset and classes 1, 2 and 3 being less distinctive for the Mauritian dataset. Here, none of the 202 cases for classes 1 and 4 in the original APTOS dataset were correctly identified. A closer look at the class-wise results shows that most of the images from class 1 were wrongly classified as class 2, and a few were classified as classes 0 and 3. Similarly, for class 4, we found that most of the images from class 4 were wrongly classified as class 2 and the rest as class 3. Based on these results, we concluded that for the APTOS dataset, classes 1 and 3 were the most difficult to learn.

Although none of the models had been trained with the data from Mauritius, the ResNet50 model achieved quite good results on this blind test dataset, achieving accuracies of 60% and above. It also obtained the best results compared to the other two models. This can be explained by the fact that the Densenet169 has more layers and may be overlearning and therefore generalizing less. Resnet50 has residual connections between layers, meaning that the output of a layer is a convolution of its input plus its input. It is also deeper than VGG16 with fewer parameters and is better able to identify the features to distinguish between the different classes of diabetic retinopathy. Moreover, although ResNet is much deeper than VGG16, the model size is substantially smaller due to the use of global average pooling rather than fully connected layers. Based on the results of the ResNet50 model, the results were further investigated, and a confusion matrix of the predicted vs. actual results was plotted, as shown in Figure 12.

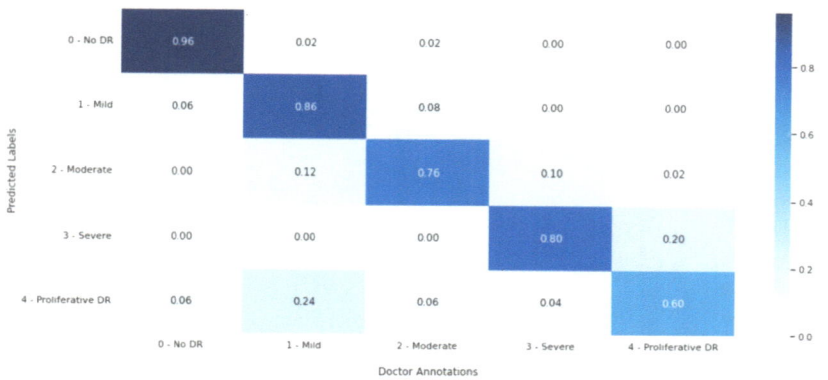

Figure 12. Confusion matrix for Mauritian data classified by ResNet50.

The precision, recall and F1 score were computed for each individual class and are displayed in Table 2. Additionally, the weighted average was also calculated.

Table 2. Performance metrics for Mauritian data classified by ResNet50.

	Precision	Recall	F1 Score
Class 0	0.9600	0.8889	0.9231
Class 1	0.8600	0.6935	0.7679
Class 2	0.7551	0.8222	0.7872
Class 3	0.7778	0.5000	0.6087
Class 4	0.6000	0.9091	0.7229
Weighted Average	0.8165	0.7933	0.7945

From the confusion matrix, we found that very good accuracies were achieved for all the classes, and the cases that were wrongly classified were close to the diagonal, being either from the class just before or just after. Thus, for class 0, the cases that were wrongly classified were actually from classes 1 and 2, for class 1, they were from classes 0 and 2, for class 2, they were from classes 1 and 3 with few cases from class 4, and for class 3, they were from class 4. This behaviour is not followed by class 4, where the wrongly classified classes were from all classes with the majority from class 2, which is quite far from class 4. Class 4 is of interest and requires further investigation. A comparison of our proposed model with the other available works in DR classification is given in Table 3.

Table 3. Comparison table of similar work.

Authors	Techniques Used	Discussions
Dai et al. [23]	Model: deep model based on ResNet Dataset: Shanghai Integrated Diabetes Prevention and Care System (Shanghai Integration Model, SIM) between 2014 and 2017 Number of images: 666,383 images	Pre-trained models (ResNet and R-CNN) were used. ROC was used to evaluate performance. Performance: AUC scores of 0.943, 0.955, 0.960 and 0.972 for mild, moderate, severe and proliferative cases were achieved, showing good performance using transfer learning
Masood et al. [11]	Model: pre-trained Inception V3 model Dataset: Eye-PACS dataset Number of images: 3908 images (800 from each class except 708 from class 4)	Performance: accuracy—48.2%, limitations: low accuracy
Li et al. [12]	Model: different pre-trained networks such as AlexNet, VGG-S, VGG16 and VGG19 Dataset: the Messidor and DR1 datasets Number of images: 1014 images (DR1), 1200 images (Messidor)	Performance: best area under the curve (AUC) (VGG-S)—98.34% (Messidor dataset), 97.86% (DR1 dataset) Limitations: number of classes is limited to DR and No DR only
Challa et al. [13]	Model: developed a deep All-CNN architecture Dataset: Eye-PACS dataset Number of images: 35,126 images	Performance: accuracy—86.64%, loss—0.46, average F1 score—0.6318 Limitation: no detailed information on overfitting
Khalifa et al. [16]	Model: AlexNet, Res-Net18, SqueezeNet and GoogleNet Dataset: APTOS dataset Number of images: 3662 images	Performance: best accuracy (AlexNet)—97.9% Limitation: high computational power needed (Intel Xeon E5-2620 processor (2 GHz), 96 GB of RAM) since the model needed to train on 14,648 images. Additionally, no detailed information was given for model overfitting during the training phase. The only method used to counter overfitting was data augmentation, which takes place before the model training phase.

Table 3. *Cont.*

Authors	Techniques Used	Discussions
Hagos et al. [17]	Model: pre-trained Inception V3 model Dataset: APTOS dataset Number of images: 2500 images (1250 for NoDR and 1250 for DR)	Performance: accuracy—90.9%, loss—3.94% Limitation: number of classes is limited to DR and No DR only
Gangwar et al. [22]	Model: deep learning hybrid model with pre-trained Inception-ResNet-v2 as a base model Dataset: Messidor-1 and APTOS dataset Number of images: 1200 images (Messidor-1), 3662 images (APTOS)	Performance: accuracy—72.33% (Messidor-1), 82.18% (APTOS dataset) Limitation: did not check whether model was overfitting
Benson et al. [24]	Model: pre-trained Inception V3 model Dataset: DR dataset obtained from VisionQuest Biomedical database Number of images: 6805 images	Performance: sensitivity—90%, specificity—90%, AUC—95% Limitation: results for No DR, MildDR, Moderate DR were 47%, 50% and 35%
Thota et al. [21]	Model: Fine-tuned and pre-trained VGG16 model Dataset: Eye-PACS dataset Number of images: 34,126 images	Performance: accuracy—74%, sensitivity—80%, a specificity—65%, AUC—80%Limitation: low accuracy compared to similar experimentations
Our proposed Model	Model: Fine-tuned and pre-trained ResNet50, VGG16, DenseNet169 models Dataset: APTOS dataset, Mauritian dataset Number of images: 3662 images (APTOS), 208 images (Mauritius)	Performance: accuracy (ResNet50)—82% (APTOS dataset), 79% (Mauritian dataset) Novelty: performed multiclass classification (5 different classes) for Mauritian dataset

Compared to similar work carried out in the field of DR classification, our proposed enhanced model was able to classify the different stages of diabetic retinopathy for a Mauritian dataset. The enhanced model was trained using the APTOS augmented dataset, and this model was used to classify the Mauritian dataset images with an overall accuracy of 79%. Furthermore, it can be said that our proposed model can be used for early detection of DR compared to Benson et al. [24], where the proposed model had a low accuracy for the early stages of DR. Meanwhile, Li et al. [12] and Hagos et al. [17] applied transfer learning for a binary classification, namely images having DR or No DR, whereas our model was used to classify all 5 stages of DR both for the APTOS and Mauritian dataset. In this paper, we have reported the use of several parameters to address overfitting of the models compared to the work of Gangwar et al. [22] and Challa et al. [13]. Finally, our model outperforms Thota et al. [21] and Masood et al. [11] in terms of accuracy.

4. Conclusions

In this work, transfer learning was applied at multiple levels with the aim of training multiple models to classify diabetic retinopathy for a completely blind dataset, the Mauritian cohort. At the initial stage, transfer learning was performed with three general pre-trained models, VGG16, ResNet50 and DenseNet169, using the APTOS dataset for diabetic retinopathy. Even after fine-tuning the three models, some classes were not being classified, and accuracies were not very high. This could be due to the dataset being highly imbalanced with almost 50% of the dataset belonging to "No DR" cases and the remaining 50% being distributed amongst the four DR classes. Hence, the dataset was augmented to achieve a comparable number of cases in each of the classes. Transfer learning was performed on the augmented APTOS dataset, and a better performance was achieved in the various experiments. It was found that the ResNet50 model produced equivalent or better results for all the classes compared to the VGG16 and DenseNet169 models. These trained enhanced models were then applied to the blind Mauritian dataset, and the results obtained are compared to the annotated local images. Again, the ResNet50, given its architecture, achieved the best results amongst the three models, and the accuracies obtained were very good. Class 0 achieved accuracies of 98%, 95% and 96% for the original APTOS dataset, the augmented APTOS dataset and the Mauritian dataset, respectively, clearly indicating that the model is able to easily distinguish this class from the other classes, thus confirming the potential of training a precursor model for class 0 versus others. It was observed that some classes performed much better than others, and this needs to be further investigated. Classes 1, 2 and 3 achieved acceptable performances while class 4 was the most difficult to classify. The diabetic retinopathy expert observed that class 3 was graded more precisely. Moreover, retinal images with pronounced choroidal fronds seemed to be identified as class 4 by the software, which clinically rates as normal variants. This is an unexpected behaviour of class 4, representing a major difference between the training APTOS data and the Mauritius data. This can be solved by further transfer learning (or fine-tuning) from the APTOS-based model to a Mauritian-specific model.

In the future, more data, such as patient demographics, can be included to ensure clinical correlation. In addition, the Mauritian cohort can be analyzed to determine whether the data are demographically representative of the population and also the extent to which they are similar to those of the APTOS cohort. Our research shows the need for a precursor software to identify normal retinal images.

Author Contributions: Conceptualization, Z.M.-D., M.H.-M.K. and C.P.-R.; methodology, Z.M.-D., M.H.-M.K. and C.P.-R.; software, K.J., Z.M.-D., M.H.-M.K. and C.P.-R.; validation, B.N.B., N.Z.M.; formal analysis, K.J., Z.M.-D., M.H.-M.K., B.N.B., N.Z.M. and C.P.-R.; investigation, K.J., Z.M.-D., M.H.-M.K., B.N.B., N.Z.M. and C.P.-R.; resources, N.Z.M. and B.N.B.; data curation, K.J.; writing—original draft preparation, K.J., Z.M.-D., M.H.-M.K. and C.P.-R.; writing—review and editing, K.J., Z.M.-D., M.H.-M.K., B.N.B., N.Z.M. and C.P.-R.; funding acquisition, Z.M.-D., M.H.-M.K., N.Z.M. and C.P.-R. All authors have read and agreed to the published version of the manuscript.

Funding: This research was funded by the Higher Education Commission (HEC) under grant number T0714 and the H3ABioNet. H3ABioNet is supported by the National Institutes of Health Common Fund under grant number U41HG006941. The content of this publication is solely the responsibility of the authors and does not necessarily represent the official views of the National Institutes of Health and of the Higher Education Commission.

Institutional Review Board Statement: Ethical clearance to collect existing fundus images from local hospitals was obtained from the Ministry of Health and Wellness of Mauritius.

Informed Consent Statement: Informed consent was obtained from all subjects involved.

Data Availability Statement: Due to the confidentiality of the data, the dataset has not been made publicly available.

Acknowledgments: Authors acknowledge the support of the Ministry of Health and Wellness, the University of Mauritius, the H3ABioNet and Sherali Zeadally.

Conflicts of Interest: The authors declare no conflict of interest. The funders had no role in the design of the study; in the collection, analyses, or interpretation of data; in the writing of the manuscript; or in the decision to publish the results.

References

1. GBD 2019 Blindness and Vision Impairment Collaborators; Vision Loss Expert Group of the Global Burden of Disease Study. Causes of blindness and vision impairment in 2020 and trends over 30 years, and prevalence of avoidable blindness in relation to VISION 2020: The Right to Sight: An analysis for the Global Burden of Disease Study. *Lancet Glob. Health* **2021**, *9*, E144–E160. [CrossRef]
2. Saeedi, P.; Petersohn, I.; Salpea, P.; Malanda, B.; Karuranga, S.; Unwin, N.; Colagiuri, S.; Guariguata, L.; Motala, A.A.; Ogurtsova, K.; et al. Global and regional diabetes prevalence estimates for 2019 and projections for 2030 and 2045: Results from the International Diabetes Federation Diabetes Atlas, 9th edition. *Diabetes Res. Clin. Pract.* **2019**, *157*, 107843. [CrossRef] [PubMed]
3. Shah, A.R.; Gardner, T.W. Diabetic retinopathy: Research to clinical practice. *Clin. Diabetes Endocrinol.* **2017**, *3*, 9. [CrossRef] [PubMed]
4. Lam, C.; Yi, D.; Guo, M.; Lindsey, T. Automated Detection of Diabetic Retinopathy using Deep Learning. *AMIA Jt. Summits Transl. Sci. Proc.* **2018**, *2018*, 147–155.
5. Oltu, B.; Karaca, B.K.; Erdem, H.; Özgür, A. A systematic review of transfer learning based approaches for diabetic retinopathy detection. *arXiv* **2021**, arXiv:2105.13793.

6. Alyoubi, W.L.; Shalash, W.M.; Abulkhair, M.F. Diabetic retinopathy detection through deep learning techniques: A review. *Inform. Med. Unlocked* **2020**, *20*, 100377. [CrossRef]
7. Kandel, I.; Castelli, M. Transfer Learning with Convolutional Neural Networks for Diabetic Retinopathy Image Classification. A Review. *Appl. Sci.* **2020**, *10*, 2021. [CrossRef]
8. Simonyan, K.; Zisserman, A. Very Deep Convolutional Networks for Large-Scale Image Recognition. *arXiv* **2014**, arXiv:1409.1556.
9. He, K.; Zhang, X.; Ren, S.; Sun, J. Deep residual learning for image recognition. In Proceedings of the 2016 IEEE Conference on Computer Vision and Pattern Recognition (CVPR), Las Vegas, NV, USA, 27–30 June 2016; pp. 770–778.
10. Zoph, B.; Vasudevan, V.; Shlens, J.; Le, Q.V. Learning transferable architectures for scalable image recognition. In Proceedings of the 2018 IEEE/CVF Conference on Computer Vision and Pattern Recognition, Salt Lake City, UT, USA, 18–23 June 2018; pp. 8697–8710.
11. Masood, S.; Luthra, T.; Sundriyal, H.; Ahmed, M. Identification of diabetic retinopathy in eye images using transfer learning. In Proceedings of the 2017 International Conference on Computing, Communication and Automation (ICCCA), Greater Noida, India, 5–6 May 2017; pp. 1183–1187.
12. Li, X.; Pang, T.; Xiong, B.; Liu, W.; Liang, P.; Wang, T. Convolutional neural networks based transfer learning for diabetic retinopathy fundus image classification. In Proceedings of the 2017 10th International Congress on Image and Signal Processing, BioMedical Engineering and Informatics (CISP-BMEI), Shanghai, China, 14–16 October 2017; pp. 1–11.
13. Challa, U.K.; Yellamraju, P.; Bhatt, J.S. A Multi-class Deep All-CNN for Detection of Diabetic Retinopathy Using Retinal Fundus Images. In *Pattern Recognition and Machine Intelligence: 8th International Conference, PReMI 2019, Tezpur, India, December 17–20, 2019, Proceedings, Part I*; Deka, B., Maji, P., Mitra, S., Bhattacharyya, D.K., Bora, P.K., Pal, S.K., Eds.; Lecture Notes in Computer Science; Springer International Publishing: Cham, Switzerland, 2019; Volume 11941, pp. 191–199.
14. Kaggle. APTOS 2019 Blindness Detection | Kaggle. Available online: https://www.kaggle.com/c/aptos2019-blindness-detection/ (accessed on 15 February 2022).
15. Kassani, S.H.; Kassani, P.H.; Khazaeinezhad, R.; Wesolowski, M.J.; Schneider, K.A.; Deters, R. Diabetic retinopathy classification using a modified xception architecture. In Proceedings of the 2019 IEEE International Symposium on Signal Processing and Information Technology (ISSPIT), Ajman, United Arab Emirates, 10–12 December 2019; pp. 1–6.
16. Khalifa, N.E.M.; Loey, M.; Taha, M.H.N.; Mohamed, H.N.E.T. Deep transfer learning models for medical diabetic retinopathy detection. *Acta Inform. Med.* **2019**, *27*, 327–332. [CrossRef]
17. Hagos, M.T.; Kant, S. Transfer Learning based Detection of Diabetic Retinopathy from Small Dataset. *arXiv* **2019**, arXiv:1905.07203. [CrossRef]
18. Sikder, N.; Chowdhury, M.S.; Shamim Mohammad Arif, A.; Nahid, A.-A. Early blindness detection based on retinal images using ensemble learning. In Proceedings of the 2019 22nd International Conference on Computer and Information Technology (ICCIT), Dhaka, Bangladesh, 18–20 December 2019; pp. 1–6.
19. Shaban, M.; Ogur, Z.; Mahmoud, A.; Switala, A.; Shalaby, A.; Abu Khalifeh, H.; Ghazal, M.; Fraiwan, L.; Giridharan, G.; Sandhu, H.; et al. A convolutional neural network for the screening and staging of diabetic retinopathy. *PLoS ONE* **2020**, *15*, e0233514. [CrossRef] [PubMed]
20. Mushtaq, G.; Siddiqui, F. Detection of diabetic retinopathy using deep learning methodology. *IOP Conf. Ser. Mater. Sci. Eng.* **2021**, *1070*, 012049. [CrossRef]
21. Thota, N.B.; Umma Reddy, D. Improving the Accuracy of Diabetic Retinopathy Severity Classification with Transfer Learning. In Proceedings of the 2020 IEEE 63rd International Midwest Symposium on Circuits and Systems (MWSCAS), Springfield, MA, USA, 9–12 August 2020; pp. 1003–1006.
22. Gangwar, A.K.; Ravi, V. Diabetic retinopathy detection using transfer learning and deep learning. In *Evolution in Computational Intelligence: Frontiers in Intelligent Computing: Theory and Applications (FICTA 2020), Volume 1*; Bhateja, V., Peng, S.-L., Satapathy, S.C., Zhang, Y.-D., Eds.; Advances in Intelligent Systems and Computing; Springer: Singapore, 2021; Volume 1176, pp. 679–689.
23. Dai, L.; Wu, L.; Li, H.; Cai, C.; Wu, Q.; Kong, H.; Liu, R.; Wang, X.; Hou, X.; Liu, Y.; et al. A deep learning system for detecting diabetic retinopathy across the disease spectrum. *Nat. Commun.* **2021**, *12*, 3242. [CrossRef] [PubMed]
24. Benson, J.; Maynard, J.; Zamora, G.; Carrillo, H.; Wigdahl, J.; Nemeth, S.; Barriga, S.; Estrada, T.; Soliz, P. Transfer learning for diabetic retinopathy. In *Medical Imaging 2018: Image Processing*; Angelini, E.D., Landman, B.A., Eds.; SPIE: Bellingham, WA, USA, 2018; p. 70.
25. Söderberg, S.; Zimmet, P.; Tuomilehto, J.; de Courten, M.; Dowse, G.K.; Chitson, P.; Gareeboo, H.; Alberti, K.G.M.M.; Shaw, J.E. Increasing prevalence of Type 2 diabetes mellitus in all ethnic groups in Mauritius. *Diabet. Med.* **2005**, *22*, 61–68. [CrossRef]
26. Housing and Population Census. Available online: https://web.archive.org/web/20121114114018/http://www.gov.mu/portal/goc/cso/file/2011VolIIPC.pdf (accessed on 15 February 2022).
27. Sudre, C.H.; Li, W.; Vercauteren, T.; Ourselin, S.; Jorge Cardoso, M. Generalised dice overlap as a deep learning loss function for highly unbalanced segmentations. In *Deep Learning in Medical Image Analysis and Multimodal Learning for Clinical Decision Support*; Cardoso, M.J., Arbel, T., Carneiro, G., Syeda-Mahmood, T., Tavares, J.M.R.S., Moradi, M., Bradley, A., Greenspan, H., Papa, J.P., Madabhushi, A., et al., Eds.; Lecture Notes in Computer Science; Springer International Publishing: Cham, Switzerland, 2017; Volume 10553, pp. 240–248.
28. Lopez-Nava, I.H.; Valentín-Coronado, L.M.; Garcia-Constantino, M.; Favela, J. Gait Activity Classification on Unbalanced Data from Inertial Sensors Using Shallow and Deep Learning. *Sensors* **2020**, *20*, 4756. [CrossRef]

29. Zhou, Y.; Wang, B.; He, X.; Cui, S.; Shao, L. DR-GAN: Conditional Generative Adversarial Network for Fine-Grained Lesion Synthesis on Diabetic Retinopathy Images. *IEEE J. Biomed. Health Inform.* **2020**, *26*, 56–66. [CrossRef]
30. Agustin, T.; Utami, E.; Fatta, H.A. Implementation of data augmentation to improve performance CNN method for detecting diabetic retinopathy. In Proceedings of the 2020 3rd International Conference on Information and Communications Technology (ICOIACT), Yogyakarta, Indonesia, 24–25 November 2020; pp. 83–88.
31. Tan, C.; Sun, F.; Kong, T.; Zhang, W.; Yang, C.; Liu, C. A survey on deep transfer learning. In *Artificial Neural Networks and Machine Learning—ICANN 2018: 27th International Conference on Artificial Neural Networks, Rhodes, Greece, October 4–7, 2018, Proceedings, Part III*; Kůrková, V., Manolopoulos, Y., Hammer, B., Iliadis, L., Maglogiannis, I., Eds.; Lecture Notes in Computer Science; Springer International Publishing: Cham, Switzerland, 2018; Volume 11141, pp. 270–279.
32. Da Rocha, D.A.; Ferreira, F.M.F.; Peixoto, Z.M.A. Diabetic retinopathy classification using VGG16 neural network. *Res. Biomed. Eng.* **2022**. [CrossRef]
33. Mule, N.; Thakare, A.; Kadam, A. Comparative analysis of various deep learning algorithms for diabetic retinopathy images. In *Health Informatics: A Computational Perspective in Healthcare*; Patgiri, R., Biswas, A., Roy, P., Eds.; Studies in Computational Intelligence; Springer: Singapore, 2021; Volume 932, pp. 97–106.
34. Huang, G.; Liu, Z.; van der Maaten, L.; Weinberger, K.Q. DenseNet Densely connected convolutional networks. In Proceedings of the 2017 IEEE Conference on Computer Vision and Pattern Recognition (CVPR), Honolulu, HI, USA, 21–26 July 2017; pp. 2261–2269.
35. Taormina, V.; Cascio, D.; Abbene, L.; Raso, G. Performance of Fine-Tuning Convolutional Neural Networks for HEp-2 Image Classification. *Appl. Sci.* **2020**, *10*, 6940. [CrossRef]
36. Zhang, C.-L.; Luo, J.-H.; Wei, X.-S.; Wu, J. In defense of fully connected layers in visual representation transfer. In *Advances in Multimedia Information Processing—PCM 2017*; Zeng, B., Huang, Q., El Saddik, A., Li, H., Jiang, S., Fan, X., Eds.; Lecture Notes in Computer Science; Springer International Publishing: Cham, Switzerland, 2018; Volume 10736, pp. 807–817.
37. ElBedwehy, M.N.; Behery, G.M.; Elbarougy, R. Face recognition based on relative gradient magnitude strength. *Arab. J. Sci. Eng.* **2020**, *45*, 9925–9937. [CrossRef]
38. Shorten, C.; Khoshgoftaar, T.M. A survey on Image Data Augmentation for Deep Learning. *J. Big Data* **2019**, *6*, 60. [CrossRef]
39. Khan, A.H.; Cao, X.; Li, S.; Katsikis, V.N.; Liao, L. BAS-ADAM: An ADAM based approach to improve the performance of beetle antennae search optimizer. *IEEE/CAA J. Autom. Sinica* **2020**, *7*, 461–471. [CrossRef]
40. Keras Transfer Learning & Fine-Tuning. Available online: https://keras.io/guides/transfer_learning/ (accessed on 2 March 2022).
41. Peng, P.; Wang, J. How to fine-tune deep neural networks in few-shot learning? *arXiv* **2020**, arXiv:2012.00204.
42. Ismail, A. View of Improving Convolutional Neural Network (CNN) Architecture (MiniVGGNet) with Batch Normalization and Learning Rate Decay Factor for Image Classification. Available online: https://publisher.uthm.edu.my/ojs/index.php/ijie/article/view/4558/2976 (accessed on 29 March 2022).
43. Usha Ruby, A. Binary cross entropy with deep learning technique for Image classification. *Int. J. Adv. Trends Comput. Sci. Eng.* **2020**, *9*, 5393–5397. [CrossRef]
44. Song, H.; Kim, M.; Park, D.; Lee, J.-G. How does Early Stopping Help Generalization against Label Noise? *arXiv* **2019**, arXiv:1911.08059. [CrossRef]
45. Jayalakshmi, G.S.; Kumar, V.S. Performance analysis of Convolutional Neural Network (CNN) based Cancerous Skin Lesion Detection System. In Proceedings of the 2019 International Conference on Computational Intelligence in Data Science (ICCIDS), Chennai, India, 21–23 February 2019; pp. 1–6.

Article

Real-Time Multi-Label Upper Gastrointestinal Anatomy Recognition from Gastroscope Videos

Tao Yu [1], Huiyi Hu [1], Xinsen Zhang [1], Honglin Lei [1], Jiquan Liu [1,*], Weiling Hu [2,3,*], Huilong Duan [1] and Jianmin Si [2,3]

[1] Key Laboratory for Biomedical Engineering of Ministry of Education, College of Biomedical Engineering and Instrument Science, Zhejiang University, Hangzhou 310027, China; yutao11615043@zju.edu.cn (T.Y.); 21815035@zju.edu.cn (H.H.); zxs1997@zju.edu.cn (X.Z.); 21915040@zju.edu.cn (H.L.); duanhl@zju.edu.cn (H.D.)

[2] Department of Gastroenterology, Sir Run Run Shaw Hospital, Medical School, Zhejiang University, Hangzhou 310027, China; jianmin_si@zju.edu.cn

[3] Institute of Gastroenterology, Zhejiang University (IGZJU), Hangzhou 310027, China

* Correspondence: liujq@zju.edu.cn (J.L.); huweiling@zju.edu.cn (W.H.); Tel.: +86-1358-889-9165 (J.L.)

Citation: Yu, T.; Hu, H.; Zhang, X.; Lei, H.; Liu, J.; Hu, W.; Duan, H.; Si, J. Real-Time Multi-Label Upper Gastrointestinal Anatomy Recognition from Gastroscope Videos. *Appl. Sci.* **2022**, *12*, 3306. https://doi.org/10.3390/app12073306

Academic Editors: Lucian Mihai Itu, Constantin Suciu and Anamaria Vizitiu

Received: 4 March 2022
Accepted: 20 March 2022
Published: 24 March 2022

Publisher's Note: MDPI stays neutral with regard to jurisdictional claims in published maps and institutional affiliations.

Copyright: © 2022 by the authors. Licensee MDPI, Basel, Switzerland. This article is an open access article distributed under the terms and conditions of the Creative Commons Attribution (CC BY) license (https://creativecommons.org/licenses/by/4.0/).

Abstract: Esophagogastroduodenoscopy (EGD) is a critical step in the diagnosis of upper gastrointestinal disorders. However, due to inexperience or high workload, there is a wide variation in EGD performance by endoscopists. Variations in performance may result in exams that do not completely cover all anatomical locations of the stomach, leading to a potential risk of missed diagnosis of gastric diseases. Numerous guidelines or expert consensus have been proposed to assess and optimize the quality of endoscopy. However, there is a lack of mature and robust methods to accurately apply to real clinical real-time video environments. In this paper, we innovatively define the problem of recognizing anatomical locations in videos as a multi-label recognition task. This can be more consistent with the model learning of image-to-label mapping relationships. We propose a combined structure of a deep learning model (GL-Net) that combines a graph convolutional network (GCN) with long short-term memory (LSTM) networks to both extract label features and correlate temporal dependencies for accurate real-time anatomical locations identification in gastroscopy videos. Our methodological evaluation dataset is based on complete videos of real clinical examinations. A total of 29,269 images from 49 videos were collected as a dataset for model training and validation. Another 1736 clinical videos were retrospectively analyzed and evaluated for the application of the proposed model. Our method achieves 97.1% mean accuracy (mAP), 95.5% mean per-class accuracy and 93.7% average overall accuracy in a multi-label classification task, and is able to process these videos in real-time at 29.9 FPS. In addition, based on our approach, we designed a system to monitor routine EGD videos in detail and perform statistical analysis of the operating habits of endoscopists, which can be a useful tool to improve the quality of clinical endoscopy.

Keywords: anatomy recognition; deep learning; endoscopy; multi-label; video analysis

1. Introduction

Gastric cancer [1] is the second leading cause of cancer-related deaths [2]. In clinical practice, Esophagogastroduodenoscopy (EGD) is a key step in the diagnosis of upper gastrointestinal tract disease. However, the rate of misdiagnosis and underdiagnosis of gastric diseases is high, reducing the detection of precancerous lesions and gastric cancer. This is because there is a great variation in EGD performed by endoscopists with different qualifications. On one hand, some inexperienced physicians may miss some critical areas and blind corners during the examination. On the other hand, physicians in densely populated areas face long examinations every day, which may lead to missed examinations and errors due to subjective mental or physical fatigue. This may result in the endoscopist not being able to comprehensively cover all anatomical locations throughout the stomach during the examination. Studies have shown that high-quality endoscopy can lead to more accurate diagnostic results [3], and it is crucial to further expand endoscopic techniques and improve routine endoscopy coverage and examination quality. Many authorities have now proposed clinical examination guidelines with corresponding expert consensus to evaluate and optimize the quality of endoscopy. The American Society of Gastrointestinal Endoscopy (ASGE) and the American College of Gastroenterology (ACG) have developed and published quality metrics common to all endoscopic procedures in EGD. The European Society of Gastroenterology (ESGE) systematically surveyed the available evidence and developed the first evidence-based performance measures for EGD (procedural integrity, examination time, etc.) in 2015 [4,5]. However, the lack of practical tools for rigorous monitoring and evaluation makes it difficult to apply many quantitative quality control indicators [6] (e.g., whether comprehensive coverage of anatomical location examination is achieved) in practice, which is a major constraint to quality control efforts.

The quality standard of GI endoscopy can be defined as: when doctors do endoscopy, they need to ensure that all key parts of the GI tract are within the scope of the examination, and maintain an appropriate observation duration, leaving no blind spots and avoiding the lens moving too fast or missing the observation of key areas. In recent years, advances in deep learning-based artificial intelligence technologies have continued to soar, with significant progress in the field of medical image recognition. Quality control of gastrointestinal endoscopy is the basis for the application of AI technology in endoscopic imaging and the prerequisite for applying artificial intelligence technology to disease screening and supplementary diagnosis. Advances have been made in the identification of gastric diseases [7,8], precancerous lesions [9–14] and gastric cancer [15–21]. It is important to use artificial intelligence systems to monitor the indicators in the quality control of gastrointestinal endoscopy in real time. However, previous studies have mainly focused on the intelligent auxiliary diagnosis of GI lesions. Due to the lack of relevant datasets for anatomical structures and the more complex and large data annotation efforts for this type of task, only a few studies were devoted to quality monitoring of routine endoscopy. Wu et al. [22] divided the anatomical location of the stomach into 10 and subdivided it into 26. DCNN was applied for anatomy classification. The final accuracy rates were 90% and 65.9%, respectively. Based on DCNN and reinforcement learning, 26 gastric anatomical locations were classified [23], and blind spots in EGD videos were monitored with an accuracy of 90.02%, which served to monitor the quality of real-time examinations. Ting et al. [24] proposed a deep ensemble feature network to combine the features extracted by multiple CNNs, to boost the recognition of three anatomic sites and two image modals with an accuracy of 96.9% and 23.8 frames per second(FPS). He et al. [25] divided the anatomical structure of endoscopy to 11 sites, and achieved 91.11% accuracy by using DenseNet121 [26]. The model was used to assist physicians in avoiding examination blind spots during examinations and to achieve comprehensive coverage of endoscopy.

Despite the good results of the above-mentioned studies on quality control of gastrointestinal endoscopy, some problems and challenges remain. First, all the above-mentioned work on anatomical location identification models is based on single-label multi-class classification, which deviates from the reality of actual clinical examinations. Multiple related anatomical sites are usually present simultaneously in the same image. When the ratio of multiple anatomical locations in the field of view is equal, a single label is not sufficient to accurately describe the currently examined location, which, in turn, increases the bias of model feature learning. Multi-label classification learning is more accurate in this application compared to multi-class image single-label classification [27], but it is challenging to further exploit this a priori relationship to improve the model accuracy due to the spatial correlation between anatomical locations, which leads to dependencies between labels. Second, all of the above work is based on anatomical location recognition models, which are trained based on static image data rather than real-time video data. It is not sufficient to identify anatomical locations under videos based on static image datasets alone. While consecutive video frames are highly similar, the dynamics of the scene cannot be expressed in static images, and this dynamically changing data is important for the application of the model in real video scenes. Although the dynamic ones can produce severe scene blurring [28] and thus affect judgment due to camera motion and gases generated during surgery, etc., the impact of such blurred data can be mitigated by some means. In conclusion, spatial and temporal factors are strong priors for the anatomical relationships within the endoscope and between consecutive frames, and are key to further improving the performance of the recognition model.

In this paper, we present a novel combined structure of deep learning models to process EGD videos to accurately identify anatomical structures of the gastrointestinal tract on real-time white-light upper gastrointestinal tract. The task consists of classifying each single frame of an EGD image sequence into a number of anatomical structures in 25 sites. Our model is built on a combination of a graph convolutional network (GCN) and a long short-term memory (LSTM) network, where the GCN is used to capture label dependencies, and the LSTM is used to extract inter-frame temporal dependencies. Specifically, we train them jointly in an end-to-end way to relate coded label interdependencies and extract high-level features of visual and temporal information of consecutive video frames. The combined features learned by our method can correlate different anatomical structures under endoscopy and are sensitive to camera movements in the video, allowing accurate identification of all anatomical structures contained in each frame of a continuous video, especially the transition frames between different anatomical locations.

The main contributions of this paper are summarized as follows: (1) Unlike previous single-label multi-class studies, we define anatomical recognition as a multi-label classification task. This setting is more in line with clinical needs and real-time video-based examination. (2) GCN-based multi-label classification algorithm. In this paper, graph structure is introduced to learn domain prior knowledge, i.e., topological interdependencies between anatomical structure labels. A ResNet-GCN model is then constructed to implement multi-label classification. (3) Fusion of ResNet-GCN and LSTM modules. Due to the complexity of EGD endoscopy scenes, it is very difficult to classify the anatomical structures of each frame accurately. Considering that EGD videos have temporal continuity and anatomical structures have spatial continuity in the video sequence, we use LSTM to learn the temporal information and spatial continuity features of anatomical structures in EGD videos based on the ResNet-GCN model. Then, we fuse the ResNet-GCN module and the LSTM module to implement an end-to-end framework, called GL-Net, for the accurate identification of UGI anatomical structures. The model fully reflects the topological dependence of labels and the continuity of anatomical structures in time and space. (4) Retrospective analysis of EGD video quality based on the GL-Net model. The quality of 1736 real EGD videos was statistically analyzed in terms of the coverage of 25 anatomy sites observed, the total examination time generated by the endoscopists, the examination time of each specific site, and the ratio of valid to invalid frames according to the endoscopic guidelines and expert consensus. The statistical analysis of the indicators gives a quantitative evaluation of the quality of the endoscopists, indicating the practical feasibility of using AI technology to ensure the quality of EGD following clinical guidelines.

The rest of this paper is organized as follows. Section 1 describes the datasets and introduces our proposed method in detail. Section 2 demonstrates the experimental results, which are discussed in Section 3. Section 4 is the conclusion of our work.

2. Materials and Methods

An overview of our proposed approach is presented in Figure 1. We used a backbone CNN model to extract visual features from static images and a GCN classification network to learn the relationship between the labels. The LSTM structure was used to model the temporal association of consecutive frames and focus on the invariant target features in the spatio-temporal information to obtain more accurate recognition.

To better exploit the correlation between labels, recurrent neural networks [29,30], attention mechanisms [31], and probabilistic graphical models [32,33] are widely used.

Wang et al. [30] used RNNs to convert labels into embedding vectors to model the correlation between each label. Zhu et al. [34] proposed a spatial regularization network (SRN) with only image-level supervised learning of the spatial regularization between labels. Recently, Chen et al. [35] proposed a multi-label image recognition model based on GCN which can capture global correlations between labels and infer knowledge from beyond a single image, and achieved good results. Inspired by Chen's work, we use graph structures to explore the dependencies between labels. Specifically, GCNs are used to disperse information from multiple labels so as to learn associative and dependent classifiers for each anatomically located label. These classifiers are further fused to image features to predict the correct outcome with label associations.

In the work of incorporating time series into deep learning models, many approaches based on dynamic time warping [36], conditional random fields [37], and hidden Markov models (HMMs) [38] have been proposed. However, the existing methods have some problems and challenges. For example, when exploring temporal correlation, these methods mostly focus on linear statistical models, which cannot accurately represent the complex temporal information during endoscopy. Second, it is difficult for these methods to accurately analyze transitional video frames where multiple targets are present at the same times, which are important for the accurate identification of anatomical locations. Several methods have been proposed to process sequential data by nonlinear modeling of temporal dependencies, such as LSTM, and have been successfully applied to many challenging tasks [28,39,40]. To address the problem of surgical procedure identification similar to EGD inspection, Jin Y et al. [28] introduced LSTM to learn temporal dependencies, and trained it in combination with convolutional neural networks. The learned temporal features are very sensitive to the changes of the surgical procedure, and can accurately identify the phase transition frames. Receiving inspiration from this approach, we proposed an LSTM fused with a GCN-based multi-label classification model for end-to-end training.

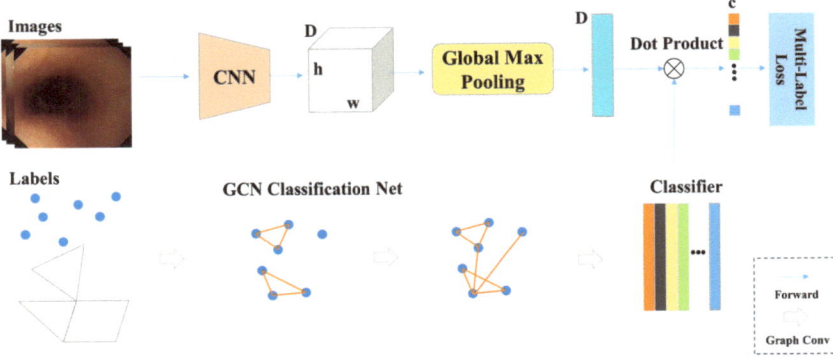

Figure 1. Overall framework of our ResNet-GCN model for multi-label image recognition. D denotes the dimension of the feature maps, h and w denote height and width. c denotes the number of category classes. The blue dots indicate all class labels.

2.1. DataSets

Following the guidance of ESGE [41] and the Japanese systematic screening protocol [42], three experts were invited to label the EGD images into 25 different anatomy sites. Representative images are shown in Figure 2. Since real endoscopy is performed under videos, severe noise (e.g., blood, bubbles, defocusing, artifacts, etc.) is generated. It is challenging to identify each image frame purely using video scenes alone. To improve the generalization ability of the dataset, 49 endoscopy videos were collected from Sir Run Run Shaw Hospital in this study. These videos were divided into a training set (39 videos) and a test set (10 videos), ensuring that images of the same case were not divided into both the training and test sets. We then split the videos into frames based on a sampling rate of 5 Hz, ensuring that the video clips contained temporal information while introducing as little redundant information as possible. This is the offset adjusted according to experience [28]. The larger the span of frames, the greater the temporal variation, so adapting the model to this variation facilitates the establishment of inter-frame relationships and the removal of invalid frames (see Figure 3). After splitting and labeling, we have 23,471 training images and 5798 test images with multi-label annotations. In the training phase, we divided the training process into two stages. In the first ResNet-GCN phase, we put all qualified images in the video together to train the gastric part classification network (GCN). In the second training data preparation phase, we took ten consecutive frames as a segment and input them together into the LSTM network. All EGD videos were captured in white light endoscopy with an OLYMPUS EVIS LUCERA ELITE CLV-290SL at FPS of 25 and resolution of 1920 × 1080 per frame. Inspection of personal information (such as date of inspection and patient name) is removed to ensure privacy and security.

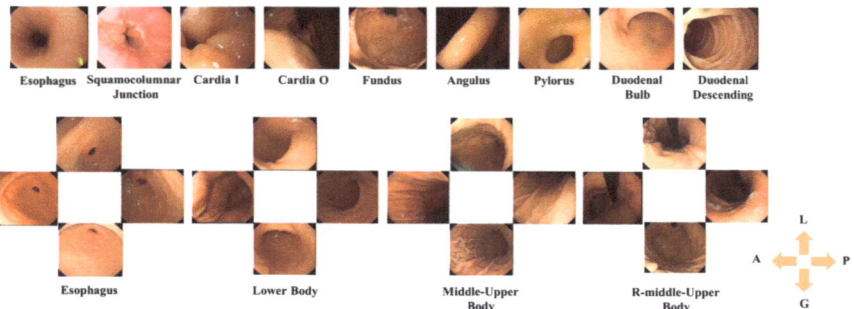

Figure 2. The representative images of UGI anatomical structures predicted by our model for multi-label classification in this paper. Specially, antrum, lower body, middle-upper body, and r-middle-upper body are further divided into four parts. Clockwise from the left side are the Anterior wall (A), Lesser curvature (L), Posterior wall (P), and Greater curvature (G). In addition, r stands for retroflex view.

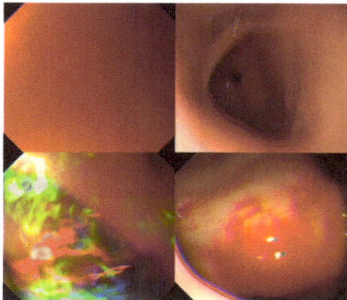

Figure 3. Invalid frame diagram in EGD video scene. The upper left figure shows the defocusing caused by the lens being too close to the gastric mucosa, the upper right figure shows the motion artifact caused by the rapid movement of the lens, the lower left figure shows the light reflection of gastric mucus under light, and the lower right figure shows the large area of blood covering the gastric mucosa.

2.2. Backbone Structure

Many innovative model design and training techniques have emerged at this stage, including the attention mechanism [43], Transformer [44], and the excellent NAS-based EfficientNet [45]. Considering the universality, stability and generality of the methods, we selected ResNet [46] as the backbone network. The residual structure allows the model capacity to vary in a flexible range, so that the model with the ResNet block as the basic unit can be built deep enough to complete the model convergence well. We use ResNet-50 [46], which was pre-trained on ImageNet [47], as the backbone network for feature extraction. Generally, the deeper the layers in the model, the larger the sensory field of the feature map and the higher the level of abstraction of the image features. Therefore, the proposed model structure attempts to extract features from one of the deepest convolutional layers of the backbone network for the construction of an attention map combining feature maps and association labels.

Let I denote an input static image or one of consecutive frames with ground-truth multi labels $y = [y^1, y^2, \ldots, y^C]$, where C is the count of all anatomical locations. The feature extraction process is expressed as

$$x = f_{GMP}(f_{Backbone}(I; \theta_{Backbone})) \in \mathbb{R}^{2048 \times 7 \times 7}, \quad (1)$$

where $f_{GAP}(\cdot)$ present the operation of global max pooling, $f_{Backbone}(\cdot)$ denotes the feature extraction from backbone structure. x is the compressed feature and contains the feature expressions in the image associated with the classification labels, which will be fused with the correlations between the labels in a matrix multiplication manner.

2.3. GCN Structure

In multi-label classification, multiple recognition targets usually appear together in an image. In some cases they must appear simultaneously, and in some cases they absolutely cannot appear at the same time. We need to efficiently establish the dependencies between targets to accurately establish feature representations in images, and correlations between multiple anatomical locations.

Since objects usually appear simultaneously in video scenes, the key to multi-label image recognition is to model the label dependencies, as shown in Figure 4. Inspired by Chen et al. [35], we model the interdependencies between anatomical locations using a graph structure where each node is a word embedding of an anatomical location, and that embedding feature is mapped to a set of classifiers constructed using GCN for image feature attention feature combinations. Thus, the approach preserves the semantic structure in the feature space and models label dependencies.

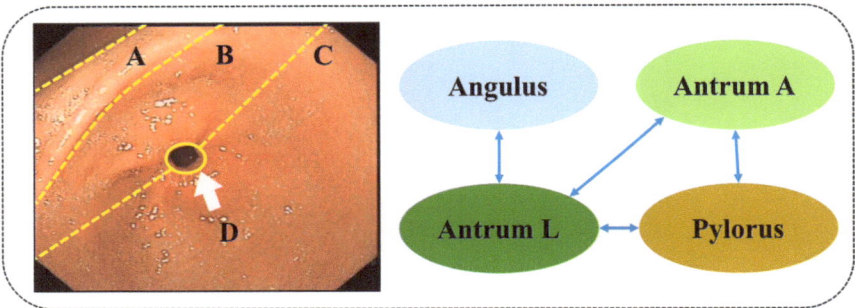

Figure 4. Multi-label images and their directed graphs. Within a single image, multiple anatomical locations appear and the image is divided into multiple regions, (A) Angulus, (B) Antrum L, (C) Antrum A and (D) Pylorus (the region indicated by the white arrow). At the right site, directed graphs are used to model label dependencies. The bidirectional arrows indicate that they are related and are likely to appear within the same view at the same time.

GCN is the operation on the graph structure. The structure uses the feature map and the corresponding correlation matrix as the input, and then updates the node features. The GCN structure can be written as follows:

$$H^{l+1} = h\left(\hat{A} H^l W^l\right),\tag{2}$$

where $h(\cdot)$ represents a non-linear mapping, $\hat{A}$ is the normalized matrix A and W^l is the transformation weight. H^{l+1} and H^l present the updated and current graph node representation.

The graph node presentation is then incorporated into the model output feature expression in the form of matrix multiplication, and the information is combined so that feature representations and labels are weighted and associated. The loss function, multi-label classification loss (e.g., binary cross entropy loss), is defined as follows:

$$L = \sum_{c=1}^{C} y^c \log(\sigma(\hat{y}^c)) + (1 - y^c)\log(1 - \sigma(\hat{y}^c))),\tag{3}$$

where $\sigma(\cdot)$ is the activation function of sigmoid.

2.4. LSTM Structure

After the above structure is trained to process video frames based on static images, the final prediction results may fluctuate due to the presence of some poor quality frames in the video. Due to the continuity of video data, temporal information provides background information for each frame identification. At the same time, individual frames may have similar appearance under the same endoscopic anatomy and scene, or they may be slightly blurred, making it difficult to distinguish them purely by their visual appearance. In contrast, the phase identification of the current frame would be more accurate if we could take into account the dependence of the current frame on the adjacent past frames. Therefore, time series information is introduced in this study to improve the stability of the model.

Temporal information modeling. In our GL-Net, we input the image features extracted from the ResNet backbone network into the LSTM network, and use the memory units of the LSTM network to correlate current frame and past frame information for improved identification using temporal dependence.

Figure 5 demonstrates the fundamental LSTM [48] units used in GL-Net. Each LSTM cell is equipped with three gates: i_t denotes input gate, f_t denotes forget gate and o_t denotes output gate. Three units are used to regulate the interaction between memory cells c_t. At timestep t, given input x_t, hidden state before h_{t-1}, and memory cell before c_{t-1}, LSTM structural units are learned and updated in the following manner:

$$i_t = \sigma(W_{xi}x_t + W_{hi}h_{t-1} + b_i),\tag{4}$$

$$f_t = \sigma(W_{xf}x_t + W_{hf}h_{t-1} + b_f),\tag{5}$$

$$o_t = \sigma(W_{xo}x_t + W_{ho}h_{t-1} + b_o),\tag{6}$$

$$g_t = \tanh(W_{xc}x_t + W_{hc}h_{t-1} + b_c),\tag{7}$$

$$c_t = f_t \odot c_{t-1} + i \odot g_t,\tag{8}$$

$$h_t = o_t \odot \tanh(c_t),\tag{9}$$

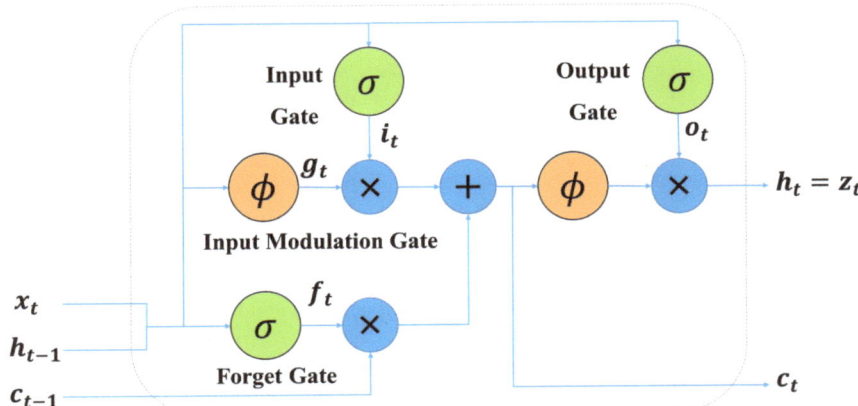

Figure 5. The structure of LSTM storage unit [49]. The arrows indicate the path of data forward propagation.

In order to fully exploit both the label association and temporal information, we propose a new recursive convolutional network, GL-Net, as shown in Figure 6. GL-Net integrates ResNet-GCN for visual descriptor extraction with label-dependent association, and the LSTM network for temporal dynamic modeling. It outperforms existing methods for independent learning of visual and temporal features. We train GL-Net end-to-end, where the parameters of the ResNet structure and the LSTM structure are co-optimized to achieve better anatomical location recognition.

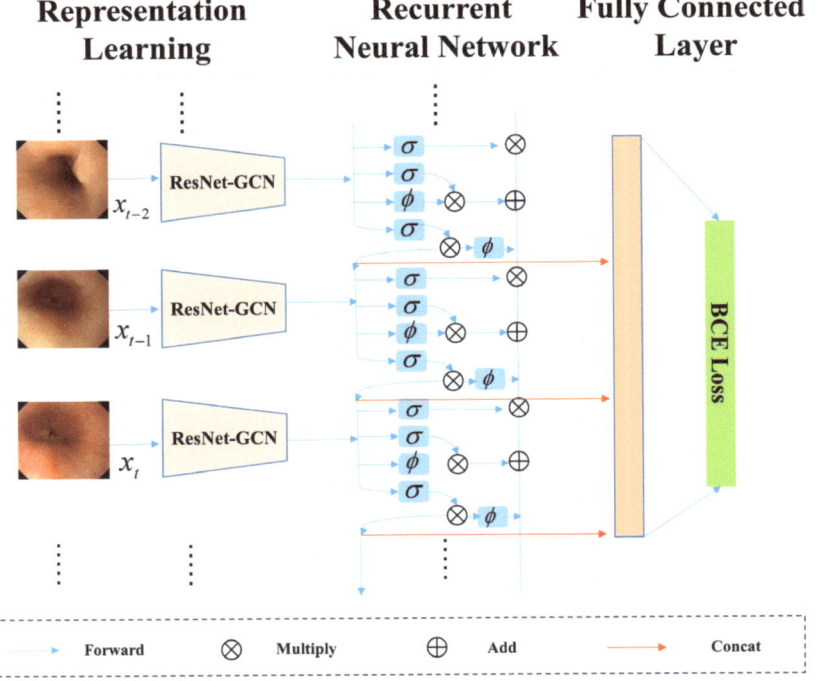

Figure 6. Overall framework of GL-Net model.

In detail, to identify the frames at time t, we extract the video clip containing a set of current frames. The sequence of frames in the video clip is represented by $x = \{x_{t'}, \ldots, x_{t-1}, x_t\}$. We use f_j to denote the representative image features of each single frame x_j. The image features $f = \{f_{t'}, \ldots, f_{t-1}, f_t\}$ of the video clips are sequentially put into an LSTM network, which is denoted by U_θ with parameters θ. With the input x_t and the previous hidden state h_{t-1}, the LSTM calculates the output o_t and the updated current hidden state h_t as $o_t = h_t = U_\theta(x_t, h_{t-1})$. Finally, the prediction of frame x_t is generated by feeding the output o_t into the softmax:

$$\hat{p}_t^i = sigmoid(W_z o_t + b_z), \tag{10}$$

where W_z and b_z respectively denote the weight and bias term, $\hat{P}_t \notin \mathbb{R}^C$ is the predicted vector and C denotes the number of classes.

Let $\hat{P}_t^i$ be the i-th element of $\hat{P}_t$, which denotes the prediction probability of frame x_t and it belongs to the class i, l_t denotes the ground truth of frame x_t, the negative log-likelihood of the frame of time t can be caculated as:

$$\ell(x_t) = -\log \hat{p}_t^{i=l_t}(U_\theta(x)). \tag{11}$$

2.5. Experimental Setups

To efficiently train the proposed model structure, we train the ResNet-GCN network first in order to subsequently initialize the entire network, considering that the parameter size of the ResNet-GCN network is larger than that of the LSTM structural units. During the training process, the images are augmented with random horizontal flips.

After training the ResNet model, we trained the GL-Net, integrating visual, labeling, and temporal information to converge. At this point, the pre-trained ResNet parameters were initialized as the parameters of its backbone model, and the parameters of the LSTM structural unit were initialized using xavier, and, empirically, the learning rate of the LSTM was set to 10 times that of the ResNet-GCN.

Our proposed model is implemented based on the Pytorch [50] framework, using a TITAN V GPU. For the first stage, our proposed structure uses two connected GCN modules with dimensions of 1024 and 2048, respectively. In the image representation learning branch, we adopt ResNet-50 as the backbone of feature extraction, which is pretrained on ImageNet. For label representations, 25-dim one-hot word embedding is adopted. SGD is employed for training, with a batch size of 16 and momentum of 0.9, with a weight decay of 5×10^{-3}. The initial learning rate is set to 0.01, and decreased to 1/10 every 10 epochs, until 1×10^{-5}.

In the end-to-end training stage, we use three LSTM layers. SGD is used as optimizer, with a batch size of 8, a momentum of 0.9 with weight decay of 1×10^{-2}, a dropout rate of 0.5, and we adopt LeakyReLU [51] as activation function. The learning rates are initially set as 1×10^{-4} for ResNet and 1×10^{-3} for LSTM, and are divided by a factor of 10 every 5 epochs. A total of 100 epochs were trained in the model.

3. Results

3.1. Evaluation Metrics

The evaluation metrics adopted in this paper are consistent with [30,52]. We compute the overall precision, recall, F1 (OP, OR, OF1) and per-class precision, recall, F1 (CP, CR, CF1). For each image, the labels are predicted as positive if their confidences are greater than the threshold (i.e., 0.5 in experience). Following [48,53], we computed the average precision (AP) for each individual class, and the average precision (mAP) for all classes.

3.2. Experimental Results

3.2.1. GCN Structure

The statistical results are presented in Table 1, and we compared them with related current spatial-temporal methods, including CNN-RNN [30], RNN-Attention [31], etc.

The models involved in the comparison all used the same set of training and test data. The benchmark backbone was kept uniform for a fair comparison. It is clear to see that the GCN-based approach obtains the best classification performance, which is due to capturing the dependencies between the labels. Compared to advanced methods for capturing dependencies of frames, our method achieves better performance in almost all metrics, which proves the effectiveness of GCN. Specifically, the proposed GCN scheme obtained 93.1% of mAP, which is 21.1% higher than their method. Even using the ResNet-50 model as the backbone model, we could still achieve better results (+17.1%). This suggests that there are strong dependencies and correlations between anatomical location labels in full-coverage examinations in white-light endoscopy scenarios, and the basic backbone of CNN together with a GCN structure can capture them well.

Table 1. Comparison of average assessment results of anatomy multi-label identification with other methods. Bold indicates the maximum value.

Method	mAP	CP	CR	CF1	OP	OR	OF1
CNN-RNN [30]	0.720	0.772	0.772	0.772	0.785	0.770	0.777
RNN-Attention [31]	0.649	0.665	0.578	0.618	0.610	0.660	0.634
ResNet-50	0.891	0.936	0.777	0.849	0.912	0.734	0.813
ResNet-GCN	0.931	0.941	0.855	0.896	0.929	0.820	0.871
GL-Net	**0.971**	**0.955**	**0.959**	**0.957**	**0.937**	**0.954**	**0.945**

We further use the heatmaps to explain the model. By weighting and summing the class activation maps [54] of the final convolutional layer, the attention map can accurately highlight the areas of the image that have a high weight on recognition, thus revealing the network's implicit attention to the image and intercepting the learning information of the network [54]. The attention maps of models is shown in Figure 7.

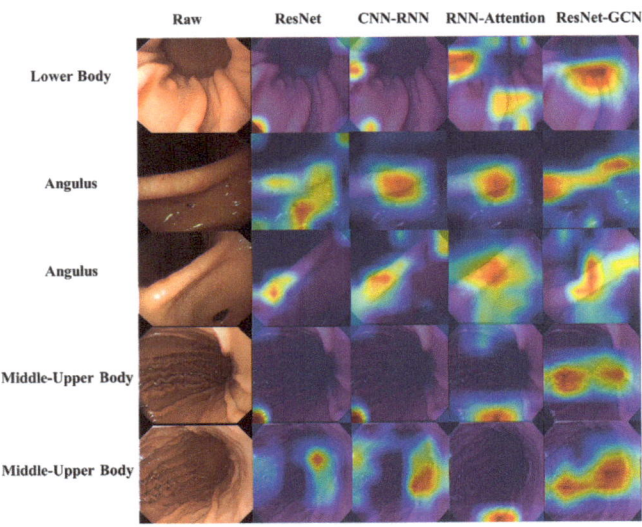

Figure 7. Visualization of the attention maps. The label on the left indicates the label for visual gradient back propagation and also the ground truth annotation. The upper label indicates the model name of the visualization. The visualization region is approximately close to red, indicating that more of the model inference weight tends to be in that region.

As illustrated in Figure 7, both in the middle and upper part of the stomach body, the GCN-based model was the best in terms of visual representation. For the other models, it is easy to randomly place weights at some locations in the image without constructing label associations. It is not possible to correctly distinguish between these anatomical structures that do not differ much between classes (greater curvature, posterior wall, anterior wall and less curvature of the body are difficult to distinguish). In contrast, the GCN-based model can pay more attention to feature regions in the image where texture features are prominent and responsive to the class. For the gastric angulus, because the structure of the angulus is prominent in the visual field, the general model can pay attention relatively accurately at this location. Compared to other models, the GCN-based model's weights are able to provide more comprehensive and complete coverage at this location, including the lesser curvature of the antral.

3.2.2. GCN with LSTM Structure

To demonstrate the importance of combining label association and temporal features for this task, we carried out a series of experiments by combining ResNet-50 with different modeling approaches, namely (1) ResNet-50 with GCN, and (2) ResNet-50 with GCN followed by LSTM.

The experimental results are listed in Table 1. The scheme with LSTM achieved better results, demonstrating the importance of temporal correlation for more accurate identification. A total of 97.1% of mAP, 95.7% CF1 and 94.5% OF1 can be seen from the GL net method proposed in this paper. Specifically, compared with ResNet-GCN, our end-to-end trainable GL-Net improves the mAP, CF1 and OF1 by 4.0%, 6.1% and 7.4%, respectively. Similarly, we compared the average accuracy of the two schemes on each anatomical structure (see in Table 2). Compared with the ResNet-GCN model without the LSTM module, the accuracy of GL-Net in the anterior wall of middle-upper body, the lesser curvature of the lower body, the posterior wall of the middle-upper body, the large curvature of the middle-upper body and angulus were improved by 22.5%, 17.6%, 11.2%, 8.8% and 8.3%, respectively.

Table 2. Comparison of average precision of each individual anatomy identification with other methods. Bold indicates the maximum value.

Anatomy	CNN-RNN	RNN-Attention	ResNet50	ResNet-GCN	GL-Net
esophagus	0.903	0.821	0.998	0.999	**1.0**
Squamocolumnar juction	0.880	0.748	0.986	0.994	**0.999**
Cardia I	0.867	0.715	0.965	0.967	**0.998**
Cardia O	0.873	0.658	0.994	**1.0**	**1.0**
Fundus	0.855	0.757	0.978	**0.979**	0.902
Middle-upper body A	0.604	0.595	0.809	0.906	**0.994**
Middle-upper body L	0.585	0.594	0.813	0.872	**0.984**
Middle-upper body P	0.546	0.562	0.680	0.745	**0.970**
Middle-upper body G	0.586	0.518	0.601	0.885	**0.904**
Lower body A	0.595	0.548	0.841	**0.918**	**0.918**
Lower body L	0.575	0.517	0.713	0.802	**0.821**
Lower body P	0.585	0.454	0.816	0.876	**0.936**
Lower body G	0.541	0.568	0.731	0.812	**0.988**
Antrum A	0.746	0.727	0.977	0.979	**0.999**
Antrum L	0.732	0.710	0.987	0.984	**0.997**
Antrum P	0.755	0.722	0.983	0.985	**0.995**
Antrum G	0.706	0.730	0.976	0.981	**0.997**
Angulus G	0.793	0.756	0.898	0.905	**0.988**
R-middle-upper body A	0.621	0.534	0.855	0.900	**0.980**
R-middle-upper body L	0.648	0.542	0.892	0.925	**0.959**
R-middle-upper body P	0.770	0.651	0.919	0.951	**0.956**
R-middle-upper body G	0.766	0.636	0.908	0.945	**0.997**
Duodenal bulb	0.751	0.623	0.997	0.998	**0.999**
Duodenal descending	0.906	0.827	0.998	**1.0**	**1.0**
Pylorus	0.753	0.722	0.969	0.980	**0.997**

By introducing label association and temporal information, our GL-Net can learn features that are more discriminative than those produced by traditional CNNs that consider only visual information. Figure 8 shows the comparison of the prediction results of the two models in the video clips. It can be seen that due to the shooting angle, bubble reflection and other reasons, the variance between some classes is small and difficult to distinguish, or sometimes features are almost completely covered. Therefore, the ResNet-GCN network, which only depends on the features of a single frame image, cannot classify correctly, while GL-Net can avoid the error by considering time dependence, and each frame is identified accurately. In addition, some frames in the video have no classification results, that is, the confidence of all the predicted results does not exceed the set threshold, which may be related to the noise in the video. GL-Net can also accurately recognize each frame in this case, which indicates that GL-Net considering the temporal information can improve the performance of UGI anatomical structure recognition in EGD video. In addition, GL-Net can process these videos in real time at 29.9 FPS, a processing speed that has great potential for application in real-time clinical scenarios.

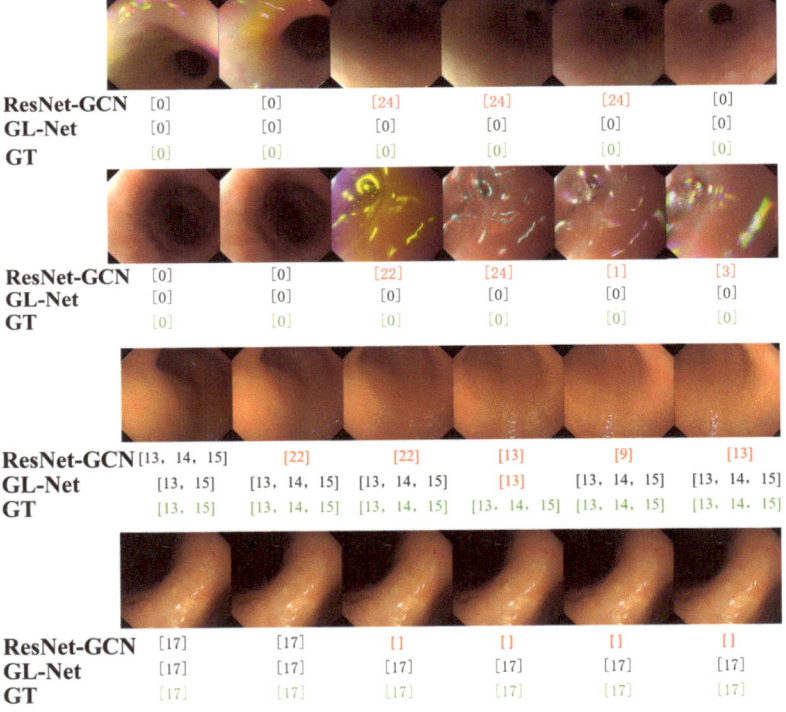

Figure 8. The prediction results on the video clips of ResNet-GCN and GL-Net. The bottom of each line corresponds to the results of ResNet-GCN, GL-Net and Ground Truth (GT) annotations separately. Specifically, 0-esophagus, 1-squamocolumnar juction, 3-outside of cardia, 9-greater curvature of lower body, 13-greater curvature of Antrum, 14-posterior wall of Antrum, 15-anterior wall of antrum, 17-angulus, 22-duodenal bulb, 24-pylorus.

3.2.3. Retrospective Analysis of EGD Videos

Based on the methodology proposed in this paper, we designed a framework for statistical analysis of the examination quality of real EGD videos in hospitals according to quality monitoring guidelines. We collected a total of 1736 EGD videos, all of which were captured with an OLYMPUS EVIS LUCERA ELITE CLV-290SL at 25 FPS and operated by expert physicians. In addition to the anatomical position identification model proposed in this paper, our system uses an invalid frame filtering model [55] to ensure that our statistical results are performed on clear and valid images.

The outputs of the proposed system are: (1) coverage statistics of the 25 sites observed; (2) total examination time; (3) examination time for each specific site; (4) the ratio of valid frames versus invalid frames.

The average coverage of anatomical structures during the EGD produced by the endoscopist was 85.81%, but only 19.28% of the patients were not blinded. In addition, the rate of misses for each anatomical structure (total number of videos with undetected anatomical structures/total number of videos) can be seen in Table 3. It can be clearly seen that most of the anatomical structures had a probability of being missed, except for the esophagus. Among them, the lower body lesser curvature had the highest miss rate of 52.41%, indicating that this area tends to be a blind area in EGD surgery. In addition, the small curvature of the middle and upper body, the descending duodenum, the posterior wall of the lower body, and the large curvature of the middle and upper body also had a blind spot rate of more than 20% in the retrograde view.

Table 3. The average miss rate of each UGI anatomical structure during EGD. Bold indicates the maximum value.

Anatomy	Miss Rate (%)
esophagus	0.00
Squamocolumnar juction	0.60
Cardia I	10.24
Cardia O	0.6
Fundus	9.04
Middle-upper body A	7.83
Middle-upper body L	10.24
Middle-upper body P	15.06
Middle-upper body G	43.98
Lower body A	13.86
Lower body L	23.49
Lower body P	18.67
Lower body G	**52.41**
Antrum A	9.64
Antrum L	13.25
Antrum P	11.45
Antrum G	9.64
Angulus	6.05
R-middle-upper body A	22.89
R-middle-upper body L	13.86
R-middle-upper body P	10.24
R-middle-upper body G	15.06
Duodenal bulb	5.42
Duodenal descending	24.70
Pylorus	6.63

As shown in Table 4, the mean examination time for all the videos was 6.572 min, but the variance is quite different, which may be because some videos take biopsies or make abnormal findings. Considering that there were some blind spots in the process of EGD, we further analysed the inspection time when 25 sites were completely observed. As can be seen, it takes 7.37 min for endoscopists to check all the anatomical structures.

Table 4. Statistics of inspection time during EGD.

Inspection Type	Mean (min)
Regular endoscopy	6.57
Coverage of all anatomy	7.37

Table 5 shows the examination time of each specific anatomical structure. Obviously, the most time-consuming site is the esophagus, which takes 85.8 s, far more than the other 24 sites. However, the endoscopists spend the least time in the lower gastric body, and the average observation time of the lesser curvature of lower body is only 1.8 s.

Table 5. Inspection time of each UGI anatomical structures during EGD. Bold indicates the maximum value.

Anatomy	Inspection Time (s)
esophagus	**85.8**
Squamocolumnar juction	15.6
Cardia I	31.2
Cardia O	26.4
Fundus	44.4
Middle-upper body A	39
Middle-upper body L	19.2
Middle-upper body P	7.8
Middle-upper body G	2.4
Lower body A	22.2
Lower body L	8.4
Lower body P	10.8
Lower body G	1.8
Antrum A	31.2
Antrum L	30
Antrum P	22.8
Antrum G	42
Angulus	45
R-middle-upper body A	15
R-middle-upper body L	16.8
R-middle-upper body P	34.2
R-middle-upper body G	25.8
Duodenal bulb	57.6
Duodenal descending	13.8
Pylorus	39.6

In addition, although no studies have clearly defined the effective time of endoscopists in operation, the visibility of mucosa has become an important indicator in the quality control guidelines of colonoscopy. Therefore, we believe that the proportion of invalid frames (including blood, bubbles, defocusing or artifacts) in the process of EGD also reflects the EGD quality. Based on this, we analysed the proportion of effective frames and invalid frames in the duration. According to the results in Table 6, the average ratio is about 2:7.

Table 6. The ratio of effective frame and invalid frame during EGD.

/	Ratio (%)
Effective Frames	22.68
Invalid Frames	77.32

4. Discussion

In this study, we used actual clinical EGD videos for real-time identification of gastric anatomical structures and quality control of computer-aided gastroscopy. We designed an efficient algorithm that integrates ResNet, GCN and LSTM networks to form the proposed GL-Net. The model achieves 97.1% mAP. Compared with previous works [23,24,56], we have the following advantages: (1) we propose a multi-label video frame-level gastric anatomical location identification method that can more accurately describe the physician's current examination location with considerable clinical significance. (2) Our model can accurately identify anatomical locations in video frames and transition frames by learning label associations and spatio-temporal features correlation of images. (3) We conducted a quantitative statistical analysis of real EGD videos to summarize existing physicians' operating habits and deficiencies, and to provide a quantitative analysis tool for effective implementation of examination quality control guidelines.

4.1. Recognition Evaluation

The purpose of this study was to use artificial intelligence to alleviate the problem that EGD quality control guidelines are not easily carried out and implemented in the clinic. In cases where the level of gastroscopy varies between physicians, there is a risk of missing a diagnosis if the entire endoscopy is not covered in that particular examination. Although similar work using CNN to assist in EGD quality control has been done in previous studies, there are several shortcomings. First, the use of a single label to represent the image is inaccurate, especially when the area occupied by different anatomical locations of the image in adjacent transition frames is large. Second, previous studies have mainly trained models on discrete still image data, which is insufficient in complex continuous video scenes and prone to a high number of false positive analysis values. In this study, we propose a new framework to address these problems. First, by introducing GCN in the training task to construct label associations, which, in turn, improves the accuracy of the location recognition. Second, the temporal associations in the video frames are addressed by introducing LSTM and continuous video frame datasets.

The GCN-based model outperforms other models with a uniform backbone structure, demonstrating its effectiveness in label-dependent modeling. The GCN has more advantages in multi-label modeling compared to the natural image dataset, which also indicates a strong interdependence between gastric anatomical locations. For analogs with lower scores, such as the large curvature and posterior wall in the upper middle body, and the small curvature, anterior and posterior wall in the lower body, there was a significant improvement. This is because GCN exploits the relationship between strong and weak label features extracted by models.

More importantly, we optimize the visual performance and sequential dynamics throughout the training process by introducing label associations and spatio-temporal priors. In general, the features generated by introducing more label associations and temporal feature constraints are more discriminative than those generated by traditional CNNs that consider only spatial information. In Figure 8, GL-Net can achieve accurate recognition results that conform to label association rules and correspond to image features, especially for frames with changing locations. In addition, based on LSTM, the results are more stable with fewer jumps, so the overall performance is improved, which is crucial for this task. Although there are many novel video-based 3D CNN methods proposed, we believe that compared to LSTM methods, 3D CNNs cannot provide correlations with longer connections due to the limitation of computational volume and computational speed. Therefore, we believe that using LSTM is the appropriate method for modeling temporal correlations. For the categories with relatively low scores, this may be due to the lack of distinct features and the insufficient number of datasets. However, considering the network performance, computational resources, and training difficulty, we use a 50-layer ResNet to implement GL-Net, so that the computational resources and training time can be controlled within a satisfactory range and satisfactory results can be obtained. With sufficient computational resources, we can choose a deeper CNN network to further improve the performance, or use multi-GPU distributed training.

In recent years, deep learning techniques in computer vision have made rapid progress, and representative recognition network structures such as VGGNet [57], Inceptions [58], ResNet [46], DenseNet [26], MobileNet [59], EfficientNet [45], and RegNet [60] have been expanding the accuracy, effectiveness, scale, and real-time performance of the networks. The Transformer [44], a self-attentive mechanism structure extending from the field of natural language processing (NLP), has given a trend to unify and combine image and text data. The reliance on data-driven deep learning models makes it easy for researchers to overlook the important role played by clinical priors in the application of medical image perception techniques; clinical tasks do not exist in isolation and data distributions are not independent of each other. Relationships between lesions and data feature distribution relationships have not been applied to the model design process. The research in this paper is inspired by the combination of clinical priori knowledge and deep learning methods. The major difference between our proposed method and previous single-label static frame methods is that the correlations between anatomical locations and the spatio-temporal relationship between consecutive frames are introduced into model design as constraints. This allows us to achieve better evaluation of our model under the same feature extraction backbone structure with the relational constraints introduced by GCN and LSTM.

4.2. Clinical Retrospective Analysis

Observing the integrity of all 25 locations is of paramount importance, however, we found that only 19.28% of patients were observed at all locations and nearly five locations had a leak rate of more than 20%. This suggests that the quality of endoscopy needs to be improved.

Studies have shown that spending more time on EGD improves the diagnostic rate, so we recorded the total procedure time during EGD and counted the procedure time per anatomical location based on model analysis. This helps the endoscopist to control the duration of each examination procedure, thereby reducing variability in the level of examination due to competent factors such as experience and fatigue examinations. This study concluded that "slow" endoscopists (who take on average more than 7 min to perform a normal endoscopy) are more likely, or even up to two times more likely, to detect high-risk gastric lesions [61]. However, in a retrospective analysis of experimental data results, the total time of the procedure was lower than the recommended time. Therefore, we recommend that endoscopists be able to increase the examination time further.

Among the various sites, the esophagus was the only one that was not missed in all videos, and had the longest examination time. On the one hand, this is because the esophagus has a certain length in space and is the entrance to the EGD. On the other hand, we have patients with Barrett's esophagus [62] in our videos. Studies have shown that the examination time of Barrett's esophagus is related to the detection rate of the associated tumor [63]. The less time spent on the lower body curvature also contributes to its high rate of missed diagnoses. The effective examination time is only 23%, so the mucosal visibility of UGI is not high enough during most EGD examinations, which is due to invalid frames when the endoscopist performs operations such as flushing and insufflation, or when the lens shakes and fails to focus. This value can be used as a reference indicator. For endoscopists with a high percentage of invalid frames, further demands can be made on the operation level.

With these data, we can clearly see the behavioral habits of Chinese doctors in gastroscopy and the possible blind spots. It is beneficial for the system to achieve quality monitoring, improve the quality of gastroscopy, and further improve the detection rate of diseases. All the indicators mentioned in this paper can reflect the details of the gastroscopy process to some extent. These indicators prove that our model has great potential value for application to improve the quality of examination.

5. Conclusions

In this paper, we propose a novel and effective recursive convolutional neural network, GL-Net, for automatic recognition of the anatomical location of the stomach in EGD videos. GL-Net consists of two partial structures, namely GCN and LSTM, which are used to extract label-dependent and time-dependent features, respectively.

The GCN part of our method is able to extract the label dependency of multi-label image recognition, compared to the currently related study of static image based recognition methods for single-label multi-class anatomical location recognition. Meanwhile, the spatial-temporal features extracted by the LSTM part are able to identify adjacent similar frames more accurately.

In addition, we designed a real-time system based on the GL-Net method to automatically monitor detailed metrics during EGD (e.g., anatomical examination coverage, effective observation frame statistics, observation statistics of each anatomical site, etc.) and perform statistical analysis on the quality of EGD examinations. A quantitative assessment of the quality of the endoscopist's examination is performed to demonstrate the professional operating habits of the endoscopist and the presence of potential accidents and problems. It also demonstrates the feasibility of implementing endoscopic quality control guidelines using artificial intelligence technology. It can effectively mitigate the subjective and empirical differences among endoscopists, improve the quality of routine endoscopy, and provide a reference for writing endoscopy reports and performing clinical procedures in real time with anatomical positions. In the future, the combination of anatomical position identification results and endoscopic mucosal health condition for comprehensive analysis is expected to further improve the quality control of computer-assisted endoscopy and assist in lesion diagnosis.

We believe that computer-aided detection and artificial intelligence techniques will play an increasing role. The rapid changes in model structure in recent years have allowed us to use increasingly advanced approaches to clinical data. However, the characteristics of the data distribution should be considered more in studies, such as the multi-label classification in this paper, which is more clinically realistic than single-label classification, and the potential associations within clinical prior knowledge and tasks, such as the construction of inter-label associations with the spatio-temporal association in this paper. Incorporating researchers' or clinicians' prior knowledge into the model training process is a more specific, accurate, and reliable solution to obtain practical solutions. We believe that in the future development of deep learning medical imaging research work, AI technology and medical knowledge will be further integrated to obtain further technical breakthroughs, as well as playing a greater role in the clinic and being more easily accepted by the public.

Author Contributions: Conceptualization, T.Y. and H.H.; methodology, T.Y. and H.H.; software, X.Z.; validation, H.L.; formal analysis, T.Y.; investigation, J.L.; resources, J.L.; data curation, W.H.; writing—original draft preparation, T.Y. and H.H.; writing—review and editing, J.L.; visualization, X.Z.; supervision, H.D. and J.S.; project administration, J.L.; funding acquisition, J.L. All authors have read and agreed to the published version of the manuscript.

Funding: This work was supported in part by the National Key Research and Development Program of Zhejiang, China (No.2021C03111) and the National Natural Science Foundation of China (No.81827804).

Institutional Review Board Statement: Ethical review and approval were waived for this study, due to the retrospective design of the study and the fact that all data used were from existing and anonymized clinical datasets.

Informed Consent Statement: Informed consent was obtained from all subjects involved in the study. Written informed consent has been obtained from the patient(s) to publish this paper.

Data Availability Statement: Not applicable.

Conflicts of Interest: The authors declare no conflict of interest.

Abbreviations

The following abbreviations are used in this manuscript:

EGD   Esophagogastroduodenoscopy
GCN   graph convolutional network
LSTM long short-term memory
ASGE The American Society of Gastrointestinal Endoscopy
ACG   The American College of Gastroenterology
ESGE The European Society of Gastroenterology
HMMs hidden Markov models

References

1. Zeiler, M.D.; Fergus, R. Visualizing and Understanding Convolutional Networks. In Proceedings of the Computer Vision—ECCV 2014, Zurich, Switzerland, 6–12 September 2014; Fleet, D., Pajdla, T., Schiele, B., Tuytelaars, T., Eds.; Springer International Publishing: Cham, Switzerland, 2014; pp. 818–833.
2. Ang, T.L.; Fock, K.M. Clinical epidemiology of gastric cancer. *Singap. Med. J.* **2014**, *55*, 621–628. [CrossRef]
3. Rutter, M.D.; Rees, C.J. Quality in gastrointestinal endoscopy. *Endoscopy* **2014**, *46*, 526–528. [CrossRef] [PubMed]
4. Cohen, J.; Safdi, M.A.; Deal, S.E.; Baron, T.H.; Chak, A.; Hoffman, B.; Jacobson, B.C.; Mergener, K.; Petersen, B.T.; Petrini, J.L.; et al. Quality indicators for esophagogastroduodenoscopy. *Gastrointest. Endosc.* **2006**, *63*, S10–S15. [CrossRef] [PubMed]
5. Park, W.G.; Cohen, J. Quality measurement and improvement in upper endoscopy. *Tech. Gastrointest. Endosc.* **2012**, *14*, 13–20.
6. Bretthauer, M.; Aabakken, L.; Dekker, E.; Kaminski, M.F.; Roesch, T.; Hultcrantz, R.; Suchanek, S.; Jover, R.; Kuipers, E.J.; Bisschops, R.; et al. Requirements and standards facilitating quality improvement for reporting systems in gastrointestinal endoscopy: European Society of Gastrointestinal Endoscopy (ESGE) Position Statement. *Endoscopy* **2016**, *48*, 291–294. [CrossRef]
7. Nayyar, Z.; Khan, M.; Alhussein, M.; Nazir, M.; Aurangzeb, K.; Nam, Y.; Kadry, S.; Haider, S. Gastric Tract Disease Recognition Using Optimized Deep Learning Features. *CMC-Comput. Mater. Contin.* **2021**, *68*, 2041–2056. [CrossRef]
8. Zhang, X.; Chen, F.; Yu, T.; An, J.; Huang, Z.; Liu, J.; Hu, W.; Wang, L.; Duan, H.; Si, J. Real-time gastric polyp detection using convolutional neural networks. *PLoS ONE* **2019**, *14*, e0214133. [CrossRef]
9. Guimares, P.; Keller, A.; Fehlmann, T.; Lammert, F.; Casper, M. Deep-learning based detection of gastric precancerous conditions. *Gut* **2020**, *69*, 4–6. [CrossRef]
10. Wang, C.; Li, Y.; Yao, J.; Chen, B.; Song, J.; Yang, X. Localizing and Identifying Intestinal Metaplasia Based on Deep Learning in Oesophagoscope. In Proceedings of the 2019 8th International Symposium on Next Generation Electronics (ISNE), Zhengzhou, China, 9–10 October 2019; pp. 1–4. [CrossRef]
11. Yan, T.; Wong, P.K.; Choi, I.C.; Vong, C.M.; Yu, H.H. Intelligent diagnosis of gastric intestinal metaplasia based on convolutional neural network and limited number of endoscopic images. *Comput. Biol. Med.* **2020**, *126*, 104026. [CrossRef]
12. Zheng, W.; Zhang, X.; Kim, J.; Zhu, X.; Ye, G.; Ye, B.; Wang, J.; Luo, S.; Li, J.; Yu, T.; et al. High Accuracy of Convolutional Neural Network for Evaluation of Helicobacter pylori Infection Based on Endoscopic Images: Preliminary Experience. *Clin. Transl. Gastroenterol.* **2019**, *10*, e00109. [CrossRef]
13. Itoh, T.; Kawahira, H.; Nakashima, H.; Yata, N. Deep learning analyzes Helicobacter pylori infection by upper gastrointestinal endoscopy images. *Endosc. Int. Open* **2018**, *6*, E139–E144. [CrossRef]
14. Lin, N.; Yu, T.; Zheng, W.; Hu, H.; Xiang, L.; Ye, G.; Zhong, X.; Ye, B.; Wang, R.; Deng, W.; et al. Simultaneous Recognition of Atrophic Gastritis and Intestinal Metaplasia on White Light Endoscopic Images Based on Convolutional Neural Networks: A Multicenter Study. *Clin. Transl. Gastroenterol.* **2021**, *12*, e00385. [CrossRef]
15. Lee, J.H.; Kim, Y.J.; Kim, Y.W.; Park, S.; Choi, Y.i.; Kim, Y.J.; Park, D.K.; Kim, K.G.; Chung, J.W. Spotting malignancies from gastric endoscopic images using deep learning. *Surg. Endosc. Other Interv. Tech.* **2019**, *33*, 3790–3797. [CrossRef]
16. Zhu, Y.; Wang, Q.C.; Xu, M.D.; Zhang, Z.; Cheng, J.; Zhong, Y.S.; Zhang, Y.Q.; Chen, W.F.; Yao, L.Q.; Zhou, P.H.; et al. Application of convolutional neural network in the diagnosis of the invasion depth of gastric cancer based on conventional endoscopy. *Gastrointest. Endosc.* **2019**, *89*, 806–815.e1. [CrossRef]
17. Ikenoyama, Y.; Hirasawa, T.; Ishioka, M.; Namikawa, K.; Yoshimizu, S.; Horiuchi, Y.; Ishiyama, A.; Yoshio, T.; Tsuchida, T.; Takeuchi, Y.; et al. Detecting early gastric cancer: Comparison between the diagnostic ability of convolutional neural networks and endoscopists. *Dig. Endosc.* **2021**, *33*, 141–150. [CrossRef]
18. Ueyama, H.; Kato, Y.; Akazawa, Y.; Yatagai, N.; Komori, H.; Takeda, T.; Matsumoto, K.; Ueda, K.; Matsumoto, K.; Hojo, M.; et al. Application of artificial intelligence using a convolutional neural network for diagnosis of early gastric cancer based on magnifying endoscopy with narrow-band imaging. *J. Gastroenterol. Hepatol.* **2021**, *36*, 482–489. [CrossRef]
19. Ling, T.; Wu, L.; Fu, Y.; Xu, Q.; An, P.; Zhang, J.; Hu, S.; Chen, Y.; He, X.; Wang, J.; et al. A deep learning-based system for identifying differentiation status and delineating the margins of early gastric cancer in magnifying narrow-band imaging endoscopy. *Endoscopy* **2021**, *53*, 469–477. [CrossRef]
20. Saito, H.; Aoki, T.; Aoyama, K.; Kato, Y.; Tsuboi, A.; Yamada, A.; Fujishiro, M.; Oka, S.; Ishihara, S.; Matsuda, T.; et al. Automatic detection and classification of protruding lesions in wireless capsule endoscopy images based on a deep convolutional neural network. *Gastrointest. Endosc.* **2020**, *92*, 144–151.e1. [CrossRef]

21. Hu, H.; Gong, L.; Dong, D.; Zhu, L.; Wang, M.; He, J.; Shu, L.; Cai, Y.; Cai, S.; Su, W.; et al. Identifying early gastric cancer under magnifying narrow-band images with deep learning: A multicenter study. *Gastrointest. Endosc.* **2021**, *93*, 1333–1341.e3. [CrossRef]
22. Wu, L.; Zhou, W.; Wan, X.; Zhang, J.; Shen, L.; Hu, S.; Ding, Q.; Mu, G.; Yin, A.; Huang, X.; et al. A deep neural network improves endoscopic detection of early gastric cancer without blind spots. *Endoscopy* **2019**, *51*, 522–531. [CrossRef]
23. Wu, L.; Zhang, J.; Zhou, W.; An, P.; Shen, L.; Liu, J.; Jiang, X.; Huang, X.; Mu, G.; Wan, X.; et al. Randomised Controlled Trial of WISENSE, a Real-Time Quality Improving System for Monitoring Blind Spots during Esophagogastroduodenoscopy. *Gut* **2019**, *68*, 2161–2169. [CrossRef] [PubMed]
24. Lin, T.H.; Jhang, J.Y.; Huang, C.R.; Tsai, Y.C.; Cheng, H.C.; Sheu, B.S. Deep Ensemble Feature Network for Gastric Section Classification. *IEEE J. Biomed. Health Inform.* **2021**, *25*, 77–87. [CrossRef] [PubMed]
25. He, Q.; Bano, S.; Ahmad, O.F.; Yang, B.; Chen, X.; Valdastri, P.; Lovat, L.B.; Stoyanov, D.; Zuo, S. Deep learning-based anatomical site classification for upper gastrointestinal endoscopy. *Int. Comput. Assist. Radiol. Surg.* **2020**, *15*, 1085–1094. [CrossRef]
26. Huang, G.; Liu, Z.; van der Maaten, L.; Weinberger, K.Q. Densely Connected Convolutional Networks. *arXiv* **2016**, arXiv:1608.06993.
27. Liu, L.; Wang, P.; Shen, C.; Wang, L.; Van Den Hengel, A.; Wang, C.; Shen, H.T. Compositional Model Based Fisher Vector Coding for Image Classification. *IEEE Trans. Pattern Anal. Mach. Intell.* **2017**, *39*, 2335–2348. [CrossRef] [PubMed]
28. Jin, Y.; Dou, Q.; Chen, H.; Yu, L.; Qin, J.; Fu, C.; Heng, P. SV-RCNet: Workflow Recognition From Surgical Videos Using Recurrent Convolutional Network. *IEEE Trans. Med. Imaging* **2018**, *37*, 1114–1126. [CrossRef] [PubMed]
29. Chen, S.F.; Chen, Y.C.; Yeh, C.K.; Wang, Y.C.F. Order-Free RNN with Visual Attention for Multi-Label Classification. *arXiv* **2017**, arXiv:1707.05495.
30. Wang, J.; Yang, Y.; Mao, J.; Huang, Z.; Huang, C.; Xu, W. CNN-RNN: A Unified Framework for Multi-label Image Classification. *arXiv* **2016**, arXiv:1604.04573.
31. Wang, Z.; Chen, T.; Li, G.; Xu, R.; Lin, L. Multi-label Image Recognition by Recurrently Discovering Attentional Regions. *arXiv* **2017**, arXiv:1711.02816.
32. Li, Q.; Qiao, M.; Bian, W.; Tao, D. Conditional Graphical Lasso for Multi-label Image Classification. In Proceedings of the 2016 IEEE Conference on Computer Vision and Pattern Recognition (CVPR), Las Vegas, NV, USA, 27–30 June 2016; pp. 2977–2986.
33. Li, X.; Zhao, F.; Guo, Y. Multi-Label Image Classification with a Probabilistic Label Enhancement Model. In Proceedings of the 30th Conference on Uncertainty in Artificial Intelligence, UAI'14, Quebec City, QC, Canada, 23–27 July 2014; AUAI Press: Arlington, VA, USA, 2014; pp. 430–439.
34. Zhu, F.; Li, H.; Ouyang, W.; Yu, N.; Wang, X. Learning Spatial Regularization with Image-Level Supervisions for Multi-label Image Classification. In Proceedings of the 2017 IEEE Conference on Computer Vision and Pattern Recognition (CVPR), Honolulu, HI, USA, 21–26 July 2017; IEEE Computer Society: Los Alamitos, CA, USA, 2017; pp. 2027–2036. [CrossRef]
35. Chen, Z.M.; Wei, X.S.; Wang, P.; Guo, Y. Multi-Label Image Recognition with Graph Convolutional Networks. *arXiv* **2019**, arXiv:1904.03582.
36. Padoy, N.; Blum, T.; Ahmadi, S.A.; Feussner, H.; Berger, M.O.; Navab, N. Statistical modeling and recognition of surgical workflow. *Med. Image Anal.* **2012**, *16*, 632–641. [CrossRef]
37. Tao, L.; Zappella, L.; Hager, G.D.; Vidal, R. Surgical Gesture Segmentation and Recognition. *Med. Image Comput. Comput. Assist. Interv.* **2013**, *16*, 339–346. [CrossRef]
38. Lalys, F.; Riffaud, L.; Morandi, X.; Jannin, P. Surgical Phases Detection from Microscope Videos by Combining SVM and HMM. In *Medical Computer Vision. Recognition Techniques and Applications in Medical Imaging*; Menze, B., Langs, G., Tu, Z., Criminisi, A., Eds.; Springer: Berlin/Heidelberg, Germany, 2011; pp. 54–62.
39. Gers, F.A.; Eck, D.; Schmidhuber, J. Applying LSTM to Time Series Predictable through Time-Window Approaches. In Proceedings of the Artificial Neural Networks—ICANN 2001, Vienna, Austria, 21–25 August 2001; Dorffner, G., Bischof, H., Hornik, K., Eds.; Springer: Berlin/Heidelberg, Germany, 2001; pp. 669–676.
40. Zeng, T.; Wu, B.; Zhou, J.; Davidson, I.; Ji, S. Recurrent Encoder-Decoder Networks for Time-Varying Dense Prediction. In Proceedings of the 2017 IEEE International Conference on Data Mining (ICDM), New Orleans, LA, USA, 18–21 November 2017; pp. 1165–1170. [CrossRef]
41. Bisschops, R.; Areia, M.; Coron, E.; Dobru, D.; Kaskas, B.; Kuvaev, R.; Pech, O.; Ragunath, K.; Weusten, B.; Familiari, P.; et al. Performance measures for upper gastrointestinal endoscopy: A European Society of Gastrointestinal Endoscopy (ESGE) Quality Improvement Initiative. *Endoscopy* **2016**, *48*, 843–864. [CrossRef]
42. Yao, K.; Uedo, N.; Kamada, T.; Hirasawa, T.; Nagahama, T.; Yoshinaga, S.; Oka, M.; Inoue, K.; Mabe, K.; Yao, T.; et al. Guidelines for endoscopic diagnosis of early gastric cancer. *Dig. Endosc.* **2020**, *32*, 663–698. [CrossRef]
43. Hu, J.; Shen, L.; Albanie, S.; Sun, G.; Wu, E. Squeeze-and-Excitation Networks. *arXiv* **2017**, arXiv:1709.01507.
44. Dosovitskiy, A.; Beyer, L.; Kolesnikov, A.; Weissenborn, D.; Zhai, X.; Unterthiner, T.; Dehghani, M.; Minderer, M.; Heigold, G.; Gelly, S.; et al. An Image is Worth 16x16 Words: Transformers for Image Recognition at Scale. *arXiv* **2020**, arXiv:2010.11929.
45. Tan, M.; Le, Q.V. EfficientNet: Rethinking Model Scaling for Convolutional Neural Networks. *arXiv* **2019**, arXiv:1905.11946.
46. He, K.; Zhang, X.; Ren, S.; Sun, J. Deep Residual Learning for Image Recognition. *arXiv* **2015**, arXiv:1512.03385.
47. Krizhevsky, A.; Sutskever, I.; Hinton, G.E. ImageNet Classification with Deep Convolutional Neural Networks. *Commun. ACM* **2017**, *60*, 84–90. [CrossRef]

48. Dong, J.; Xia, W.; Chen, Q.; Feng, J.; Huang, Z.; Yan, S. Subcategory-Aware Object Classification. In Proceedings of the 2013 IEEE Conference on Computer Vision and Pattern Recognition, Portland, OR, USA, 23–28 June 2013; pp. 827–834. [CrossRef]
49. Hochreiter, S.; Schmidhuber, J. Long short-term memory. *Neural Comput.* **1997**, *9*, 1735–1780. [CrossRef]
50. Paszke, A.; Gross, S.; Massa, F.; Lerer, A.; Bradbury, J.; Chanan, G.; Killeen, T.; Lin, Z.; Gimelshein, N.; Antiga, L.; et al. PyTorch: An Imperative Style, High-Performance Deep Learning Library. *arXiv* **2019**, arXiv:1912.01703.
51. Xu, B.; Wang, N.; Chen, T.; Li, M. Empirical Evaluation of Rectified Activations in Convolutional Network. *arXiv* **2015**, arXiv:1505.00853.
52. Ge, W.; Yang, S.; Yu, Y. Multi-Evidence Filtering and Fusion for Multi-Label Classification, Object Detection and Semantic Segmentation Based on Weakly Supervised Learning. *arXiv* **2018**, arXiv:1802.09129.
53. Wei, Y.; Xia, W.; Lin, M.; Huang, J.; Ni, B.; Dong, J.; Zhao, Y.; Yan, S. HCP: A Flexible CNN Framework for Multi-Label Image Classification. *IEEE Trans. Pattern Anal. Mach. Intell.* **2016**, *38*, 1901–1907. [CrossRef]
54. Zhou, B.; Khosla, A.; Lapedriza, A.; Oliva, A.; Torralba, A. Learning Deep Features for Discriminative Localization. *arXiv* **2015**, arXiv:1512.04150.
55. Xu, Z.; Tao, Y.; Wenfang, Z.; Ne, L.; Zhengxing, H.; Jiquan, L.; Weiling, H.; Huilong, D.; Jianmin, S. Upper gastrointestinal anatomy detection with multi-task convolutional neural networks. *Healthc. Technol. Lett.* **2019**, *6*, 176–180.
56. Chang, Y.Y.; Li, P.C.; Chang, R.F.; Yao, C.D.; Chen, Y.Y.; Chang, W.Y.; Yen, H.H. Deep learning-based endoscopic anatomy classification: An accelerated approach for data preparation and model validation. *Surg. Endosc.* **2021**, 1–11. [CrossRef]
57. Simonyan, K.; Zisserman, A. Two-Stream Convolutional Networks for Action Recognition in Videos. In Proceedings of the Neural Information Processing Systems (NIPS'14), Montreal, QC, Canada, 8–13 December 2014; Ghahramani, Z., Welling, M., Cortes, C., Lawrence, N., Weinberger, K.Q., Eds.; Curran Associates, Inc.: Red Hook, NY, USA, 2014; MIT Press: Cambridge, MA, USA, 2014; Volume 1, pp. 568–576.
58. Szegedy, C.; Liu, W.; Jia, Y.; Sermanet, P.; Reed, S.; Anguelov, D.; Erhan, D.; Vanhoucke, V.; Rabinovich, A. Going Deeper with Convolutions. *arXiv* **2014**, arXiv:1409.4842.
59. Howard, A.G.; Zhu, M.; Chen, B.; Kalenichenko, D.; Wang, W.; Weyand, T.; Andreetto, M.; Adam, H. MobileNets: Efficient Convolutional Neural Networks for Mobile Vision Applications. *arXiv* **2017**, arXiv:1704.04861.
60. Radosavovic, I.; Prateek Kosaraju, R.; Girshick, R.; He, K.; Dollár, P. Designing Network Design Spaces. *arXiv* **2020**, arXiv:2003.13678.
61. Teh, J.L.; Tan, J.R.; Lau, L.J.F.; Saxena, N.; Salim, A.; Tay, A.; Shabbir, A.; Chung, S.; Hartman, M.; Bok-Yan So, J. Longer Examination Time Improves Detection of Gastric Cancer During Diagnostic Upper Gastrointestinal Endoscopy. *Clin. Gastroenterol. Hepatol.* **2015**, *13*, 480–487.e2. [CrossRef] [PubMed]
62. Conio, M.; Filiberti, R.; Blanchi, S.; Ferraris, R.; Marchi, S.; Ravelli, P.; Lapertosa, G.; Iaquinto, G.; Sablich, R.; Gusmaroli, R.; et al. Risk factors for Barrett's esophagus: A case-control study. *Int. J. Cancer* **2002**, *97*, 225–229. [CrossRef] [PubMed]
63. Gupta, N.; Gaddam, S.; Wani, S.B.; Bansal, A.; Rastogi, A.; Sharma, P. Longer inspection time is associated with increased detection of high-grade dysplasia and esophageal adenocarcinoma in Barrett's esophagus. *Gastrointest. Endosc.* **2012**, *76*, 531–538. [CrossRef] [PubMed]

Article

CoroNet: Deep Neural Network-Based End-to-End Training for Breast Cancer Diagnosis

Nada Mobark [1,*], Safwat Hamad [2] and S. Z. Rida [3]

1. Faculty of Computer and Information, South Valley University, Qena 83523, Egypt
2. Faculty of Computer and Information Sciences, Ain Shams University, Cairo 11566, Egypt; shamad@cis.asu.edu.eg
3. Faculty of Science, South Valley University, Qena 83523, Egypt; szagloul1000@gmail.com
* Correspondence: nada_elshreif@yahoo.com

Abstract: In 2020, according to the publications of both the Global Cancer Observatory (GCO) and the World Health Organization (WHO), breast cancer (BC) represents one of the highest prevalent cancers in women worldwide. Almost 47% of the world's 100,000 people are diagnosed with breast cancer, among females. Moreover, BC prevails among 38.8% of Egyptian women having cancer. Current deep learning developments have shown the common usage of deep convolutional neural networks (CNNs) for analyzing medical images. Unlike the randomly initialized ones, pre-trained natural image database (ImageNet)-based CNN models may become successfully fine-tuned to obtain improved findings. To conduct the automatic detection of BC by the CBIS-DDSM dataset, a CNN model, namely CoroNet, is proposed. It relies on the Xception architecture, which has been pre-trained on the ImageNet dataset and has been fully trained on whole-image BC according to mammograms. The convolutional design method is used in this paper, since it performs better than the other methods. On the prepared dataset, CoroNet was trained and tested. Experiments show that in a four-class classification, it may attain an overall accuracy of 94.92% (benign mass vs. malignant mass) and (benign calcification vs. malignant calcification). CoroNet has a classification accuracy of 88.67% for the two-class cases (calcifications and masses). The paper concluded that there are promising outcomes that could be improved because more training data are available.

Keywords: breast cancer; mammogram; coronet; deep learning; convolutional neural network; transfer learning

1. Introduction

Cancer ranks a significant obstacle to rising life expectancy, and is a leading cause of death worldwide. In 2019, WHO reported that the first or second major reason for death earlier than the age of 70 is cancer, in 112 of 183 nations. It is ranked third or fourth in the other 23 countries [1]. It causes an irregular growth of cells and is frequently named depending on the part of the body in which it occurs. Cancer usually spreads out rapidly throughout the body tissues [2]. It starts in cells, the smallest units of body tissues and organs, e.g., in the breast. Mostly, cancer results from mutations, anarchic division, and multiplication or abnormal changes in the cells. New cells usually replace the old or damaged cells that die. This process occasionally fails, and the cell can keep up uncontrollable or orderless division, creating more cells similar to it and causing a tumor.

A tumor is divided into benign (uncancerous) or malignant (cancerous). Benign tumors are not dangerous, because they do not cause cancer: their cells appear close to normal, grow slowly, and do not attack near tissues or harm other body parts. In contrast, malignant tumors are dangerous. If they are not checked, they ultimately exceed the original tumor and attack other body parts.

Cases and deaths are broken down by global region and type of cancer. In 2020, 19.3 million new cases of cancer (18.1 million excluding NMSC, excluding basal cell carcinoma)

as well as 10 million deaths (9.9 million excluding NMSC, excluding basal cell carcinoma) occurred in various countries of the world (Table 1). Figure 1 depicts the global distribution of new cases and fatalities for the 10 most common types of cancer among females worldwide in 2020 [3].

Table 1. New cases and deaths for 10 cancer types in 2020.

Location of Cancer	Number of New Cases (% of All Locations)		Number of New Deaths (% of All Locations)	
Brain, nervous system	(1.6)	308,102	(2.5)	251,329
Colon	(6.0)	1,148,515	(5.8)	576,858
Female breast	**(11.7)**	**2,261,419**	**(6.9)**	**684,996**
Leukemia	(2.5)	474,519	(3.1)	311,594
Liver	(4.7)	905,677	(8.3)	830,180
Lung	(11.4)	2,206,771	(18.0)	1,796,144
Nonmelanoma of skin	(6.2)	1,198,073	(0.6)	63,731
Ovary	(1.6)	313.959	(2.1)	207.252
Prostate	(7.3)	1,414,259	(3.8)	375,304
Stomach	(5.6)	1,089,103	(7.7)	768,793

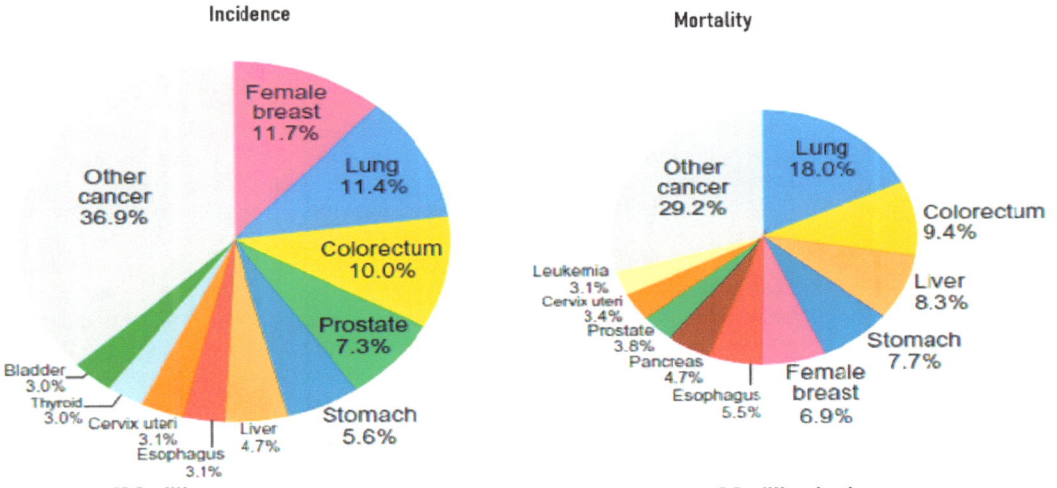

Figure 1. Distributing the Cases and Deaths of the 10 Prevalent Cancers in 2020 for Both Sexes [3].

To detect the presence of cancer in the body, effective techniques are available. In the early stages, Breast Cancer (BC) can be detected through screening; hence, the treatment can be more effective. Several methods are available, including ultrasound, magnetic resonance imaging (MRI), CT, tomosynthesis, and mammography, as well as molecular breast imaging. Because mammography is cheap and available, it is considered the most widely adopted screening method. When examining a human breast, mammography employs low-dose X-rays. Notably, mammography is a simple and affordable method by specialists. Actually, it is considered the gold standard method of detecting the early stages of BC before the lesions turn into something clinically tangible. Its images show cancerous masses and calcium deposits more brightly. As a result, the death rate decreased by 25 to 30%. A specialist receives two views of the breast, producing two images, namely MLO (Medio Lateral Oblique) and CC (Cranio Caudal) views [4,5].

Specialists have accomplished results of cancer detection that have varied broadly. Even the performances of top clinicians pave the way for further improvements [6,7]. Although mammography is used extensively, interpreting its images has challenged specialists. For instance, false positives may cause patient anxiety [8] and unimportant follow-ups, as well as invasive diagnostic proceedings. The types of cancer that are not identified at screening may be unidentifiable until the advanced stages, when they are hard to treat [9].

In the 1990s, mammography had computer-aided detection (CAD). Since then, many assisting tools have been adopted for medical purposes [10]. Although they have been thought of as promising [11,12], this generation of software did not succeed in obtaining a better performance compared to readers in actual settings [6,12,13]. Lately, several developments have resulted in the reissuance of the field, because of the successful attempts of deep learning. Scholars and researchers employed several machine learning methods to detect BC using mammograms [14].

The Digital Database for Mammography Screening (DDSM) [15] represents the highest generally utilized databases of the public mammogram. Several papers utilized the traditional techniques of automatic, not manual, extraction of features, including fractional Fourier transform, Gray Level Co-Occurrence Matrix (GLCM), and Gabor filter, in order to secure features, followed by applying SVM or further classifiers to conduct the classification [16,17]. Furthermore, neural networks were utilized as classifiers [18,19]. Recently, several papers have employed CNN for feature generations, using mammograms [20,21]. Some authors utilized pre-trained CNN as transfer learning uses. Lévy et al. [22] surpassed human performance in the classification of DDSM images using CNN, exploiting transfer learning on pretrained models such as AlexNet, the ImageNet Large Scale Visual Recognition Challenge's winning network in 2012 (ILSVRC), and GoogLeNet, which won the 2014 edition of the same competition [23,24]. Guan [25] only used one Convolutional CNN, with the front convolutional layers being responsible for feature generation and the back fully connected (FC) layers acting as the classifier. Therefore, our CNN uses mammographic images as the input, and the (predicted) label as the output. With no evident overfitting, the average validation accuracy for abnormal vs. normal cases converged at around 0.905. In 2018, Xi et al. used VGGNet, the winner of the ImageNet challenge in 2014, to achieve a 92.53% classification accuracy [26,27]. The same authors exploited ResNet to localize the abnormalities within the full mammography images [28]. Recently, Ragab et al. extracted ROIs from mammography, both manually and with threshold-based techniques, then classified them using AlexNet chained with SVM [29]. On the CBIS DDSM dataset, they claimed an accuracy of 87.2% with a 0.94 AUC. Shen et al. further extended these studies by comparing the findings of several state-of-the-art architectures; when averaging the top four models, they were able to obtain a 0.91 AUC [30].

In 2020, an important article was published in Nature [31], in which the authors trained an ensemble of three models on more than 28,000 mammogram images. Then, they compared its predictions with the decisions of radiologists. The actual labels were determined using follow-up exams or biopsies. It turns out that AI beats humans in terms of sensitivity and specificity.

Some scholars have addressed the scarcity of images in the DDSM dataset by proposing data augmentation techniques. Hussain et al. [32] compared different transformations, proving that using augmentation functions that preserve a high amount of information (i.e., not too disruptive) helps to increase the classification accuracy. Similar results were obtained by Costa et al. in a less extensive study [33].

In this study, we aim to perform abnormality classification in mammography using CNNs. The dataset of interest is the CBIS DDSM. The mammogram images feature two kinds of breast abnormalities: mass and calcification, which can be either benign or malignant. In supplementary, we display the advances of the CAD methods utilized in detecting and diagnosing BC, using mammograms that encompass pre-processing, feature selection, features extraction, and contrast enhancement, as well as methods of classification.

In this paper, Section 2 is dedicated to the Materials and Methods, whereas Section 3 is devoted to the methodology and pre-trained models. Section 4 explores the discussion of classification through classifiers and combined classifiers. Section 5 covers the concluding remarks.

2. Materials and Procedures

2.1. Materials

The mammogram is one the most important methods on the effectiveness and sensitivity of the screening modality [34].

2.1.1. Mammography Datasets

Various datasets are publicly accessible. They differ in terms of size, image format, image type, and resolution, etc., such as DDSM and DDSM's Curated Breast Imaging Subset (CBIS-DDSM) as show (Table 2).

One of the most significant characteristics of a mammogram is the utilization of low-energy X-rays, to screen and diagnose the human breast. Two master views are introduced for acquiring the X-ray images: CC and MLO (Figure 2). Mammography mainly aims to detect BC early [35,36], ordinarily by detecting abnormal regions or masses in the images of the X-ray. These masses are often highlighted by a physician or an expert radiologist.

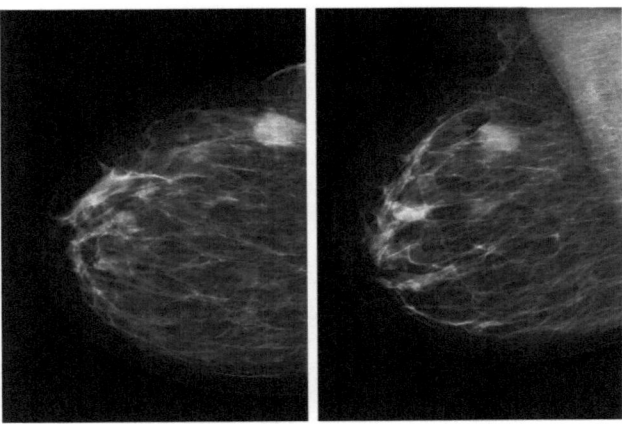

Figure 2. Mammography in CC and MLO views, respectively.

Table 2. Mammography datasets for breast cancer [1,37,38].

Dataset	Type	# of Images	View	Format	Classes	Resolution (Bit/Pixel)	Availability
DDSM [37]	Digital Mammogram (DM)	10,480	MLO/CC	LJPEG	Normal, benign and malignant	8–16	Publicly available
CIBS-DDSM [38]	Digital Mammogram (DM)	10,239	MLO/CC	DICOM	Benign and malignant	10	Publicly available

The paper applied mammographic images from databases. As a dataset, DDSM was first assembled and became available online in 2007 by South Florida University. It contains 2620 scanned film mammographic images of normal, benign, and malignant cases, all stored in Lossless Joint Photographic Experts Group format (LJPEG) with altered sizes and resolutions [37].

DDSM is employed to conduct research in the systems of detecting and classifying BC. It shows real breast data with a resolution of 42 microns, 16 bits, and an average size of 3000 × 4800 pixels. It [15,36,39] holds 2620 scanned film mammography studies distributed in 43 volumes. DDSM database holds 695 normal cases and 1925 abnormal cases (914 malignant/cancer cases and 870 benign cases, as well as 141 benign without callback), specifying the boundaries and locations of the abnormal cases. For every case, four images can be found to represent the left and right breasts in the MLO and CC views (Figure 2) [34]. An experienced radiologist can recognize malignant and benign masses in all mammograms. CBIS-DDSM Dataset: CBIS-DDSM is a developed and united edition of DDSM. Table 3 displays the distribution of data.

Table 3. Distribution of data.

Type	Normal	Abnormal		Total
		Benign	Malignancy	
Train	1190	688	719	2597
Test	128	64	64	256
Total	1318	752	783	2853

2.1.2. Data Pre-Processing

The dataset is provided as a set of numpy arrays, containing the images and labels to use for training and testing. Before these data can be actually used as input for the NN models, a few pre-processing steps are necessary. Depending on the specific classification task (e.g., mass/calcification, benign/malignant, ...), the actions to perform can be slightly different. The following list describes the whole sequence for preparing the data:

1. Import the training and testing data as numpy arrays from shared npy files.
2. When the baseline patches are not needed, remove them and the corresponding labels from the arrays (even indices).
3. Remap the labels, depending on how many, and which classes are involved in the specific classification. If the task is to only distinguish between the masses and calcification, only two labels (0–1) are needed. Conversely, four labels (03) are required when it is also important to discriminate benign abnormalities from malignant ones.
4. Normalize the pixel values to be in a range that is compatible with the chosen model. Scratch CNN models using input in the range (0, 1) floating point, while VGGNet and other pretrained models are designed to work with images in (0, 255) that are further pre-processed with custom transformations (channel swapping, mean subtraction, ...).
5. Shuffle the training set and corresponding labels accordingly.
6. Distribute the training data to "validation" and "training" subsets. The former will be used to compute the loss function exploited by the optimizer, where the actual performance is monitored on an independent group during training, using a validation set.
7. Instantiate Keras generators as data sources for the network. Data augmentation settings can be specified at this stage.

At the end of the pipeline, one or more of the resulting samples are effectively visualized to verify that:

- The data are formatted as expected (size, range ...)
- The images content is still meaningful and was not accidentally corrupted during the process.

2.2. Methodology

2.2.1. Pre-Trained Models

CNNs have grown deeper in the past few years, because they have shown great performance; with the state-of-the-art networks going from 7 layers to 1000 layers. In this paper, we use some of these state-of-art architectures, pre-trained on ImageNet, for transfer learning from natural images to breast cancer images.

2.2.2. Pre-Trained VGG Architecture

A very deep convolutional network has many versions (VGG) [27], and has been published by researchers from Oxford University as one of the best networks; it is known as simple. Its architecture is very easy and deep; the convolution layers and dropout layers are basically switched between. To replicate the influence of bigger receptive fields, the first step is to use numerous small 3 × 3 filters in each convolutional layer and to merge them in a sequence (VGG).

Despite the simple architecture, the network is costly regarding the cost of the computation and memory, because the dramatically rising kernels cause more computational time and a bigger sized model. The applied VGG16 architecture includes 13 convolutional layers and five pooling layers, and attains 9.9% top-5 error on ImageNet. Its immense size makes the training an extremely cumbersome process; notwithstanding, VGG16 is often used for transfer learning, thanks to its flexibility.

2.2.3. Pre-Trained ResNet50 Architecture

Microsoft Research team introduced the ResNet50 for Image Recognition [28]; a deep residual learning model. Notably, it is one of the best developed models. Due to the novel concept of residual layers, some levels are bypassed to prevent a vanishing gradient. The authors developed an elegant, simple, and straightforward idea by gathering a standard deep CNN and adding shortcut connections that avoid limited convolutional layers simultaneously. These connections generate residual blocks, as the convolutional layer's output is prompted by the block's input tensor. The ResNet50 model, for example, is made up of 50 layers of similar blocks connected by shortcuts. These connections keep the computation time to a minimum, and provide a rich combination of features at the same time; see Figure 3.

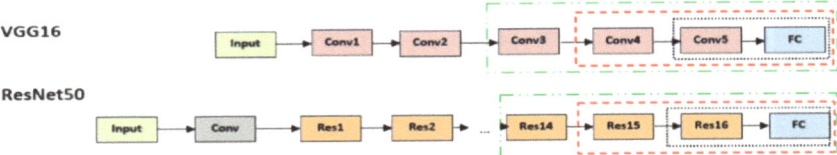

Figure 3. The architecture of VGG16 and ResNet50.

2.2.4. Pre-Trained MobileNet Architecture

MobileNet [40] is a scooped architecture proposed by Google to operate via mobile phones and embedded devices or systems that do not have computational power. Its architecture employed depthwise separable convolutions to radically decrease the sum of trainable parameters, in rapprochement with regular CNNs with corresponding depth. Both the spatial and depth dimensions are handled by the depthwise separable convolution (number of channels). It breaks up the kernel into two parts, one for depthwise convolution, and the other for pointwise convolution. The cost of calculation is considerably reduced when kernels are used. MobileNet provides findings of rapprochement with AlexNet, while drastically reducing the trainable parameters. Table 4 shows Summary of the architectures of CNN.

Table 4. Summary of the architectures of CNN.

Model	Main Finding	Depth	Dataset	Error Rate	Input Size	Year
AlexNet	Utilizes Dropout and ReLU	8	ImageNet	16.4	$227 \times 227 \times 3$	2012
VGG	Increased depth, small filter size	16, 19	ImageNet	7.3	$224 \times 224 \times 3$	2014
ResNet	Robust against overfitting because of symmetry mapping-based skip links	50,152	ImageNet	3.57	$224 \times 224 \times 3$	2016
Xception	A depthwise convolution followed by a pointwise convolution	71	ImageNet	0.055	$229 \times 229 \times 3$	2017
MobileNet-v2	Inverted residual structure	53	ImageNet	-	$224 \times 224 \times 3$	2018

3. New Method

The present part tackles the work method of the suggested methods.

3.1. Convolutional Neural Network (CNN)

Deep CNN represents one of the distinctive types of neural networks that have found major and popular use in machine learning and computer-aided detection applications [41] for better performance and efficiency. The CNN has demonstrated extraordinary performance in several competitions regarding image processing and computer vision. The fantastic uses of CNN involve speech recognition, natural language processing, video processing, and object detection, as well as image classification and segmentation.

CNN is a mathematical structure, which usually includes three types of building blocks:

✓ Convolution layers;
✓ Pooling layers;
✓ Fully connected layers.

Convolution and pooling building blocks perform feature extraction, while the third charts the extracted features into a final output, such as classification. A convolution layer has an interesting part of CNN that is made of many mathematical operations, like convolution, which represents a specialized type of linear operation.

The strong learning ability of the deep CNN network is firstly due to it using several feature extraction phases that can acquire representations based on data automatically. There has been an acceleration in the CNN network by research, due to the large amounts of available data and hardware improvements. Researchers have reported exciting deep CNN architectures. Many inspirational ideas have been discovered for achieving developments in CNN networks, including the use of several activation and loss functions, architectural innovations, regularization, and parameter optimization. They are achieved through architectural innovations and important developments in the representation capacity of CNN deep networks.

3.2. Architecture and Development of the Model

The CNN model, i.e., CoroNet, was proposed to automatically detect BC from mammogram images according to Xception CNN architecture [42,43]. Xception Extreme Inception architecture represents the major feature of Xception (the predecessor model). In addition, it consists of a 71-layer deep CNN architecture pre-trained on an ImageNet dataset. The major conception behind Xception is its depthwise separable convolution. Using this method, the operations' number is decreased using a factor proportional to $1/k$. Xception employs depthwise separable convolution layers with residual connections instead of traditional convolutions. Separable in-depth Convolution replaces the traditional $n \times n \times k$ convolution with a $1 \times 1 \times k$ point-wise convolution followed by a channel-wise $n \times n$ spatial convolution.

Residual connections represent "skip connections" whose authorized gradients flux directly via a network, without travelling via non-linear functions of activation; consequently, disappearing gradients are avoided. In the case of residual connections, the output of a weight layer series is combined with the original input and passed via a non-linear activation function.

Out of the 33,969,964 parameters in CoroNet, 54,528 are non-trainable, and the other 33,969,964 are trainable. Xception represents the base model of CoroNet while adding a dropout layer, and two completely connected layers, ultimately. In Table 5, CoroNet's architecture, layer-wise parameters, and output shape are all depicted. In order to specify the overfitting problem, we used Transfer Learning to initialize the model's parameters.

Table 5. CoroNet Architecture Details.

Layer (Type)	Output Shape	No of Parameters
Xception (Model)	$5 \times 5 \times 2048$	20,861,480
flatten (Flatten)	51,200	0
dropout (Dropout)	51,200	0
dense (Dense)	256	13,107,456
dense_1 (Dense)	4	1028
	Total parameters: 33,969,964	
	Trainable parameters:33,915,436	
	Non-trainable parameters: 54,528	

4. Results and Discussion

The authors performed two scenarios for CoroNet, for the detection of BC from mammogram images. The first model was the major multi-class model (two-class CoroNet), trained to categorize mammogram images into two groups: masses and calcifications. The other was the four-class CoroNet (malignant mass vs. benign mass and malignant calcification vs. benign calcification).

CoroNet, the proposed model, was implemented in Keras on top of Tensorflow 2.0. It was pre-trained on the ImageNet dataset before being retrained end-to-end on the prepared dataset using the Adam optimizer with a learning rate of 0.0001, a batch size of 128, and an epoch value of 200. The data were shuffled before each epoch was activated, which was known as data shuffling. Google Colab was used to perform all of the experiments and training attempts.

The adopted models' training and performance were evaluated with reference to significant parameters, namely, validation loss, training loss, validation accuracy, and training accuracy, at various epochs. Table 6 shows these parameters' results. The parameters were considered to estimate the trained models' under-fitting and over-fitting. The graphs of training loss vs. validation loss and training accuracy vs. validation accuracy of each model were presented (Figures 4–7). In sum, CoroNet demonstrates the minimum training and validation loss, and shows the best accuracies of training and validation.

Table 6. Training performance of the CNN models in the present paper.

Models	Epoch Stop	Validation Accuracy	Training Accuracy	Validation Loss	Testing Loss
VGG 16	13	86.54	68.90	0.2886	0.4320
CoroNet	84	94.73	99.73	0.6079	0.0069
MobileNet	29	68.41	70.24	0.5759	0.6054
ResNet50	12	72.15	74.40	0.5457	0.5948

Figure 4. VGG16 Feature Extraction.

Figure 5. ResNet50 Feature Extraction.

Figure 6. MobileNet Feature Extraction.

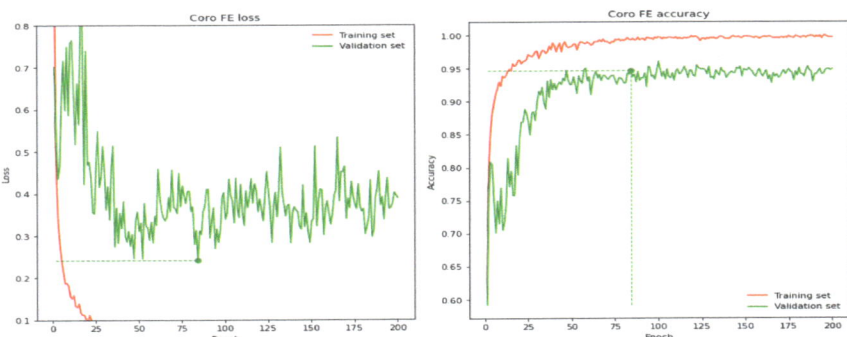

Figure 7. CoroNet Feature Extraction.

5. Conclusions

Deep convolutional neural networks (CNNs) are frequently used for medical image analysis, Unlike the randomly initialized ones, pre-trained natural image database (ImageNet)-based CNN models have a better chance of being successfully fine-tuned to produce better results than those that are randomly initialized. A CNN model called CoroNet is suggested to perform the automatic detection of BC by the CBIS-DDSM dataset. It leverages the Xception architecture, which was completely trained on whole-image BC based on mammograms, and pre-trained on the ImageNet dataset. This paper proved that the convolutional design method "CoroNet" outperforms its alternative networks. In four-class classification, experiments demonstrate that it can achieve an overall accuracy of 94.92 percent (benign mass vs. malignant mass and benign calcification vs. malignant calcification). For the two-class examples, CoroNet has a classification accuracy of 86.67%. (calcifications and masses).

High-resolution mammography handling is seen as a significant difficulty. In order to see the fine features contained in these high-resolution mammograms, the models must also be updated. Although there are various imaging modalities that can be employed, such as MRI and ultrasound, the majority of the current CAD relies on X-ray mammography. The use of 3D mammograms for diagnosis rather than 2D mammograms is another difficult issue that necessitated research in order to make the most of the 3D property, and to improve detection and classification performance.

Author Contributions: Conceptualization, N.M. and S.H.; methodology, N.M.; software, N.M.; validation, S.H. and N.M.; formal analysis, N.M.; investigation, N.M.; resources, N.M.; data curation, S.H.; writing—original draft preparation, N.M.; writing—review and editing, S.H. and N.M.; visualization, N.M.; supervision, S.Z.R.; project administration, S.H.; funding acquisition, N.M. All authors have read and agreed to the published version of the manuscript.

Funding: This research received no external funding.

Institutional Review Board Statement: Not applicable.

Informed Consent Statement: Not applicable.

Data Availability Statement: Not applicable.

Conflicts of Interest: The authors declare no conflict of interest.

References

1. Hassan, N.M.; Hamad, S.; Mahar, K. Mammogram breast cancer CAD systems for mass detection and classification: A review. *Multimed. Tools Appl.* **2022**, *81*, 20043–20075. [CrossRef]
2. Ponraj, D.N.; Jenifer, M.E.; Poongodi, P.; Manoharan, J.S. A survey on the preprocessing techniques of mammogram for the detection of breast cancer. *J. Emerg. Trends Comput. Inf. Sci.* **2011**, *2*, 656–664.

3. Sung, H.; Ferlay, J.; Siegel, R.L.; Laversanne, M.; Soerjomataram, I.; Jemal, A.; Bray, F. Global Cancer Statistics 2020: GLOBOCAN Estimates of Incidence and Mortality Worldwide for 36 Cancers in 185 Countries. *CA Cancer J. Clin.* **2021**, *71*, 209–249. [CrossRef] [PubMed]
4. Saira Charan, S.; Khan, M.J.; Khurshid, K. Breast Cancer Detection in Mammograms using Convolution Neural Network. In Proceedings of the 2018 International Conference on Computing, Mathematics and Engineering Technologies (iCoMET), Sukkur, Pakistan, 3–4 March 2018; Volume 1, pp. 2–6.
5. Omonigho, E.L.; David, M.; Adejo, A.; Aliyu, S. Breast Cancer: Tumor Detection in Mammogram Images Using Modified AlexNet Deep Convolution Neural Network. In Proceedings of the 2020 International Conference in Mathematics, Computer Engineering and Computer Science (ICMCECS), Ayobo, Nigeria, 18–21 March 2020; pp. 1–6. [CrossRef]
6. Elmore, J.G.; Jackson, S.L.; Abraham, L.; Miglioretti, D.L.; Carney, P.A.; Geller, B.M.; Yankaskas, B.C.; Kerlikowske, K.; Onega, T.; Rosenberg, R.D.; et al. Variability in Interpretive Performance at Screening Mammography and Radiologists' Characteristics Associated with Accuracy. *Radiology* **2009**, *253*, 641–651. [CrossRef]
7. Lehman, C.D.; Wellman, R.D.; Buist, D.S.M.; Kerlikowske, K.; Tosteson, A.N.A.; Miglioretti, D.L. Diagnostic Accuracy of Digital Screening Mammography With and Without Computer-Aided Detection. *JAMA Intern. Med.* **2015**, *175*, 1828–1837. [CrossRef] [PubMed]
8. Tosteson, A.N.A.; Fryback, D.G.; Hammond, C.S.; Hanna, L.G.; Grove, M.R.; Brown, M.; Wang, Q.; Lindfors, K.; Pisano, E.D. Consequences of False-Positive Screening Mammograms. *JAMA Intern. Med.* **2014**, *174*, 954–961. [CrossRef]
9. Houssami, N.; Hunter, K. The epidemiology, radiology and biological characteristics of interval breast cancers in population mammography screening. *NPJ Breast Cancer* **2017**, *3*, 12.[CrossRef]
10. Gilbert, F.J.; Astley, S.M.; Gillan, M.G.; Agbaje, O.F.; Wallis, M.G.; James, J.; Boggis, C.R.; Duffy, S.W. Single Reading with Computer-Aided Detection for Screening Mammography. *N. Engl. J. Med.* **2008**, *359*, 1675–1684. [CrossRef]
11. Giger, M.L.; Chan, H.-P.; Boone, J. Anniversary paper: History and status of CAD and quantitative image analysis: The role of Medical Physics and AAPM. *Med. Phys.* **2008**, *35*, 5799–5820. [CrossRef]
12. Fenton, J.J.; Taplin, S.H.; Carney, P.A.; Abraham, L.; Sickles, E.A.; D'Orsi, C.; Berns, E.A.; Cutter, G.; Hendrick, R.E.; Barlow, W.E.; et al. Influence of Computer-Aided Detection on Performance of Screening Mammography. *N. Engl. J. Med.* **2007**, *356*, 1399–1409. [CrossRef]
13. Kohli, A.; Jha, S. Why CAD Failed in Mammography. *J. Am. Coll. Radiol.* **2018**, *15*, 535–537. [CrossRef] [PubMed]
14. Ganesan, K.; Acharya, U.R.; Chua, C.K.; Min, L.C.; Abraham, K.T.; Ng, K.-H. Computer-Aided Breast Cancer Detection Using Mammograms: A Review. *IEEE Rev. Biomed. Eng.* **2012**, *6*, 77–98. [CrossRef] [PubMed]
15. Heath, M.; Bowyer, K.; Kopans, D.; Moore, R.; Kegelmeyer, W.P. The digital database for screening mammography. In Proceedings of the 5th International Workshop on Digital Mammography, Toronto, Canada, 1–14 June 2000; pp. 212–218.
16. Khan, S.; Hussain, M.; Aboalsamh, H.; Bebis, G. A comparison of different Gabor feature extraction approaches for mass classification in mammography. *Multimed. Tools Appl.* **2015**, *76*, 33–57. [CrossRef]
17. Narváez, F.; Alvarez, J.; Garcia-Arteaga, J.D.; Tarquino, J.; Romero, E. Characterizing Architectural Distortion in Mammograms by Linear Saliency. *J. Med. Syst.* **2016**, *41*, 26. [CrossRef] [PubMed]
18. Wang, S.; Rao, R.V.; Chen, P.; Zhang, Y.; Liu, A.; Wei, L. Abnormal Breast Detection in Mammogram Images by Feed-forward Neural Network Trained by Jaya Algorithm. *Fundam. Inform.* **2017**, *151*, 191–211. [CrossRef]
19. Nithya, R.; Santhi, B. Classification of Normal and Abnormal Patterns in Digital Mammograms for Diagnosis of Breast Cancer. *Int. J. Comput. Appl.* **2011**, *28*, 21–25. [CrossRef]
20. Zhu, W.; Lou, Q.; Vang, Y.S.; Xie, X. Deep multi-instance networks with sparse label assignment for whole mammogram classification. In *International Conference on Medical Image Computing and Computer-Assisted Intervention*; Springer: Cham, Switzerland, 2017; pp. 603–611.
21. Sampaio, W.B.; Diniz, E.M.; Silva, A.C.; de Paiva, A.C.; Gattass, M. Detection of masses in mammogram images using CNN, geostatistic functions and SVM. *Comput. Biol. Med.* **2011**, *41*, 653–664. [CrossRef]
22. Lévy, D.; Jain, A. Breast mass classification from mammograms using deep convolutional neural networks. *arXiv* **2016**, arXiv:1612.00542.
23. Krizhevsky, A.; Sutskever, I.; Hinton, G.E. Imagenet classification with deep convolutional neural networks. *NIPS* **2012**, *60*, 84–90. [CrossRef]
24. Szegedy, C.; Liu, W.; Jia, Y.; Sermanet, P.; Reed, S.; Anguelov, D.; Erhan, D.; Vanhoucke, V.; Rabinovich, A. Going deeper with convolutions. In Proceedings of the IEEE Conference on Computer Vision and Pattern Recognition, Boston, MA, USA, 7–12 June 2015; pp. 1–9.
25. Guan, S.; Loew, M. Breast cancer detection using synthetic mammograms from generative adversarial networks in convolutional neural networks. *J. Med. Imaging* **2019**, *6*, 31411. [CrossRef]
26. Xi, P.; Shu, C.; Goubran, R. Abnormality detection in mammography using deep convolutional neural networks. In Proceedings of the 2018 IEEE International Symposium on Medical Measurements and Applications (MeMeA), Rome, Italy, 11–13 June 2018; pp. 1–6.
27. Simonyan, K.; Zisserman, A. Very deep convolutional networks for large-scale image recognition. *arXiv* **2014**, arXiv:1409.1556.
28. He, K.; Zhang, X.; Ren, S.; Sun, J. Deep residual learning for image recognition. In Proceedings of the IEEE Conference on Computer Vision and Pattern Recognition, Las Vegas, NV, USA, 27–30 June 2016; pp. 770–778.

29. Ragab, D.A.; Sharkas, M.; Marshall, S.; Ren, J. Breast cancer detection using deep convolutional neural networks and support vector machines. *PeerJ* **2019**, *7*, e6201. [CrossRef] [PubMed]
30. Shen, L.; Margolies, L.R.; Rothstein, J.H.; Fluder, E.; McBride, R.; Sieh, W. Deep Learning to Improve Breast Cancer Detection on Screening Mammography. *Sci. Rep.* **2019**, *9*, 12495. [CrossRef] [PubMed]
31. McKinney, S.M.; Sieniek, M.; Godbole, V.; Godwin, J.; Antropova, N.; Ashrafian, H.; Back, T.; Chesus, M.; Corrado, G.S.; Darzi, A.; et al. International evaluation of an AI system for breast cancer screening. *Nature* **2020**, *577*, 89–94. [CrossRef]
32. Hussain, Z.; Gimenez, F.; Yi, D.; Rubin, D. Differential Data Augmentation Techniques for Medical Imaging Classification Tasks. In *AMIA Annual Symposium Proceedings*; American Medical Informatics Association: Bethesda, MD, USA, 2018; Volume 2017, pp. 979–984.
33. Costa, A.C.; Oliveira, H.C.R.; Catani, J.H.; de Barros, N.; Melo, C.F.E.; Vieira, M.A.C. Data augmentation for detection of architectural distortion in digital mammography using deep learning approach. *arXiv* **2018**, arXiv:1807.03167.
34. Elmore, J.G.; Armstrong, K.; Lehman, C.D.; Fletcher, S.W. Screening for Breast Cancer. *JAMA J. Am. Med. Assoc.* **2005**, *293*, 1245–1256. [CrossRef]
35. Friedewald, S.M.; Rafferty, E.A.; Rose, S.L.; Durand, M.A.; Plecha, D.M.; Greenberg, J.S.; Hayes, M.K.; Copit, D.S.; Carlson, K.L.; Cink, T.M.; et al. Breast Cancer Screening Using Tomosynthesis in Combination With Digital Mammography. *JAMA J. Am. Med. Assoc.* **2014**, *311*, 2499–2507. [CrossRef]
36. Guan, S.; Loew, M. Breast Cancer Detection Using Transfer Learning in Convolutional Neural Networks. In Proceedings of the 2017 IEEE Applied Imagery Pattern Recognition Workshop (AIPR), Washington, DC, USA, 10–12 October 2017; pp. 1–8. [CrossRef]
37. Heath, M.; Bowyer, K.; Kopans, D.; Kegelmeyer, P.; Moore, R.; Chang, K.; Munishkumaran, S. Current Status of the Digital Database for Screening Mammography. In *Digital Mammography*; Springer: Berlin/Heidelberg, Germany, 1998; pp. 457–460. [CrossRef]
38. Lee, R.S.; Gimenez, F.; Hoogi, A.; Miyake, K.K.; Gorovoy, M.; Rubin, D.L. A curated mammography data set for use in computer-aided detection and diagnosis research. *Sci. Data* **2017**, *4*, 170177. [CrossRef]
39. Suckling, J. The Mammographic Image Analysis Society Digital Mammogram Database. *Digit. Mammo* **1994**, *1069*, 375–378.
40. Howard, A.G.; Zhu, M.; Chen, B.; Kalenichenko, D.; Wang, W.; Weyand, T.; Andreetto, M.; Adam, H. Mobilenets: Efficient convolutional neural networks for mobile vision applications. *arXiv* **2017**, arXiv:1704.04861.
41. Khan, A.; Sohail, A.; Zahoora, U.; Qureshi, A.S. A survey of the recent architectures of deep convolutional neural networks. *Artif. Intell. Rev.* **2020**, *53*, 5455–5516. [CrossRef]
42. Khan, A.I.; Shah, J.L.; Bhat, M.M. CoroNet: A deep neural network for detection and diagnosis of COVID-19 from chest x-ray images. *Comput. Methods Programs Biomed.* **2020**, *196*, 105581. [CrossRef] [PubMed]
43. Chollet, F. Xception: Deep learning with depthwise separable convolutions. In Proceedings of the IEEE Conference on Computer Vision and Pattern Recognition, Honolulu, HI, USA, 21–26 July 2017; pp. 1251–1258.

Article

Improving Medical X-ray Report Generation by Using Knowledge Graph

Dehai Zhang *, Anquan Ren, Jiashu Liang, Qing Liu, Haoxing Wang and Yu Ma

School of Software, Yunnan University, Kunming 650504, China
* Correspondence: dhzhang@ynu.edu.cn

Abstract: In clinical diagnosis, radiological reports are essential to guide the patient's treatment. However, writing radiology reports is a critical and time-consuming task for radiologists. Existing deep learning methods often ignore the interplay between medical findings, which may be a bottleneck limiting the quality of generated radiology reports. Our paper focuses on the automatic generation of medical reports from input chest X-ray images. In this work, we mine the associations between medical discoveries in the given texts and construct a knowledge graph based on the associations between medical discoveries. The patient's chest X-ray image and clinical history file were used as input to extract the image–text hybrid features. Then, this feature is used as the input of the adjacency matrix of the knowledge graph, and the graph neural network is used to aggregate and transfer the information between each node to generate the situational representation of the disease with prior knowledge. These disease situational representations with prior knowledge are fed into the generator for self-supervised learning to generate radiology reports. We evaluate the performance of the proposed method using metrics from natural language generation and clinical efficacy on two public datasets. Our experiments show that our method outperforms state-of-the-art methods with the help of a knowledge graph constituted by prior knowledge of the patient.

Keywords: radiology report; computer-aided diagnosis; prior knowledge; knowledge graph; deep learning

1. Introduction

Medical images are important to diagnose and detect underlying diseases, and radiological reports are essential to aid clinical decision making [1]. They describe some observations of the image such as the extent, size, and location of the disease. The physician communicates findings and diagnoses from the patient's medical scan through the medical report. This process is often laborious, taking an average of 5 to 10 min to write a medical report [2]. The daily task of a radiologist involves analyzing a large number of medical images, which helps the physician to locate the lesion more accurately. Due to the increasing demand for medical images, radiologists still have a large workload. However, the process of writing radiology reports can be time-consuming and tedious for radiologists [3], and it can also be error-prone when writing a report. In addition, the ability to automatically generate accurate reports helps radiologists and physicians to make quick and meaningful diagnoses. Its potential efficiency and benefits can be substantial, especially in critical situations such as outbreaks of COVID or similar pandemics. In order to reduce the burden on radiologists, it is important to be able to generate reports accurately and automatically. These reasons provide a good motivation for our research into the automatic generation of medical reports.

With the development of image captioning and the availability of large-scale datasets, the application of deep learning in the automatic generation of medical reports has been continuously deepened. However, how to generate accurate radiology reports is still a challenging task, because the radiology report generation task is quite different from the image captioning task. First, a radiology report is generated to output a paragraph, which is usually composed of several sentences, while image captioning generally only needs to generate one sentence. Secondly, the generation of radiology reports requires extensive domain knowledge to generate clinically coherent text and the use of medical terms to describe normal and abnormal medical observations [4]. In addition, in the image caption, the model needs to cover all the details of the image as much as possible to generate rich captions. However, in radiology report generation, the model only needs to focus on abnormal areas and infer potential diseases to generate radiology reports. The relationship between potential diseases can determine the accuracy of the generated radiology reports.

Most existing methods focus on the image-to-fluent text aspect of the medical report generation problem. These methods generally perform reasonably well in addressing verbal fluency, but their results are significantly unsatisfactory in terms of clinical accuracy. The possible reason is that their results are far from proficient in revealing the topics related to the expected diseases and symptoms in the generated texts, and they ignore the associations between these underlying diseases. Adding a knowledge graph can alleviate this problem. A knowledge graph describes the relationships between concepts, entities, and their keys in the objective world in a structured way. Knowledge graphs are also often used as prior knowledge, which can provide complementary information for accurate reporting. Medical reports typically consist of many long sentences describing various disease-related symptoms and related topics in precise and domain-specific terms, which may potentially influence each other. Deep learning methods often suffer from a lack of knowledge when not explicitly taught compared to experienced radiologists, which limits the accuracy of the generation. Modeling the associations between medical observations in the form of a knowledge graph allows us to further leverage prior knowledge to generate high-quality reports. To this end, Zhang et al. [5] and Li et al. [6] combined knowledge reasoning based on the knowledge graph with encoders and decoders for radiology report generation. However, their prior knowledge is manually predefined and, therefore, requires domain experts to be closely involved in the design and implementation of the system, which is a waste of time and effort. Due to the nature of the graph, their method can usually achieve high accuracy, but may miss some important findings. While it is feasible to manually identify and implement a high-quality knowledge graph to obtain good accuracy, it is often impractical to exhaustively encode all nodes and relationships in this way. Our work uses prior knowledge from text mining to build a generic knowledge graph to alleviate some of these concerns.

We propose an innovative framework for automatic report generation based on prior knowledge, which seamlessly integrates prior and linguistic knowledge at different levels. First, we investigate a data-driven approach to automatically obtain associations between disease labels in radiology reports. This prior knowledge is a natural extension of human-designed knowledge. Disease labels are defined as nodes in the knowledge graph, which are related and influence each other during the propagation of the graph. Secondly, we establish a graph convolutional neural network to aggregate and transmit information between each node to obtain prior knowledge [7]. Specifically, a set of multi-view chest X-ray images are sent to the convolutional neural network for image feature extraction, and then the content of the clinical document instructed by the doctor is used for text feature extraction using Transformer [8]. The two extracted features are summed and normalized to wound together to obtain a hybrid image–text feature, called a contextualized embedding. The image–text hybrid features and the adjacency matrix constructed according to the knowledge graph are transferred to the three-layer graph convolutional network, and the special features of each knowledge graph node are learned to obtain the episodic representation with prior knowledge. Then, these node features are transferred to two branches, a linear classifier for disease classification, and a generator made of a transformer to generate reports. After generating the report, the generated report is passed into the text classifier again to fine-tune the generated report. Unlike previous studies, additional text mining concepts are added to the model as labels for classification as well as nodes in the knowledge graph, and the expressive power of the model is enriched by training on the chest X-ray image dataset with structured labels.

We evaluate our proposed method on the publicly accessible Open-I [9] and MIMIC-CXR [10] datasets, where we employ Natural Language Generation (NLG) and clinical efficacy (CE) metrics to analyze the quality of clinically generated reports. The results show that the proposed method achieves good performance in both natural language generation and clinical efficacy indicators. It is also shown that the addition of prior knowledge helps to improve the quality and accuracy of the automatic generation of radiology reports.

Our contributions are outlined below:

1. We mine and model the text, and according to the mined information, we use prior knowledge to build the knowledge graph and construct the corresponding adjacency matrix (Section 3.1.1).
2. We combine text–image hybrid features with knowledge reasoning based on the knowledge graph to improve the quality and accuracy of radiology report generation (Section 3.1.4).
3. Our experiments show that our proposed model outperforms state-of-the-art methods. The knowledge graph composed of prior knowledge of patients plays a crucial role in improving the quality of generated reports.

2. Related Work

2.1. Image-Based Captioning and Medical Report Generation

Most of the work on image-based captioning is based on the classical structure of CNN + LSTM, which aims to generate real sentences or relevant paragraphs of a topic to summarize the visual content in an image or video [11–14]. With the development of computer vision and natural language processing technology, many works combine radiology images and free text to automatically generate radiology reports to help clinical radiologists make a quick and meaningful diagnosis [15]. Radiology report generation takes X-ray images as input and generates descriptive reports to support inference of better diagnostic conclusions beyond disease labels. Many radiology report generation methods follow the practice of image captioning models [16–18]. For example, ref. [19] adopted an encoder–decoder architecture and proposed a hierarchical generator and attention mechanism to generate long reports. Xue et al. [20] fused the visual features and semantic features of the previous sentence through the attention mechanism, used the fused features to generate the next sentence, and then generated the whole report in a loop. Wang et al. [21] proposed an embedding network with text and image as input to jointly learn text and image information and train the CNN-LSTM architecture end-to-end, which was then combined with a multi-level attention model to generate a chest X-ray report. Chen et al. [22] recorded important information during the generation process and then further assisted the generation of radiology reports by providing memory-driven transformers. Jing et al. [23] used reinforcement learning to exploit structural information between and within reports to generate high-quality radiology reports. Liu et al. [24] combined self-key sequence training and reinforcement learning to optimize the emergence of disease keywords in radiology reports. Shin et al. [25] adopted the CNN-RNN framework to generate radiological reports describing detected diseases based on visual features on chest X-ray image datasets.

Our work is mainly similar to that of Hoang et al. [26], who proposed a fully distinguishable end-to-end structural model, which mainly consists of three complementary modules for classifier, generator, and interpreter, which increase the linguistic fluency and clinical accuracy of generated reports. However, it does not add the association between disease labels, and lacks the association between disease labels in the classification. We use the association between labels learned from the text knowledge base to promote the semantic alignment between disease labels and images, which can better show the correlation between disease labels in classification, improve the accuracy of label classification, and further improve the accuracy of report generation.

2.2. Knowledge Graph

The concept of the knowledge graph was formally proposed by Google in 2012 to achieve a more intelligent search engine. Since 2013, it has gained popularity in academia and industry and plays an important role in intelligent question answering, intelligence analysis, anti-fraud, and other applications. A knowledge graph is essentially a knowledge base known as a semantic network, which is a knowledge base with a directed graph structure where the nodes of the graph represent entities or concepts. The edges of the graph represent semantic relations between entities/concepts, such as similarity relations between two entities, or syntactical correspondences.

In our knowledge graph, we use the tool SentencePiece [27] to obtain entities and then determine the relationship between entities based on the number of co-occurrences to build a knowledge graph, which is a structured way to represent knowledge graphically. In a knowledge graph, information is represented as a set of nodes, which are connected by a set of labeled directed lines to represent the relationship between nodes. A knowledge graph can well represent the relationship between nodes.

2.3. Transformer

Transformer was first introduced in the context of machine translation with the aim of speeding up training and improving remote dependency modeling. It is implemented by parallel processing of sequential data and an attention mechanism, which consists of a multi-head self-attention module and a feed-forward layer. By considering multi-head self-attention mechanisms and graph attention networks [28], recent transformer-based models have shown considerable progress in many difficult tasks, such as image generation [29], question answering, and linguistic reasoning [30]. Radiology reports are usually composed of several long sentences. As the traditional RNN is not suitable for generating long sentences and paragraphs, Chang et al. [31] designed a hierarchical RNN architecture as the decoder to generate long sentences, but the effect was not satisfactory. The recently emerged Transformer architecture can alleviate this problem. Therefore, we mainly use Transformers to compose our text codec in our work.

3. Our Approach

Our framework consists of a classification module, a generation module, and an interpretation module, as shown in Figure 1. The classification module consists of a multi-view image encoder, a text encoder, and a graph convolutional network based on a knowledge graph. We first build a knowledge graph in a data-driven manner (Section 3.1.1), then use a multi-view image encoder to read multiple chest X-ray images and extract the global visual feature representation, which is passed to a fully connected layer to decouple the global visual feature representation into multiple low-dimensional visual embedding (Section 3.1.2). At the same time, the text encoder reads the clinical documents and summarizes the content into text summary embedding, and then uses the "Add &LayerNorm" operation to wrap the visual embedding and text summary embedding together to obtain the image–text hybrid features, which are referred to as context-related embedding of the disease topic (Section 3.1.3). The episodic embedding is passed to a graph convolutional network (GCN) based on a knowledge graph that propagates semantic correlations between disease topics based on the knowledge graph to inject prior knowledge into concept representation learning (Section 3.1.4). The generation module takes as initial input the rich disease embedding that passes through the graph convolutional network to generate the text (Section 3.2). Finally, the generated text is sent to the interpreter for fine-tuning to align with the disease-related topics of the classification module (Section 3.3). In what follows, we elaborate on these three modules.

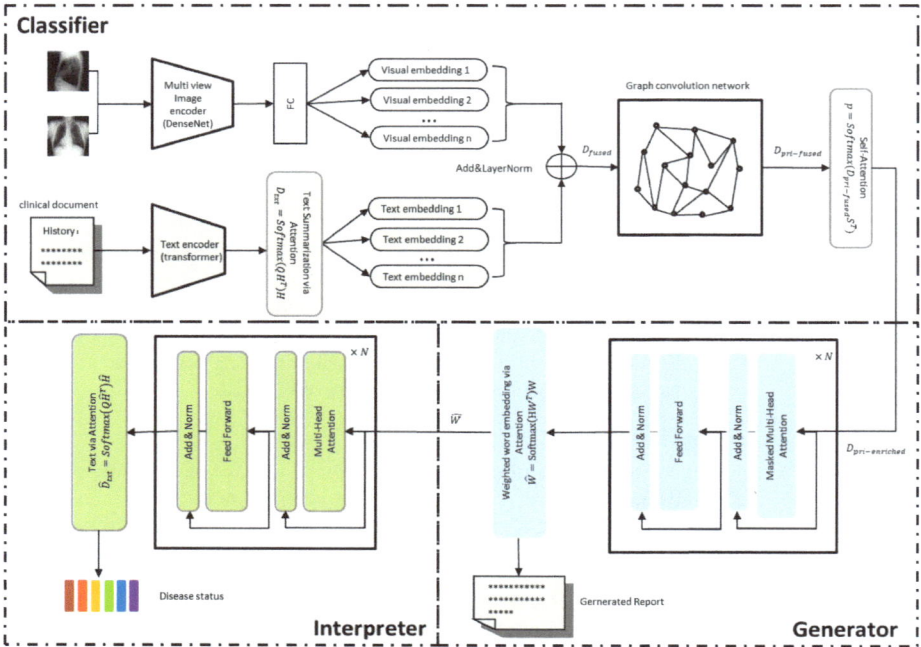

Figure 1. Our model mainly includes three parts. The classifier is used to read chest X-ray images and clinical history to extract visual features and text features, and then the image–text hybrid features and the adjacency matrix of the knowledge graph are passed into the graph convolutional network to obtain the features containing prior knowledge. The obtained features are combined with topic features and state features and fed into the generator based on the visual self-attention model to generate the radiology report. The generated report is then passed into the interpreter to adjust the generated radiology report.

3.1. Classifier

3.1.1. Construction of Knowledge Graph

In our study, the nodes of the knowledge graph are disease-related topics. Edges are semantic associations between concepts. Our knowledge graph consists of two parts. The first part is defined by the CheXpert [32] tagger, a rule-based system that extracts and classifies medical reports into 14 common diseases. The label of each disease has four states, namely positive, negative, uncertain, or unmentioned. The MIMIC-CXR dataset has been annotated by the CheXpert tagger. The second part consists of supplementary concepts and their interrelationships, mined from radiology reports in a data-driven manner. Specifically, we count nouns in radiology reports using the SentencePiece tool, which is an unsupervised text tagger and de-tagger, and then we select nouns with top-k occurrences as additional disease labels, if they are not included in the fourteen disease labels defined by the CheXpert tagger. We establish the knowledge graph according to the co-occurrence of labels in radiology reports (Figure 2), and then construct the incidence matrix and binarize the matrix to form the adjacency matrix of the knowledge graph [33]. Specifically, we build a $n \times n$ matrix, where each row or column represents a label, and then calculate the values in the matrix based on the number of co-occurrences of the labels in the radiology report. If the number of co-occurrences between two labels is greater than the average number of co-occurrences, the two labels are considered to be associated, and the corresponding value of the matrix is assigned a value of 1. On the contrary, if the number of co-occurrences between two labels is less than the average number of co-occurrences, the two labels are regarded as not associated, and the corresponding value of the matrix is assigned a value of 0. This matrix is then regarded as the adjacency matrix of the knowledge graph.

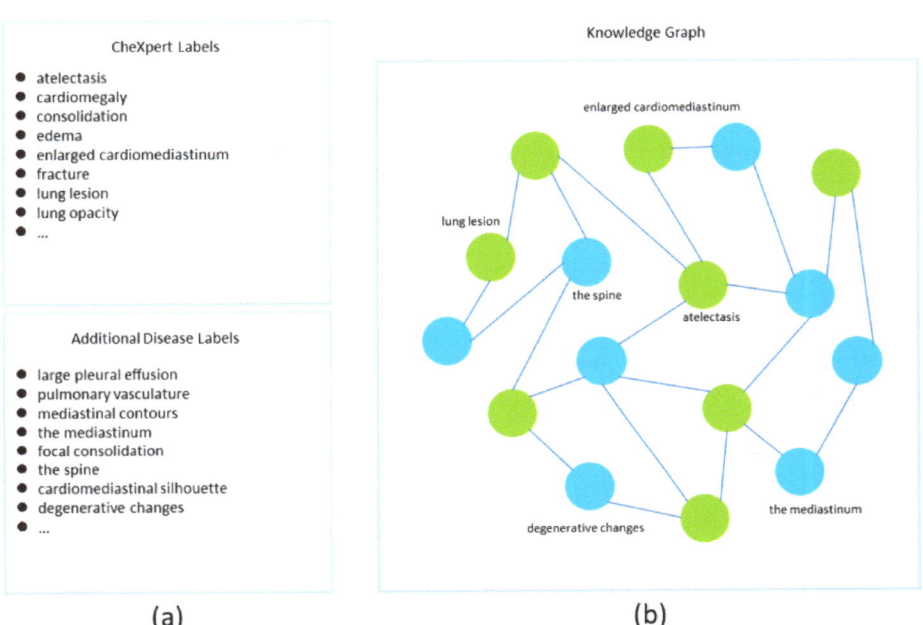

Figure 2. Construction of the knowledge graph. The knowledge graph is built based on the co-occurrence of the tags in the radiology report, and only the tag names of some nodes are shown. In the figure, part (**a**) represents all node names in the knowledge graph; part (**b**) represents the knowledge graph, where the green nodes represent the 14 common disease labels defined by CheXpert, and the blue nodes represent the additional disease labels defined by us.

3.1.2. Multi-View Image Encoder

Each medical study consists of m chest X-ray images $\{X_i\}_{i=1}^{m}$, as DenseNet can achieve better performance than ResNet [34] with fewer parameters and computational cost. We extract its corresponding latent features $\{\mathcal{X}_i\}_{i=1}^{m} \in \mathbb{R}^c$ via a Densenet-121 [35] image encoder with shared weights, where c is the number of features. Then, we obtain multi-view latent features $\mathcal{X} \in \mathbb{R}^c$ by combining m latent feature sets $\{\mathcal{X}_i\}_{i=1}^{m}$, referring to the method proposed by Su et al. [36]. When $m = 1$, the multi-view encoder reduces to a single-image encoder.

3.1.3. Text Encoder and Disease Contextualized Representation

Let T be a text document of length l with word embedding $\{w_1, w_2, \ldots, w_l\}$, where $w_i \in \mathbb{R}^e$ denotes the i-th word in the text and e is the embedding dimension. We use the Transformer encoder as our text feature extractor to extract a set of hidden states $H = \{h_1, h_2, \ldots, h_l\}$, where $h_i \in \mathbb{R}^e$ denotes the attention feature of the i-th word to other words in the text.

$$h_i \in Encoder(w_i | w_1, w_2, \ldots, w_l) \tag{1}$$

n disease-related topics are queried from the whole document T, which is summarized as $Q = \{q_1, q_2, \ldots, q_n\}$. We refer to this retrieval process as text summarization embedding $D_{txt} \in \mathbb{R}^{n \times e}$:

$$D_{txt} = Softmax\left(QH^T\right)H \tag{2}$$

$q_i \in \mathbb{R}^e$ is randomly initialized and updated by the attention mechanism. $Softmax(QH^T)$ represents the word-attention heat-map of n query diseases in the document.

As shown in Figure 1, we decouple the multi-view latent features $\mathcal{X} \in \mathbb{R}^C$ extracted by the image encoder into a low-dimensional disease representation $D_{img} \in \mathbb{R}^{n \times e}$, where each row is a vector $\varphi_j(x) \in \mathbb{R}^e, j = 1, 2, \ldots, n$. $\varphi_j(x)$ is defined as follows:

$$\varphi_j(x) = A_j^T x + b_j \tag{3}$$

where $A_j \in \mathbb{R}^{c \times e}$ and $b_j \in \mathbb{R}^e$ are trainable parameters for the j-th class of disease representation. n denotes the number of disease representations and e is the dimension of the embedding. Then, the visual embedding D_{img} and the text summary embedding D_{txt} are twisted together to form the disease situational representation $D_{fused} \in \mathbb{R}^{n \times e}$:

$$D_{fused} = LayerNorm(D_{img} + D_{txt}) \tag{4}$$

The fusion of visual and textual information allows our model to simulate the workflow of a hospital to screen the visual manifestations of a disease based on a patient's clinical history.

3.1.4. Graph Convolutional Networks and Contextualized Representations of Diseases with Prior Knowledge

We use the GCN to model the intrinsic association between diseases or topics, and the adjacency matrix is built based on the knowledge graph detailed above. The GCN updates its node representation via message passing, and graph convolution is represented as [37]:

$$\hat{H}^l = ReLU\left(BN\left(Conv1d\left(H^l\right)\right)\right) \tag{5}$$

$$m = ReLU\left(D^{-1/2} \hat{A} D^{-1/2} H^l W^l\right) \tag{6}$$

$$H^{l+1} = ReLU(BN(Conv1d(concat(\hat{H}^l, m)))) \tag{7}$$

where H^l is the state in layer l and H^0 is initialized using the disease contextualized representation. $\hat{A} = A + I_N$ is the adjacency matrix with self-connection, A is the adjacency matrix of the knowledge graph, I_N is the identity matrix of order N, $D = \text{diag} \sum_j A_{ij}$ is the degree matrix of the graph, BN is the batch normalization, and W^l is a trainable layer-specific weight matrix. We extract D_{fused} to the semantic information between nodes by the GCN, and obtain the disease situational representation $D_{pri\text{-}fused} \in \mathbb{R}^{n \times e}$ with prior knowledge.

3.1.5. Rich Disease Representation with Prior Knowledge

To further improve the accuracy of generated reports, we introduce rich disease representations with prior knowledge. The main idea behind rich disease representations with prior knowledge is to further encode informative attributes about the disease state, such as positive, negative, uncertain, or unmentioned. Formally, let k be the number of states and the state embedding be $S \in \mathbb{R}^{k \times e}$, then the confidence of each disease classification as one of k disease states is:

$$p = \text{Softmax}\left(D_{pri\text{-}fused} S^T\right) \tag{8}$$

S is a trainable parameter initialized randomly. D_{fused} is used as the feature of multi-label classification, and the classification loss function is as follows:

$$\mathcal{L}_{p-c} = -\frac{1}{n} \sum_{i=1}^{n} \sum_{j=1}^{k} y_{ij} \log(p_{ij}) \tag{9}$$

where $y_{ij} \in \{0,1\}$ and $p_{ij} \in [0,1]$ represent the true and predicted values of the i-th disease label, respectively. The state embedding $D_{states} \in \mathbb{R}^{n \times e}$ can be calculated as follows:

$$D_{states} \begin{cases} yS, \text{ Training stage} \\ pS, \text{ Other stages} \end{cases} \quad (10)$$

Finally, $D_{enriched} \in \mathbb{R}^{n \times e}$ is the disease representation enriched with prior knowledge, and is composed of disease embedding states, disease name topics, and episodic disease representations with prior knowledge.

$$D_{pri\text{-}enriched} = D_{states} + D_{topics} + D_{pri\text{-}fused} \quad (11)$$

where $D_{topics} \in \mathbb{R}^{n \times e}$ is randomly initialized to represent the disease or topic to be generated and is a trainable parameter. The rich disease representation with prior knowledge provides a clear and accurate disease description, which provides strong data support for the subsequent generation module.

3.2. Generator

Our report generator is derived from Transformer. As shown in Figure 1, the network consists of the masked multi-head self-attention module and the feed-forward layer superimposed on each other N times. The previous disease embedding and word embedding are then used to calculate the hidden state $h_i \in \mathbb{R}^e$ for each word of the medical reporter species, and the disease embedding is denoted as $D_{pri\text{-}enriched} = \{d_i\}_{i=1}^n$:

$$h_i = Encoder(w_i | w_1, w_2, \ldots, w_{i-1}, d_1, d_2, \ldots, d_n) \quad (12)$$

Then, we predict the possible words based on the hidden state $H = \{h_i\}_{i=1}^l \in \mathbb{R}^{l \times e}$.

$$p_{word} = Softmax\left(HW^T\right) \quad (13)$$

$W \in \mathbb{R}^{v \times e}$ is this vocabulary embedding, v is the vocabulary size, l is the length of the document, and $p_{word,ij}$ represents the confidence that the i-th position in the generated medical report selects the j-th word in the vocabulary. The loss function of the generator is the cross entropy of the real word $y_{word,ij}$ and the predicted word $p_{word,ij}$.

$$\mathcal{L}_g = -\frac{1}{l} \sum_{i=1}^{l} \sum_{j=1}^{v} y_{word,ij} \log\left(p_{word,ij}\right) \quad (14)$$

The final generated report $\hat{W} \in \mathbb{R}^{l \times e}$ is:

$$\hat{W} = p_{word} W \quad (15)$$

3.3. Interpreter

To make the generated report more consistent with the original output of the classifier, we refer to the idea of CycleGAN [38]. We build a text classifier based on the text encoder above, input the generated report into the text classifier, output the state of the disease-related topic, compare it with the original output of the classifier module, and then fine-tune the generated report by adjusting the word representation output $\hat{W}$.

First, the text encoder summarizes the current medical report $\hat{W}$ and outputs the report summary embedding of the queried disease Q.

$$\hat{D}_{txt} = Softmax\left(Q\hat{H}^T\right)\hat{H} \in \mathbb{R}^{n \times e} \quad (16)$$

where $\hat{H}$ is calculated from the medical report $\hat{W}$ using Equation (1). Each report summary embedding $\hat{d}_i \in \mathbb{R}^e$ is classified into one of k disease-related states, and $\hat{d}_i$ is the i-th line of $\hat{D}_{txt}$.

$$p_{int} = Softmax\left(\hat{D}_{txt}S^T\right) \in \mathbb{R}^{n \times k} \tag{17}$$

The loss function of the interpreter is similar to the loss function of the classifier.

$$\mathcal{L}_i = -\frac{1}{n}\sum_{i=1}^{n}\sum_{j=1}^{k} y_{ij} \log(p_{int,ij}) \tag{18}$$

Here, $y_{ij} \in \{0,1\}$ denotes the confidence of the true disease label, and $p_{int,ij}$ denotes the confidence of the predicted label. Overall, the training loss function of the whole model can be summarized as follows.

$$\mathcal{L}_{all} = \mathcal{L}_{p-c} + \mathcal{L}_g + \mathcal{L}_i \tag{19}$$

4. Experiments

The experimental section evaluates the medical report generation task from two aspects: verbal performance and clinical accuracy performance. The evaluation of the experiments is performed on two widely used chest X-ray datasets, the Open-I and MIMIC-CXR datasets.

4.1. Datasets

4.1.1. MIMIC-CXR Dataset

The MIMIC-CXR dataset is a large publicly available dataset of chest X-ray images in JPG format, containing 377,110 images and 227,835 medical reports of 65,379 patients from multiple viewpoints. Chest X-ray images from three main perspectives are reported: anterior-posterior (AP), posterior-anterior (PA), and lateral (LA). Each study included a comparison, clinical history, indication, reason for examination, impression, and findings section. In our approach, we use multi-view images and concatenate the clinical history, reason for examination, and indication sections as contextual information. For consistency, we follow the experimental setup of [39] and focus on generating the "findings" section as the corresponding radiology report.

4.1.2. Open-I Dataset

The Open-I dataset, collected at Indiana University Hospital, is a public radiology dataset containing 3955 radiology studies corresponding to 7470 frontal and lateral chest X-ray images. These radiological studies are related to one or more chest X-ray images. Each study reported impression, findings, comparison, and indication sections. Similar to the MIMIC-CXR dataset, we use multi-view chest X-ray images (frontal and lateral) and the indicator part as context input. In our approach, we follow the approach of the existing literature [2] and concatenate the impression part and the survey results part as the correct generated report.

4.2. Implementation Detail

We use Densenet-121 as the core of our CNN model, and all images are resized to 256×256. We use Transformer as the core of the text encoder. Both generators and interpreters are implemented based on Transformers and trained from scratch, all hyper-parameters are selected based on the performance on the validation set, and the number of reporting encoder layers is set to 12. We train the classification and generation reports on the Open-I and MIMIC-CXR datasets using the Adam optimizer with an initial learning rate of 3×10^{-4} and weight decay of 1×10^{-2}. For the interpreter, the Open-I dataset is trained using a learning rate of 3×10^{-5}, and the MIMIC-CXR dataset is trained using an Adam optimizer with a learning rate of 3×10^{-6} and a weight decay of 1×10^{-2}. We train the model with epochs of 50 for both Open-I and MIMIC-CXR datasets. We evaluate the proposed model on the validation set. Our experiments are trained in parallel on two RTX3090 sheets, and the experiment for each dataset is mainly divided into two parts. The first part is the training of the classifier and generator, and then the trained model is added to the interpreter for the second part of training. Each part is trained for 50 epochs in our experiments. The training time of the first and second parts is about 54 s and 59 s per round for the Open-I dataset, and about 34 min and 35 min per epoch for the MIMIC-CXR dataset, respectively.

4.3. Experimental Results

4.3.1. Language Generation Performance

We employ the widely used NLG metric to evaluate the proposed model, which includes scores from BLEU-1 to BLEU-4 [40], ROUGE-L [41], and METEOR [42]. We use the nlg-eval library [43] to calculate the BLEU-1 to BLEU-4 scores, ROUGE-L scores, and METEOR scores. In Table 1, our experimental results are compared with other state-of-the-art methods, and all metrics have a certain improvement. The scores of BLEU-1 to BLEU-4 are obtained by analyzing the sequence of consecutive words appearing in the prediction report. In our results, BLEU-1 to BLEU-4 are significantly improved, indicating that our method ignores some meaningless words and focuses more on describing diseases with long sentences. ROUGE-L and METEOR are also much better than previous excellent methods, which mean that our method can generate accurate reports and the framework is effective.

Table 1. A comparison of our method and many existing methods, using different linguistic metrics: BLEU-1 to BLEU-4, METEOR, and ROUGE-L, with the best results highlighted in bold.

Datasets	Methods	BLEU-1	BLEU-2	BLEU-3	BLEU-4	METEOR	ROUGE-L
Open-I	TieNet [21]	0.330	0.194	0.124	0.081	N/A	0.311
	Liu et al. [24]	0.359	0.237	0.164	0.113	N/A	0.354
	KERP [6]	0.482	0.325	0.226	0.162	N/A	0.339
	HRGR-Agent [44]	0.438	0.298	0.208	0.151	N/A	0.322
	SD&C [23]	0.464	0.301	0.210	0.154	N/A	0.362
	CoAtt [2]	0.455	0.288	0.205	0.154	N/A	0.369
	R2Gen [22]	0.470	0.304	0.219	0.165	0.187	0.371
	PPKED [45]	0.483	0.315	0.224	0.168	0.190	0.376
	SGF [46]	0.467	0.334	0.261	0.215	0.201	0.415
	Hoang et al. * [26]	0.490	0.362	0.286	0.233	0.213	0.440
	Ours	**0.505**	**0.379**	**0.303**	**0.251**	**0.218**	**0.446**
MIMIC-CXR	Liu et al. [24]	0.313	0.206	0.146	0.103	N/A	0.306
	R2Gen [22]	0.353	0.218	0.145	0.103	0.142	0.277
	GumbelTransformer [39]	0.415	0.272	0.193	0.146	0.159	0.318
	PPKED [45]	0.360	0.224	0.149	0.106	0.149	0.284
	Hoang et al. * [26]	0.489	0.351	0.267	0.211	0.209	0.381
	Ours	**0.491**	**0.358**	**0.278**	**0.225**	**0.215**	**0.389**

* indicates that the experimental results are reproduced in our experimental environment.

4.3.2. Clinical Accuracy Performance

We use the CheXpert [10] label as a measure to evaluate the clinical accuracy of generated reports. We compare 14 common diseases proposed in CheXpert and MIMIC-CXR based on precision, precision, recall, and F-1 metrics. We show the macro and micro scores, respectively. A high macro score indicates an improvement in the detection of all 14 diseases, while a higher micro score indicates an improvement in the impact caused by the imbalance of the dataset, such as the higher frequency of some diseases than others. The results of our comparison are shown in Table 2. Compared with other experiments, our clinical performance has improved in most of the indicators in the macro and micro scores.

Table 2. The clinical accuracy of the generated reports was quantitatively compared by evaluating 14 common diseases defined together in the CheXpert and MIMIC-CXR datasets, with the best results highlighted using bold font.

Datasets	Methods	Acc.	Macro Scores				Micro Scores			
			AUC.	F-1.	Prec.	Rec	AUC.	F-1.	Prec.	Rec
Open-I	S&T [11]	0.915	N/A	N/A	N/A	N/A	N/A	N/A	N/A	N/A
	SA&T [16]	0.908	N/A	N/A	N/A	N/A	N/A	N/A	N/A	N/A
	TieNet [21]	0.902	N/A	N/A	N/A	N/A	N/A	N/A	N/A	N/A
	Liu et al. [24]	0.918	N/A	N/A	N/A	0.190	N/A	N/A	N/A	N/A
	Hoang et al. [26]	0.937	0.702	0.152	0.142	0.173	0.877	0.626	0.604	0.649
	Ours	**0.938**	**0.749**	**0.193**	**0.246**	**0.181**	**0.925**	**0.636**	**0.614**	**0.660**
MIMIC-CXR	SA&T [16]	N/A	N/A	0.101	0.247	0.119	N/A	0.282	0.364	0.230
	AdpAtt [17]	N/A	N/A	0.163	0.341	0.166	N/A	0.347	0.417	0.298
	Liu et. al [24]	0.867	N/A	N/A	0.309	0.134	N/A	N/A	**0.586**	0.237
	GumbelTransformer [39]	N/A	N/A	0.214	0.327	0.204	N/A	0.398	0.461	0.350
	Hoang et al. [26]	0.887	0.784	0.412	0.432	0.418	0.874	0.576	0.567	0.585
	Ours	**0.890**	**0.858**	**0.560**	**0.587**	**0.593**	**0.907**	**0.640**	0.579	**0.715**

4.4. Ablation Studies

The quantitative results of our method in the Open-I dataset are shown in Table 3. Because the MIMIC-CXR dataset is too large and the effects of our method in both datasets are improved, we mainly focus on the smaller dataset Open-I when analyzing the quantitative results. By observing Table 3, it can be seen that after adding rich disease embedding containing prior knowledge to the classifier, all evaluation indicators are improved, and after adding the interpreter on this basis, the indicators are again improved to a certain extent. Compared with the model that only uses rich disease embedding and adds prior knowledge, the BLEU-1 value of the highest index is increased by 6.7% from 0.445 to 0.475. After adding rich disease embedding with prior knowledge to the classifier, our model adds an interpreter to obtain the best performance. Compared with the basic model with the interpreter, our final model also has a great improvement in various indicators. The BLEU-4 value is improved compared with the basic model without the interpreter and the model with the interpreter, and it is 7% higher than the basic model with the interpreter. It can be seen that the prior knowledge we incorporate is aided by the automatic generation of accurate radiology reports.

Table 3. Ablation studies. Base with $D_{enriched}$ refers to the model composed of the classifier and generator mentioned in this paper, and the classifier uses enriched disease embedding. Base with $D_{pri-enriched}$ refers to the model composed of the classifier and generator mentioned in this paper. Rich disease embedding with prior knowledge is used in the classifier. Interpreter is the Interpreter mentioned above. The best results are highlighted in bold.

Datasets	Methods	BLEU-1	BLEU-2	BLEU-3	BLEU-4	METEOR	ROUGE-L
Open-I	Base with $D_{enriched}$	0.445	0.328	0.256	0.206	0.198	0.434
	Base with $D_{enriched}$ + Interpreter	0.490	0.362	0.286	0.233	0.213	0.440
	Base with $D_{pri-enriched}$	0.475	0.349	0.273	0.222	0.208	0.433
	Base with $D_{pri-enriched}$ + Interpreter	**0.505**	**0.379**	**0.302**	**0.251**	**0.218**	**0.446**

4.5. Qualitative Results

In Figure 3, we visualize the radiology reports of the three sets of images. The same color in the reference report and the generated report indicates the same description of the chest X-ray image, and the underscore indicates additional findings of our model for the chest X-ray image. In the first example, the reports generated by our model are almost identical to the actual situation. In the second example, the report generated by our model completely covers the actual situation and finds an additional disease based on the input image. Although this disease may indeed not exist, it can be used as a reminder to assist the doctor to make a more comprehensive diagnosis. In the third example, our model generates a longer report based on the obtained features, and it is almost the same as the actual situation, indicating that our model also has a relatively good performance in generating complex sentences. It can be seen that our method can extract features from images and clinical documents combined with prior knowledge to generate readable and accurate reports.

	Input image	Reference report	Generated report
Example 1		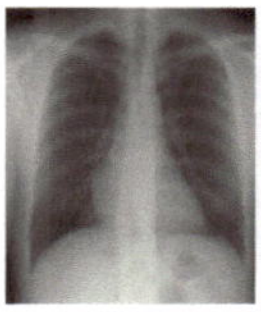 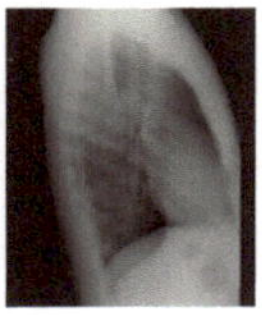 the lungs are clear . no focal air space consolidation . no pleural effusion or pneumothorax . normal cardiomediastinal silhouette	no acute cardiopulmonary abnormalities . no focal consolidation . no visualized pneumothorax . heart size and cardiomediastinal silhouette are within normal limits . lungs are grossly clear . no pleural effusions . xxxx are intact .
Example 2		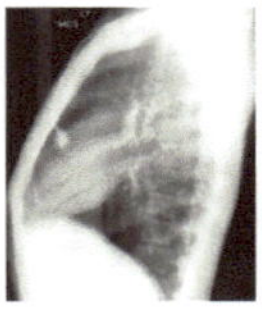 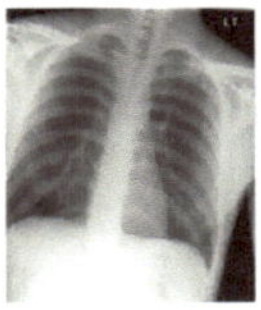 no acute cardiopulmonary findings . heart size normal . no focal airspace disease . no pneumothorax or effusions .	no acute cardiopulmonary findings . heart size and mediastinal contour within normal limits . no focal airspace disease . no pneumothorax or pleural effusion . <u>degenerative changes of the thoracic spine</u> .
Example 3		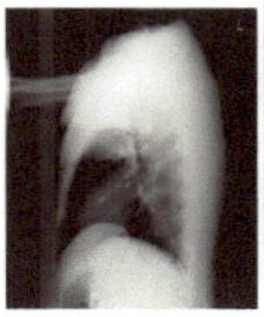 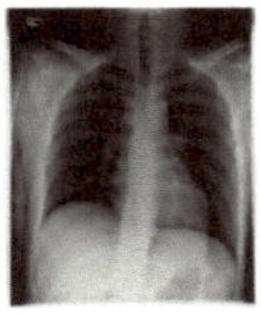 no acute cardiopulmonary abnormalities . the trachea is midline . the cardiomediastinal silhouette is normal . the lungs are clear without evidence of focal infiltrate or effusion . there is no pleural effusion or pneumothorax . the visualized bony structures reveal no acute abnormalities .	no acute cardiopulmonary process . the trachea is midline . the cardiomediastinal silhouette is within normal limits for size and contour . the lungs are normally inflated without evidence of focal airspace disease . there is no pneumothorax or effusion . bony structures reveal no acute abnormalities .

Figure 3. Examples of three visual reports selected from the Open-I dataset. The same color emphasizes the same description of the chest X-ray image. Additional findings of our model for images are highlighted by underlines.

5. Conclusions and Outlook

In this work, we propose a model to enhance the accuracy of generated medical reports based on prior knowledge. We validate the proposed model experimentally, and we validate the effectiveness of our added prior knowledge on the Open-I and MIMIC-CXR datasets. The experimental results show that our model achieves relatively excellent performance in the indicators of natural language generation and clinical efficacy. Ablation experiments show that our model can learn visual features and text features better after adding prior knowledge, so it can generate medical reports more accurately. In addition, the establishment of our knowledge graph is built according to the dataset, which does not need additional experts to build, so it can be more convenient to apply to other datasets.

In our work, we have not considered the influence of location information on the generation of radiology reports, which is important. In the future, we will explore the impact of including location information in disease classification on improving the accuracy of generated reports. Next, we will explore how to improve the accuracy of our classifier, which is related to the accuracy of our automated reports. Specifically, we will pre-train our image encoder and text encoder using a public dataset and then try to incorporate location information into the classifier.

Author Contributions: Conceptualization, D.Z., A.R., and Q.L.; methodology, A.R., Q.L., J.L., Y.M., and D.Z.; investigation, A.R. and Y.M.; visualization, H.W., A.R., and J.L.; project administration, D.Z. and Q.L.; writing—original draft preparation, A.R., D.Z., J.L., and H.W.; writing—review and editing, Q.L., D.Z., A.R., J.L., Y.M., and H.W. All authors have read and agreed to the published version of the manuscript.

Funding: This research was funded by (i) Natural Science Foundation China (NSFC) under Grant No. 61402397, 61263043, 61562093 and 61663046; (ii) Open Foundation of Key Laboratory in Media Convergence of Yunnan Province under Grant No. 220225201. (iii) Open Foundation of Key Laboratory in Software Engineering of Yunnan Province: 2020SE304. (iv) Practical innovation project of Yunnan University, Project No. 2021z34, No. 2021y128 and 2021y129.

Acknowledgments: This research was supported by the Yunnan Provincial Key Laboratory of Software Engineering, the Kunming Provincial Key Laboratory of Data Science and Intelligent Computing and the Key Laboratory in Media Convergence of Yunnan Province.

Conflicts of Interest: The authors declare no conflict of interest.

References

1. *Handbook of Medical Image Computing and Computer Assisted Intervention*; Academic Press: Cambridge, MA, USA, 2019.
2. Jing, B.; Xie, P.; Xing, E. On the automatic generation of medical imaging reports. *arXiv* **2017**, arXiv:1711.08195.
3. Bruno, M.A.; Walker, E.A.; Abujudeh, H.H. Understanding and confronting our mistakes: The epidemiology of error in radiology and strategies for error reduction. *Radiographics* **2015**, *35*, 1668–1676. [CrossRef]
4. Shin, H.C.; Lu, L.; Kim, L.; Seff, A.; Yao, J.; Summers, R.M. Interleaved text/image deep mining on a very large-scale radiology database. In Proceedings of the IEEE Conference on Computer Vision and Pattern Recognition, Boston, MA, USA, 7–12 June 2015; pp. 1090–1099.
5. Zhang, Y.; Wang, X.; Xu, Z.; Yu, Q.; Yuille, A.; Xu, D. When radiology report generation meets knowledge graph. In Proceedings of the AAAI Conference on Artificial Intelligence, New York, NY, USA, 7–12 February 2020; Volume 34, pp. 12910–12917.
6. Li, C.Y.; Liang, X.; Hu, Z.; Xing, E.P. Knowledge-driven encode, retrieve, paraphrase for medical image report generation. In Proceedings of the AAAI Conference on Artificial Intelligence, Honolulu, HI, USA, 27 January–1 February 2019; Volume 33, pp. 6666–6673.
7. Yao, T.; Pan, Y.; Li, Y.; Mei, T. Exploring visual relationship for image captioning. In Proceedings of the European conference on Computer Vision (ECCV), Munich, Germany, 8–14 September 2018; pp. 684–699.
8. Vaswani, A.; Shazeer, N.; Parmar, N.; Uszkoreit, J.; Jones, L.; Gomez, A.N.; Kaiser, Ł.; Polosukhin, I. Attention is all you need. In Proceedings of the Advances in Neural Information Processing Systems 2017, Long Beach, CA, USA, 4–9 December 2017; Volume 30.
9. Demner-Fushman, D.; Kohli, M.D.; Rosenman, M.B.; Shooshan, S.E.; Rodriguez, L.; Antani, S.; Thoma, G.R.; McDonald, C.J. Preparing a collection of radiology examinations for distribution and retrieval. *J. Am. Med. Inform. Assoc.* **2016**, *23*, 304–310. [CrossRef] [PubMed]
10. Johnson, A.E.W.; Pollard, T.J.; Berkowitz, S.J.; Greenbaum, N.R.; Lungren, M.P.; Deng, C.-y.; Mark, R.G.; Horng, S. MIMIC-CXR, a de-identified publicly available database of chest radiographs with free-text reports. *Sci. Data* **2019**, *6*, 1–8. [CrossRef] [PubMed]
11. Vinyals, O.; Toshev, A.; Bengio, S.; Erhan, D. Show and tell: A neural image caption generator. In Proceedings of the IEEE Conference on Computer Vision and Pattern Recognition, Boston, MA, USA, 7–12 June 2015; pp. 3156–3164.
12. Goyal, Y.; Khot, T.; Summers-Stay, D.; Batra, D.; Parikh, D. Making the v in vqa matter: Elevating the role of image understanding in visual question answering. In Proceedings of the IEEE Conference on Computer Vision and Pattern Recognition, Honolulu, HI, USA, 21–26 July 2017; pp. 6904–6913.
13. Rennie, S.J.; Marcheret, E.; Mroueh, Y.; Ross, J.; Goel, V. Self-critical sequence training for image captioning. In Proceedings of the IEEE Conference on Computer Vision and Pattern Recognition, Honolulu, HI, USA, 21–26 July 2017; pp. 7008–7024.
14. Tran, A.; Mathews, A.; Xie, L. Transform and tell: Entity-aware news image captioning. In Proceedings of the IEEE/CVF Conference on Computer Vision and Pattern Recognition, Seattle, WA, USA, 14–19 June 2020; pp. 13035–13045.
15. Zhou, S.K.; Greenspan, H.; Davatzikos, C.; Duncan, J.S.; Van Ginneken, B.; Madabhushi, A.; Prince, J.L.; Rueckert, D.; Summers, R.M. A review of deep learning in medical imaging: Imaging traits, technology trends, case studies with progress highlights, and future promises. *Proc. IEEE* **2021**, *109*, 820–838. [CrossRef]
16. Xu, K.; Ba, J.; Kiros, R.; Cho, K.; Courville, A.; Salakhudinov, R.; Zemel, R.; Bengio, Y. Show, attend and tell: Neural image caption generation with visual attention. In Proceedings of the International Conference on Machine Learning, PMLR, Lille, France, 7–9 July 2015; pp. 2048–2057.

17. Lu, J.; Xiong, C.; Parikh, D.; Socher, R. Knowing when to look: Adaptive attention via a visual sentinel for image captioning. In Proceedings of the IEEE conference on computer vision and pattern recognition, Honolulu, HI, USA, 21–26 July 2017; pp. 375–383.
18. Anderson, P.; He, X.; Buehler, C.; Teney, D.; Johnson, M.; Gould, S.; Zhang, L. Bottom-up and top-down attention for image captioning and visual question answering. In Proceedings of the IEEE conference on computer vision and pattern recognition, Salt Lake City, UT, USA, 18–22 June 2018; pp. 6077–6086.
19. Yuan, J.; Liao, H.; Luo, R.; Luo, J. Automatic radiology report generation based on multi-view image fusion and medical concept enrichment. In *Proceedings of the International Conference on Medical Image Computing and Computer-Assisted Intervention, Shenzhen, China, 13–17 October 2019*; Springer: Cham, Switzerland, 2019; pp. 721–729.
20. Xue, Y.; Xu, T.; Rodney Long, L.; Xue, Z.; Antani, S.; Thoma, G.R.; Huang, X. Multimodal recurrent model with attention for automated radiology report generation. In *Proceedings of the International Conference on Medical Image Computing and Computer-Assisted Intervention, Granada, Spain, 16–20 September 2018*; Springer: Cham, Switzerland, 2018; pp. 457–466.
21. Wang, X.; Peng, Y.; Lu, L.; Lu, Z.; Summers, R.M. Tienet: Text-image embedding network for common thorax disease classification and reporting in chest X-rays. In Proceedings of the IEEE Conference on Computer Vision and Pattern Recognition, Salt Lake City, UT, USA, 18–22 June 2018; pp. 9049–9058.
22. Chen, Z.; Song, Y.; Chang, T.-H.; Wan, X. Generating radiology reports via memory-driven transformer. *arXiv* **2020**, arXiv:2010.16056.
23. Jing, B.; Wang, Z.; Xing, E. Show, describe and conclude: On exploiting the structure information of chest X-ray reports. *arXiv* **2020**, arXiv:2004.12274.
24. Liu, G.; Hsu TM, H.; McDermott, M.; Boag, W.; Weng, W.-H.; Szolovits, P.; Ghassemi, M. Clinically accurate chest X-ray report generation. In Proceedings of the Machine Learning for Healthcare Conference, PMLR, Ann Arbor, MI, USA, 9–10 August 2019; pp. 249–269.
25. Shin, H.C.; Roberts, K.; Lu, L.; Demner-Fushman, D.; Yao, J.; Summers, R.M. Learning to read chest X-rays: Recurrent neural cascade model for automated image annotation. In Proceedings of the IEEE Conference on Computer Vision and Pattern Recognition, Las Vegas, NV, USA, 27–30 June 2016; pp. 2497–2506.
26. Nguyen, H.; Nie, D.; Badamdorj, T.; Liu, Y.; Zhu, Y.; Truong, J.; Cheng, L. Automated generation of accurate\& fluent medical X-ray reports. *arXiv* **2021**, arXiv:2108.12126.
27. Kudo, T.; Richardson, J. Sentencepiece: A simple and language independent subword tokenizer and detokenizer for neural text processing. *arXiv* **2018**, arXiv:1808.06226.
28. Veličković, P.; Cucurull, G.; Casanova, A.; Romero, A.; Liò, P.; Bengio, Y. Graph attention networks. *arXiv* **2017**, arXiv:1710.10903.
29. Chen, M.; Radford, A.; Child, R.; Wu, J.; Jun, H.; Luan, D.; Sutskever, I. Generative pretraining from pixels. In Proceedings of the International Conference on Machine Learning, Virtual Event, 13–18 July 2020; pp. 1691–1703.
30. Devlin, J.; Chang, M.W.; Lee, K.; Toutanova, K. Bert: Pre-training of deep bidirectional transformers for language understanding. *arXiv* **2018**, arXiv:1810.04805.
31. Yin, C.; Qian, B.; Wei, J.; Li, X.; Zhang, X.; Li, Y.; Zheng, Q. Automatic generation of medical imaging diagnostic report with hierarchical recurrent neural network. In Proceedings of the 2019 IEEE International Conference on Data Mining (ICDM), Beijing, China, 8–11 November 2019; pp. 728–737.
32. Irvin, J.; Rajpurkar, P.; Ko, M.; Yu, Y.; Ciurea-Ilcus, S.; Chute, C.; Marklund, H.; Haghgoo, B.; Ball, R.; Shpanskaya, K.; et al. Chexpert: A large chest radiograph dataset with uncertainty labels and expert comparison. In Proceedings of the AAAI Conference on Artificial Intelligence, Honolulu, HI, USA, 27 January–1 February 2019; Volume 33, pp. 590–597.
33. Chen, Z.M.; Wei, X.S.; Wang, P.; Guo, P. Multi-label image recognition with graph convolutional networks. In Proceedings of the IEEE/CVF Conference on Computer Vision and Pattern Recognition, Long Beach, CA, USA, 16–20 June 2019; pp. 5177–5186.
34. He, K.; Zhang, X.; Ren, S.; Sun, J. Deep residual learning for image recognition. In Proceedings of the IEEE Conference on Computer Vision and Pattern Recognition, Las Vegas, NV, USA, 27–30 June 2016; pp. 770–778.
35. Huang, G.; Liu, Z.; Van Der Maaten, L.; Weinberger, K.Q. Densely connected convolutional networks. In Proceedings of the IEEE Conference on Computer Vision and Pattern Recognition, Honolulu, HI, USA, 21–26 July 2017; pp. 4700–4708.
36. Su, H.; Maji, S.; Kalogerakis, E.; Learned-Miller, E. Multi-view convolutional neural networks for 3d shape recognition. In Proceedings of the IEEE International Conference on Computer Vision, Santiago, Chile, 7–13 December 2015; pp. 945–953.
37. Jia, N.; Tian, X.; Zhang, Y.; Wang, F. Semi-supervised node classification with discriminable squeeze excitation graph convolutional networks. *IEEE Access* **2020**, *8*, 148226–148236. [CrossRef]
38. Zhu, J.Y.; Park, T.; Isola, P.; Efros, A.A. Unpaired image-to-image translation using cycle-consistent adversarial networks. In Proceedings of the IEEE International Conference on Computer Vision, Venice, Italy, 22–29 October 2017; pp. 2223–2232.
39. Lovelace, J.; Mortazavi, B. Learning to generate clinically coherent chest X-ray reports. In Proceedings of the Findings of the Association for Computational Linguistics: EMNLP 2020, Punta Cana, Dominican Republic, 8–12 November 2020; Volume 2020, pp. 1235–1243.
40. Papineni, K.; Roukos, S.; Ward, T.; Zhu, W.-J. Bleu: A method for automatic evaluation of machine translation. In Proceedings of the 40th Annual Meeting of the Association for Computational Linguistics, Philadelphia, PA, USA, 7–12 July 2002; pp. 311–318.
41. Lin, C.Y. *Rouge: A Package for Automatic Evaluation of Summaries*; Text Summarization Branches Out; Association for Computational Linguistics: Barcelona, Spain, 2004; pp. 74–81.

42. Banerjee, S.; Lavie, A. METEOR: An automatic metric for MT evaluation with improved correlation with human judgments. In Proceedings of the Acl Workshop on Intrinsic and Extrinsic Evaluation Measures for Machine Translation and/or Summarization, Ann Arbor, MI, USA, 29 June 2005; pp. 65–72.
43. Sharma, S.; Asri, L.E.; Schulz, H.; Zumer, J. Relevance of unsupervised metrics in task-oriented dialogue for evaluating natural language generation. *arXiv* **2017**, arXiv:1706.09799.
44. Li, Y.; Liang, X.; Hu, Z.; Xing, E.P. Hybrid retrieval-generation reinforced agent for medical image report generation. In Proceedings of the Neural Information Processing Systems 2018, held at Palais des Congres de Montreal, Montreal CANADA, 2–8 December 2018 Advances in Neural Information Processing Systems 2018, Montreal, QC, Canada, 2–8 December 2018; Volume 31.
45. Liu, F.; Wu, X.; Ge, S.; Fan, W.; Zou, Y. Exploring and distilling posterior and prior knowledge for radiology report generation. In Proceedings of the IEEE/CVF Conference on Computer Vision and Pattern Recognition, Nashville, TN, USA, 19–25 June 2021; pp. 13753–13762.
46. Li, J.; Li, S.; Hu, Y.; Tao, H. A Self-Guided Framework for Radiology Report Generation. *arXiv* **2022**, arXiv:2206.09378.

MDPI AG
Grosspeteranlage 5
4052 Basel
Switzerland
Tel.: +41 61 683 77 34

Applied Sciences Editorial Office
E-mail: applsci@mdpi.com
www.mdpi.com/journal/applsci

Disclaimer/Publisher's Note: The statements, opinions and data contained in all publications are solely those of the individual author(s) and contributor(s) and not of MDPI and/or the editor(s). MDPI and/or the editor(s) disclaim responsibility for any injury to people or property resulting from any ideas, methods, instructions or products referred to in the content.